Law and Policy of the European Gas Market

Law and Policy of the European Gas Market

Monica Waloszyk

Centre for Energy Law, University of Aberdeen, UK

Edward Elgar

Cheltenham, UK • Northampton, MA, USA

Published by
Edward Elgar Publishing Limited
The Lypiatts
15 Lansdown Road
Cheltenham
Glos GL50 2JA
UK

Edward Elgar Publishing, Inc.
William Pratt House
9 Dewey Court
Northampton
Massachusetts 01060
USA

A catalogue record for this book
is available from the British Library

Library of Congress Control Number: 2014932607

This book is available electronically in the ElgarOnline.com Law Subject Collection, E-ISBN 978 1 78347 347 2

MIX
Paper from responsible sources
FSC® C013056
FSC www.fsc.org

ISBN 978 1 78347 346 5

Typeset by Columns Design XML Ltd, Reading
Printed and bound in Great Britain by T.J. International Ltd, Padstow

Contents

Preface

This book is concerned with the law and politics of the EU gas market and particularly the regulatory and competitive choices of institutions and bodies operating in the market, in view of rendering it more integrated. The main question under discussion is how regulatory and competitive choices influence the achievement of an integrated EU gas market and the gas market design. The answers are pursued in an interdisciplinary manner. This is done with regard to the fact that legal choices and afferent institutional arrangements occur not in the protective confines of a legislative laboratory, but in a constantly evolving geopolitical perimeter, where gas relationships are frequently renegotiated. The book considers a number of legal and political science subquestions and theories, which allow an analysis of causes and facilitating factors for the choice of market regulation tools and of the legislative and institutional effects that such choices determine.

The structure is divided into seven chapters, including introduction and conclusion. These look at the factual and legal benchmarks for the reformation of the European gas market, the specificities of the natural gas sector in terms of legal and policy setting of relationships with third-country producers, the legal and geopolitical aspects contributing to the gas market segmentation, the gas market structure in the aftermath of the Energy Sector Inquiry and the adoption of the Third Energy Package, the relationship between market liberalization and market integration, and the flaws of the gas market design, deriving from the ambiguous articulation of gas policy goals at an EU level and the shabby matching of legal tools and accompanying institutional design available to their pursuit.

The conclusions attempt to contribute to the discussion regarding the reassessment of the concurrent application of competition law and regulation in the EU gas market.

Acknowledgements

The successful completion of this book has relied on the generous advice, support and patience of a number of people. I am indebted to Professor Herwig Hofmann, University of Luxembourg, supervisor of my doctoral thesis on which this book is based, for his support throughout the lengthy writing process and his encouragement in times of doubt, as well as for very useful discussion and for his ongoing support of my professional development. I also warmly thank all the members of my Thesis Committee, Professor Mark Cole, Professor Rusen Ergec and Professor Jörg Gerkrath (all of the University of Luxembourg), Professor Alexander Türk (Kings College, London) and Professor Gerard Rowe (Viadrina European University, Frankfurt (Oder)) for their interest and positive evaluation of work. I am especially grateful to Professor Rowe, who has provided me with constant inspiration, support and guidance since my time as a student at the Viadrina. I thank Blathnaid O'Hannrachain for her considerable contribution in proofreading my work. Dr Derek Averre, Nicolina Bordian, Christian Deprez, Dr Ekaterina Islentyeva, Dr Helene Ragheboom, Dr Jaroslaw Janczak, Marie Marty, Mariana Peña, Martin Petschko, Carola Rappe, Gavin Robinson, Dharamveer Singh and Willy Tadjudie all helped me through discussion, advice and simple reassurance, whether as colleagues at the University of Luxembourg or at conferences. I thank all of them. To my husband, Maciek, and to my family I owe you a great debt of gratitude for your love, support and kindness, without which I might well not have embarked on the journey of writing this work.

Abbreviations

ACER	Agency for the Cooperation of Energy Regulators
BSEP	Black Sea Environmental Program
BSREC	Black Sea Regional Energy Centre
CEER	Council of European Energy Regulators
CESR	Committee of European Securities Regulators
CJEU	Court of Justice of the European Union
CNE	Comisíon nacional de energía
COMP	Competition
DG ENER	Directorate-General Energy
DSO	Distribution System Operator
EASA	European Aviation Safety Agency
EASEE	European Association for Streamlining of Energy Exchange
EBN	Energie Beheer Nederland
EBRD	European Bank for Reconstruction and Development
EC	European Community
ECN	European Competition Network
ECT	Energy Charter Treaty
EEA	European Economic Area
EEC	European Economic Community
EFSA	European Food Safety Authority
ENEL	Ente Nazionale per l'Energia Elettrica
ENI	Ente Nazionale Idrocarburi
ENPI	European Neighbourhood and Partnership Instrument
ENTSO	European Network of Transmission System Operators
ENTSOG	European Network of Transmission System Operators for Gas
ERGEG	European Regulators' Group for Electricity and Gas
ESA	European Supervisory Authorities
EU	European Union

Euratom	European Atomic Energy Community
Eurostat	European Statistical System
EUSA	European Union Studies Association
EWM	Early Warning Mechanism
GATS	General Agreement on Trade in Services
GATT	General Agreement on Trade and Tariffs
GAZPROM	Gazovaya Promyshlennost
GDF	Gaz de France
GDP	Gross Domestic Product
GFU	Gas Negotiation Committee
GIS	Gas Insulated Switchgear
GRI	Gas Regional Initiative
IEA	International Energy Agency
INOGATE	Interstate Oil and Gas Transportation to Europe
ISO	independent system operator
ITO	independent transmission operator
LNG	liquefied natural gas
Med-TSO	Mediterranean – Transmission System Operator
MEGAL	Mittel-Europäische-Gasleitung
NATO	North Atlantic Treaty Organization
NCA	National Competition Authority
NLNG	Nigerian LNG Limited
NRA	National Regulatory Authority
OFGEM	Office of Gas and Electricity Markets
OJ	Official Journal of the European Union (formerly Communities)
PCA	Partnership and Cooperation Agreement
PCI	Projects of Common Interest
PSO	public service obligation
RWE	Rheinisch-Westfälisches Elektrizitätswerk AG
SEA	Single European Act
SPS	Sanitary and Phytosanitary Measures
TACIS	Technical Assistance for the Commonwealth of Independent States
TANAP	Trans-Anatolian gas pipeline
TBT	Technical Barriers to Trade

TEC	Treaty Establishing the European Community
TEP	Third Energy Package
TEU	Treaty on the European Union
TFEU	Treaty on the Functioning of the European Union
TRACECA	Transport Corridor Europe–Caucasus–Asia
TRIMS	Agreement on Trade Related Investment Measures
TRIPS	Agreement on Trade Related Aspects of Intellectual Property Rights
TSO	Transmission System Operator
VCLT	Vienna Convention on the Law of Treaties
WTO	World Trade Organization

Table of cases

Table of legal acts

INTERNATIONAL TREATIES AND CONVENTIONS

TREATIES AND CONVENTIONS OF THE EUROPEAN COMMUNITIES AND THE EUROPEAN UNION

INTERNATIONAL AGREEMENTS AND GUIDELINES

AGREEMENTS, COMMUNICATIONS, DECLARATIONS AND GUIDELINES OF THE EUROPEAN COMMUNITIES AND OF THE EUROPEAN UNION

NATIONAL LEGAL ACTS

Table of legal acts of the European Union

COUNCIL REGULATIONS

COUNCIL DIRECTIVES

COUNCIL DECISIONS

COMMISSION REGULATIONS

COMMISSION DECISIONS

COMMISSION DOCUMENTS

Green Papers and White Papers

Communications

Staff Working Documents, Reports and Impact Assessments

PREPARATORY ACTS OF THE EUROPEAN UNION

1. Introduction

On 3 March 2011, the Third Energy Package (TEP) entered into force in the EU. In comparison to its predecessors, the package appears to be more focused on delivering an integrated energy market,[1] in which regulatory processes across the Member States are coordinated and harmonized, the market interconnection is strengthened and the capacity of the Union[2] to react in the eventuality of a supply crisis is reinforced. In this respect, the focus of the latest stage of the European Commission's attempt to reform the European Union's (EU) electricity and gas markets stretches beyond the mere liberalization of the energy market, towards the achievement of an integrated administration to govern an integrated EU gas market.

The goal of energy market integration – at times associated with or overlapping the goal of market liberalization – has been pursued to a certain extent by the EU Commission from the beginning of the energy sector reformation. However, the clarity of its articulation has been highly influenced by both internal and external factors. Over the last two decades, European energy markets in general, and gas and electricity markets in particular, have undergone an important restructuring process.[3] During this process the role of market actors, the internal market structure and design, and the international energy context have evolved. This accumulation of economic, political and societal factors on national, regional and international levels has influenced and at times steered the direction taken by the European energy legislation.

[1] In this sense, a real cross-border energy market is expected to emerge by pursuing regulatory methods targeting the establishment of entry/exit zones, market-based balancing and binding network codes.

[2] The present text takes into consideration the changes brought about by the Lisbon Treaty to the three-pillars system of the European Union, which resulted in the European Community being absorbed by the European Union. As a consequence, the concept of "Union" or "European Union" is employed unitarily throughout the text and refers both to the present EU and to the former EC. The word "Community" is used either when it is an integral part of a quotation or a case-law, or as part of a bibliographical reference.

[3] See Haase and Bressers, 2008, p. 1; Stepina, 2008, p. 1.

In the EU market restructuring has been performed mostly by means of regulatory reforms, aimed at creating the foundations of a comprehensive common energy policy, encompassing the needs of a single energy market. Starting with the 1990s, the importance of regulation in the implementation of sector reforms has gained momentum at the expense of state intervention. Such a development has occurred in an attempt to render investments in transmission networks more efficient, to enhance the operation of energy networks despite the dismantling of traditional network governance structures and furthermore to facilitate competition in the production and supply of energy.[4] In this respect, several European Directives, Regulations,[5] green papers, Court decisions, joint declarations and joint positions have been adopted. The internal market project has been built around the idea of developing a strategy simultaneously addressing energy security, economic efficiency and environmental protection issues within the European Union. In time, this should be accompanied by a coherent and coordinated European external policy in the field of energy, translated into a unitary position of Member States towards third parties in energy exchanges.[6]

1.1 GENERAL CONTEXT

Certainly, the development of a comprehensive energy policy addressing legal, political, economic and social concerns has been neither singular nor restricted to Europe. A mere look at the energy policies of most developed countries shows important changes during the last decades, owing to climate change, energy efficiency issues and security of supply concerns.[7] Specifically, the increasing global demand for energy,[8] accompanied by a general decrease in the world's traditional energy resources, has raised the awareness of policy-makers with respect to economic and political challenges resulting from an unequal partitioning of energy

---

[4] Jamasb and Pollitt, 2008, pp. 4584–4589.

[5] The term "regulation" can refer to both a specific form of legal tool and a mode of governance of the EU gas market. In order to distinguish between the two, the study capitalizes the term when referring to the legal tool.

[6] The extent to which such endeavour has been successful is to be further discussed within the following chapters.

[7] Schaffer, 2009, p. 9.

[8] International Energy Agency, World Energy Outlook 2010 Factsheet, http://www.worldenergyoutlook.org/docs/weo2010/factsheets.pdf

resources at a global level, associated with the concentration of major volumes of such resources in the hands of very few energy suppliers.

The competition for access to energy resources has always dominated international relations, notwithstanding the variations in energy demand.[9] It could be maintained, however, that the necessity to draft an energy policy has grown as a direct consequence of an increasingly insecure energy supply. The geopolitical context serving as a background for the initiation of an energy policy at European level provides evidence in this sense. Specifically, it can be argued that the acceleration of the European processes leading to the establishment of an energy policy has been highly influenced by emerging doubts related to traditional energy resources and import dependencies on politically sensitive countries.[10] As explained below, the geopolitical context has been accompanied by legal and institutional changes regarding the modes of governance at Union level, which in themselves favoured the adoption of such a policy. Nevertheless, considering the failure of previous attempts to forge a common policy encompassing all forms of energy,[11] it could be further argued that Member States required an external motivation for pooling together their energy resources, while renouncing an important part of their sovereignty rights over them.

There is an intrinsic link between the availability of energy and the economic development of a society. This fact is perhaps best illustrated by the wealth gap between the richest and the poorest countries in the world and the scientific and societal advancements registered in the former during the last and current centuries. These could be easily connected to the availability of a multitude of energy resources and the facilitation of access to them. Paraphrasing the US Energy Secretary,[12] it appears that the use of oil, gas and electricity in our modern society has resulted in each citizen of a developed country having the equivalent of roughly 100 servants working for them.[13] It comes thus as no surprise that the guarantee of an uninterrupted access to energy resources and of the security of energy supply represents a core interest of energy policy designs all over the world.

---

[9] Haghighi, 2007, p. 1.

[10] Cronshaw *et al.*, 2008, p. 13.

[11] For example, the Treaty Establishing the European Atomic Energy Community (1957) and the Treaty Establishing the Coal and Steel Communities (1951).

[12] Dr Steven Chu, US Secretary of Energy between 21 January 2009 and 22 April 2013.

[13] Crooks, 2009b.

On the EU agenda for establishing secure and integrated energy markets, the guarantee of a secure and reliable access to natural gas resources represents an utmost concern. Among European Union energy resources, natural gas constitutes an essential component, amounting to one-quarter of the primary energy supply and "contributing mainly to electricity generation, heating, feedstock for industry and fuel for transportation".[14] During recent years, the stake of geopolitics in natural gas market arrangements has increased considerably. This is due partially to gas having grown from a marginally traded fuel, consumed within disparate town limits, to a fuel transported across great distances for the supply of a large variety of economic sectors.[15] The increased dependency of different industries on the use of gas, together with the concentration of the majority of natural gas resources within the hands of very few world suppliers, has granted the fuel political characteristics it did not possess before.[16] Closely related to the geopolitics of gas, but moving further towards the creation of an integrated European gas market and the establishment of a common set of rules that could govern the gas relationships among the EU Member States and between these and third-country suppliers, the EU has been involved in a law-making process, renewed and reviewed at regular time intervals since 1998.

It should be noted that, within the European Union, the development of a gas market is of relatively new origin, despite the use of gas for street and home lighting throughout northwestern Europe since as early as the nineteenth century.[17] There are various reasons for this late development. These range from the late breakthrough of gas field discoveries in Europe to the lack of a sustained movement towards a real European energy market and an afferent common energy policy until the end of the 1980s and the beginning of the 1990s. Indeed until that time, it can be argued that the segmentation of gas markets along national borders had raised no polemics, similar to the monopolistic arrangement of national gas industries. Certainly, reliance on the welfare state as the sole regulator of the market – responsible for ensuring fair redistribution and taxation – coupled with the natural state sovereignty upon energy matters contributed to the maintenance of such a *status-quo*. That persisted even as the patterns of governance in Europe underwent significant changes, which later on resulted in the gradual shifting of vertical decision-making from national to supranational *loci*. As a consequence, the dawning of the gas

---

14 Paragraph 1 of the Preamble of Regulation (EU) No. 994/2010.

15 Victor, 2006, p. 1.

16 For a different opinion refer to Haghighi, 2007, p. 13.

17 Spanjer, 2008, p. 2.

market liberalization movement has seen industry and industry-oriented governments manifesting strong opposition to the liberalizing trend, while appearing resolved to maintain the existent market arrangements unchanged.[18] That allowed a few national incumbents to enjoy an uncontested monopoly of energy production, transmission and distribution.

Notwithstanding an undisputed tendency to pool together essential national resources as part of the European integration, no aim of establishing a European common energy market was visible until the late 1980s. The strategic importance of energy sectors determined each Member State to guard fiercely its sovereignty in this field and to either discard with indifference or vehemently oppose any attempt of relocating decision-making related to energy policy at EU level.[19] The change occurred during the 1970s, when the role of the positive state "as planner, direct producer of goods and services, and employer of last resort"[20] began to erode as a consequence of rising unemployment and escalating inflation rates. Despite a new approach of addressing market failure through regulation, rather than state intervention, the latter was still regarded as legitimate in the field of energy until the mid-1980s. Such a legitimacy was maintained owing to security of supply anxieties generated by the two oil shocks of the previous decade.[21] By the end of the 1980s, however, the EU was facing a deepening of the economic and monetary integration, which challenged the traditional modes of governance of Member States' governments.[22]

Indeed, the adoption of the Single European Act signalled that Member States acknowledged their incapacity to face international competition alone. It also announced the extension of the EU's competences over areas previously reserved to state sovereignty. At the same time, the 1986 gas and oil price decline and the availability of substantial gas volumes in Europe, owing to increased efforts in exploration and production investments,[23] lessened the leverage of the threat of resources scarcity on the security of Member States energy supply. As a direct consequence, the legitimacy of state intervention within the regulation of national energy industries became questionable. Additionally, an attained maturity of

---

[18] Haase, 2008.
[19] Padgett, 1992, p. 55.
[20] Majone, 1997, p. 141.
[21] Spanjer, 2008, p. 9.
[22] Majone, 1997, p. 139.
[23] Cronshaw *et al.*, 2008, p. 13.

European energy markets through the achievement of previously state-coordinated substantial investments brought the monopolistic structure of the market under scrutiny.

The dismantling of monopolistic gas market structures through European-led legislative and regulatory reforms has faced both open and disguised opposition from various industry stakeholders, much like the reformation of any network industry. Whereas the former type of opposition appears to have faded away between 1998, when the phased transition process to a single gas market began, and 2003, when a second European package of measures was adopted,[24] the latter has subsisted until the present day, in a more or less pronounced form throughout the majority of EU Member States. The International Energy Agency (IEA) identified four general trends of industry stakeholders' strategies to counteract the EU regulatory reform: acquiring assets in other countries in order to balance the loss of their national monopolistic market share, merging downstream local utilities on a national or regional basis, increasing mergers and acquisitions between gas and power companies and vertical integration of newly created pan-European energy groups.[25] The effects of industry's response to liberalization upon the sector regulation process have been visible in the findings of the Energy Sector Inquiry.[26] The inquiry identified several obstacles hampering the positive effects of the first two cycles of energy reforms. These stemmed mostly from the energy market structure, differing or contradictory regulatory regimes across the Member States, the regulatory lag and poor EU energy market interconnection.

It would appear that the goal of an integrated European gas market that should have been achieved by 2007[27] still remains far from being attained. As argued in the following chapters, the partial failure of the initial legislative packages resided both in the conception and design of the energy market and afferent policy at an EU level, as well as in the response of Member States and national structures to these. Hence, it could be assumed that the potential success of the new energy package is to a great extent dependent on its capacity to address such matters. TEP seems to provide the necessary measures to correct the market failures of the past, but the effectiveness of their enforcement within the Member States is at the moment still unclear. In order to avoid a speculative

24 Cameron, 2005a, pp. 7–8.
25 Cronshaw *et al.*, 2008, pp. 29–30.
26 DG Competition Report on Energy Sector Inquiry, /*SEC/2006/1724/.
27 Cameron, 2005a, p. 8.

approach, the present study focuses on the actual degree of market integration and the legislative, institutional and policy tools used in forging it.

1.2 THE FOCUS OF THE BOOK

The book is concerned with the law and politics of the EU market of natural gas[28] and particularly with the regulatory and competitive choices of institutions and bodies operating in the market, in view of rendering it more integrated. The analysis is not limited to the development of a common gas market, mainly from a competition law point of view, but takes the position that consideration of sectoral regulation as such is indispensable when analysing the gas market design and its afferent policy. Unlike other legal scholarship, the present study questions the validity of considering regulation as a temporary resource, which should in time be completely replaced by the application and enforcement of EU competition law.[29] It maintains the basic notion of competition as being suitable for rectifying deficiencies in energy supply, but it advocates that its tools might not be sufficient to answer the specific needs of the gas sector. Accordingly, sector specific regulation might be more indicated to balance regulation-for-competition with the fulfilment of public service obligations, security of supply concerns and environmental and consumer protection, which might take place at the expense of the former. Hence, the application of both sector regulation and competition law might be necessary as long as the gas sector preserves its particularities.

The main question operated in the study is how regulatory and competitive choices influence the achievement of an integrated EU gas market and gas market design. When answering the question it has to be remembered that the application of competition law and specific sector legislation to the EU gas market, as well as their afferent institutional arrangements, does not occur in the protective confines of a legislative laboratory. Instead it takes place within a constantly evolving geopolitical perimeter, where gas supply relationships between market players are

---

[28] This refers to gas produced from conventional reservoirs, which is transported from producers to end-consumers through a pipeline grid or by means of liquefied natural gas tanks. It does not regard manufactured gas resulted from the transformation of oil or coal, or natural gas developed from unconventional reservoirs, which could themselves be the subject of distinct research.

[29] See, e.g., Geradin, 2006; Larouche, 2006.

frequently renegotiated. Hence, the purely internal analysis of EU gas legislation and Member States gas regulatory regimes isolated from the geopolitical context seems to be insufficient. Having due regard to this, the book follows an interdisciplinary approach, in which legal and political science subquestions and theories are analysed and inferred. In this way, the research combines an analysis of causes and facilitating factors of market regulation choices with an inquiry into the effects of the specific legislative and institutional arrangements that such choices determine.

To satisfy the purpose of the main research question, the study suggests the structure presented below. Following the introduction, the second chapter provides the factual and legal benchmarks for the reformation of the European gas market, which help the placement of the present market integration development into an evolving legislative and geopolitical context. The chapter details the elements that make gas different from other network industries, before dwelling upon the legislative framework applicable to the development of an integrated gas market. In this respect, it explains the transformation of EU gas market regulation as a consequence of the substantive changes that have occurred in general market governance since the Single European Act.

The present directions taken by EU gas sector regulation are explained in the third chapter by looking at the process that has led to the adoption of the TEP. Subsequently, the chapter ponders upon the legal basis for EU intervention in the gas market. This is followed by an analysis of the setting of relationships with natural gas producers. The legal background of gas supply to the EU is discussed by referring to international law and agreements, the applicability of EU competition law to foreign gas undertakings and obligations and conditions issued from specific gas regulation. The focus moves gradually to a comprehensive case study on EU–Russia gas relationships.

The fourth chapter addresses the market structure in terms of the level of market concentration and the degree of market interconnection. The question guiding the chapter is how the instruments at hand can be better adapted to facilitate an integrated market structure. In this respect, the chapter first assesses the maintained foreclosure of the EU gas market caused by traditional and newly emerged market players and the entrenched characteristics of gas supply relationships. It does that by looking at the level of competition in the supply market. Second, it turns to discussing legal aspects contributing to gas market segmentation, while pondering upon the mechanisms that could correct gaps in market interconnection.

The fifth chapter moves onto the analysis of the market design achieved through liberalization. It does this by pondering upon the possibilities available to the actors active in the market in order to attain the goals of market integration. Such possibilities derive from both competition policy and sector regulation, and their various mixings exert a different scope in influencing the shape and functioning of the gas markets.

The sixth chapter is centred on integrated institutional arrangements. It first presents the individual bodies, their competencies, functions, resources, interaction, organization schemes and accountability, and then it discusses the shared competencies in the EU gas market. To a certain extent, it deals with implementation mechanisms and procedures from administrative rules to individual decisions. In this way, it infers cooperation procedures and information sharing and moves into discussing the more complex issue of integrated administration on the EU gas market.

The seventh chapter is dedicated to concluding remarks on the future of regulation in EU network industries. It encompasses an evaluation of European gas regulation. In light of the findings of previous chapters, this chapter attempts to observe to what extent the established normative goals are achieved using the legal tools available. Additionally, it attempts to formulate proposals for the harmonization of national and EU gas regulatory levels.

2. The specifics of the EU gas market

The introduction and spread of use of natural gas in the territories of Member States presents several specificities in terms of governmental choice of energy resources mixes, exploitation development in connection to petroleum products and environmental concerns. The rise of natural gas as an essential energy resource for the achievement of national energy policy objectives has been greatly influenced by three geopolitical developments. These succeeded one another in the three decades following 1970. The first development refers to the 1970s oil shocks. This pointed out the weaknesses in EU energy policy. In the aftermath of the shocks a need for energy diversification was triggered throughout Western European economies.[1] Second, around the 1980s, major changes occurred within national patterns of governance. As a consequence, traditional employment policies, which favoured the production and use of a limited array of energy resources,[2] were challenged and dismantled. This opened the door for the development of the natural gas industry. Third, starting with the 1990s, global environmental concerns emerged on the European political agenda and played a significant role in redefining the composition of the Member States' energy mixes by making natural gas the fuel of choice for consumers interested in its low environmental impact. Protectionist approaches nurtured by some of the largest Member States towards traditional energy resources such as coal and nuclear energy were challenged by the adoption of the Kyoto Protocol in 1997 and most recently by the nuclear incident at Fukushima, in Japan.[3]

Natural gas represents the world's fastest growing primary energy resource and its consumption is supposed to double by 2030, overtaking oil, which has hitherto been the world's leading energy resource. Gas's complexity from a legal point of view is reflected in its simultaneous qualification as a commodity, a service placed under public service

---

[1] Cronshaw *et al.*, 2008, p. 7.

[2] For example, coal.

[3] http://ec.europa.eu/energy/nuclear/safety/doc/20110525_eu_stress_tests_specifications.pdf

obligations and a network industry, prone to thriving under monopoly. Understanding the complexity of this energy resource is a necessary step in comprehending the nature of regulation, competition policy and law applied to this sector and the obstacles to achieving the market liberalization goals.

2.1 NATURAL GAS: A COMPLEX ENERGY RESOURCE

During the last 30 years, natural gas has become one of the most important sources of energy in Europe owing to its extensive use in electricity production, its low environmental impact in comparison to other traditional energy resources and its relatively low production costs in contrast to those generated by newer energy trends.[4] Natural gas is a multifacetted energy resource, which can be perceived as a commodity, as a service placed under public service obligations and as a network industry. The coexistence of these aspects influences the complexity of the legal framework accompanying the EU gas market. Each feature of natural gas is regulated by different European and national law provisions, which have to be taken into consideration. There is no unitary European energy law addressing the gas sector. Single market principles, environmental provisions, international law covering trade relations and competition law and policy interact with specific sector legislation in shaping the market. Complementary to an intricate legislative framework, the understanding of gas as a commodity, service or network industry attracts a specific institutional alignment, reflected in the organization of the market's institutions at all layers of the regulatory structure.

There are different types of gas that are traded on Member States' energy markets and that can be distinguished from one another depending on their energy quality.[5] In this sense, gas as a commodity produced from different fields or wells is likely to feature different energy content, according to the types and percentages of contaminants and water that reach the gas stream. Energy quality is an important aspect to consider when addressing energy efficiency objectives, security of supply, trade and safety, but also when ensuring the interoperability of transmission and distribution systems across Member States. In order for network users to gain effective access to the transmission system, they must

---

4 For example, development of renewable resources.

5 Cameron, 2007, p. 25.

provide the responsible transmission system operator with information upon the quality of the gas they intend to supply.[6] Should the natural gas for which access is demanded fail to meet the quality parameters of the transmission system operator (TSO), the latter may refuse to transmit the gas.[7] In an attempt to foster the cross-border interoperability of systems, the European Commission has mandated the European Committee for Standardization to harmonize natural gas quality in the EU.[8] However, the process is still ongoing and the provisions of Directive 2009/73/EC remain vague as to energy quality requirements.[9]

As a service, gas is subjected to public service obligations. Although the specific sector legislation does not provide for the universality of the service of gas, the safety of its supply is essential to protect and satisfy the needs of end-consumers. Public service obligations applied to gas include security, regularity, quality, price of supplies, and the protection of vulnerable consumers and the environment.[10] Under the liberalization Directives, Member States enjoy a discretionary right of imposing public service obligations, which have to be clearly defined, transparent, non-discriminatory and verifiable. The fulfilment of such criteria is necessary, while they can constitute legitimate reasons for refusing access to the network or licences for new production and for transport capacity. The satisfaction of public service obligations appears sometimes to be at odds with demands of market competitiveness.[11]

In its perception as both a commodity and a service, natural gas is governed by the single market principle of free circulation of goods and services in the EU, sanctioned by the provisions of the Treaty on the Functioning of the European Union (TFEU) and its predecessors.[12]

---

6 Annex I.3 of Regulation (EC) No. 715/2009.

7 Article 4.6 of Operational Order of a Transmission System Operator EUSTREAM, 2007.

8 European Commission, 2007: Mandate to CEN for Standardization in the Fields of Gas Qualities.

9 Recital 41 of Directive 2009/73/EC.

10 Both telecom and electricity are associated with the concept of universal service. This concept is lacking from the specific sector legislation regulating the EU gas market, while gas is a substitutable energy resource. This does not exempt the supply of gas from public service obligations, but limits its interpretation from a universal service point of view.

11 See Section 5.1.2.2.

12 See Title II and Title IV of TFEU. As will result from the following paragraphs, this aspect could be drawn upon in order to justify increased cooperation among national regulatory authorities (NRAs) with respect to cross-border trade in national gas.

However, the double-sided nature of gas allows the emergence of legal constraints to the application of these Treaty provisions to the natural gas sector. Article 36 TFEU stipulates that the provisions related to the free movement of goods can be submitted to certain exemptions, should such be required for public security reasons. Although exemptions are generally strictly defined, Member States have been able in the past to invoke their security of supply obligations in order to elude the application of general rules on trade and competition to natural gas cases.[13] Directive 2009/73/EC attempts to limit the public service obligations related to natural gas that Member States could appeal to by imposing minimal standards and by rendering the entire process more transparent.[14] However, at the same time it introduces the concept of "energy poverty" and emphasizes the close connection between public service obligations and consumer protection, all the more so when it comes to vulnerable consumers and their access to gas resources. Hence, despite the relatively early stage of application throughout the EU of the 2009 gas Directive, one could ask whether the public service obligations' provisions of the new secondary law are not liable to perpetuate the legal constraints to free movement and free competition of natural gas in the Union.

Added to the above, a third way of looking at gas presents it as a network industry. This encompasses three complementary components: an upstream segment supplying and producing the basic products and services; a network infrastructure; and a downstream service provision activity.[15] Scarce or absent indigenous sources of production and the confinement of the transport of gas to a limited number of rigid pipeline grids have been translated into stable dependency relationships between EU Member States and a couple of producing third countries. Such an aspect is intrinsic to the gas market and has no similar equivalent in the case of the electricity, renewables or nuclear energy markets. To reach EU territory, natural gas is transported across long distances and interstate borders. This, coupled with the fact that natural gas is not exactly a fungible commodity, increases its exposure to political pressure originating either at the source or in the transit countries.[16] While the largest

---

13 Padgett, 1992, p. 56.

14 Obligation of notification article 3(4), article 3(11) of Directive 2009/73/EC.

15 Bergman *et al.*, 1998.

16 Schaffer, 2009, p. 4.

amounts of natural gas are traded via pipelines, gas markets are regionalized. Within such markets, cross-border gas transactions generally take the shape of transit or supply agreements between major gas incumbents.[17]

Natural gas has known the slowest rate of technological change among all European utilities. One of the essential factors contributing to the accelerated liberalization of the telecommunications sector has been the occurrence of important technological developments at the same time as the introduction of deregulation policies.[18] A similar technological boom has been absent from the gas sector and is also unlikely to happen in the near future. As a consequence, natural gas network infrastructure remains characterized by rigidity and assets specificity. This means that network assets may not be interchanged between different uses.

Gas transport requires the presence of a physical link between producers and consumers – often in the shape of a pipeline. The entire EU gas network may be imagined as a collection of nodes connected by transport links.[19] The transport links are provided by pipelines, whereas the nodes are represented by consumers, liquefied natural gas (LNG) facilities and storage. In what concerns the latter, natural gas can be stored according to four different methods: in underground facilities, in transmission and distribution pipelines, in LNG containers situated above the ground or by means of "line-packing".[20] This provides gas system operators with an important degree of flexibility in balancing their systems over time, unlike in the case of electricity, where owing to the lack of storage potential, balancing must occur in a matter of few seconds.[21] Access to storage facilities is an essential condition for suppliers who want to be active in gas retail markets. This is particularly true in the case of suppliers with a limited number of customers and a fluctuating demand.

---

[17] Undertakings participating directly or indirectly in the trade of gas have traditionally been large incumbents, often vertically integrated, some of which have also been historically involved in the trade of oil and other petroleum products. See also Cameron, 2007, p. 25.

[18] Nyssens and Schnichels, 2007, p. 1381.

[19] Bergman *et al.*, 1998, p. 19.

[20] Line-packing refers to "the ability of a natural gas pipeline to effectively 'store' small quantities of gas on a short-term basis by increasing the operating pressure of the pipe". See, Natural Gas Market Report Glossary, http://www.spragueenergy.com/pages/content.aspx?p=Natural%20Gas%20Marketwatch%20Glossary

[21] Cameron, 2007, pp. 22–26.

These require significantly higher storage capacities than their competitors, who benefit from a large and diversified customer base.[22]

Out of the three segments composing the EU gas network, the latter two are at least partly under the direct influence of the EU legislator. Starting with the 1990s liberalization wave – when it became apparent that monopoly was not an intrinsic aspect of the natural gas network – the regulatory design of the gas market has differentiated between competitive and monopolistic components of the gas network. In this sense production and wholesale or retail supply were considered to be competitive parts of the system exposed to free and open competition. At the same time – owing to their high fixed costs of pipeline construction and the relatively low cost of their operation – transmission and distribution continued to be regarded as monopolistic activities, which required reinforced regulation.

Despite this differentiation, the structure of the gas chain continues to entail a vertically integrated character from production to consumption. By a way of consequence, regulatory action in one segment of the chain will most likely have an impact upon the other segments.[23] For example, unbundling requirements under Directive 2009/73/EC are *prima facie* designed for the regulation of transmission system operators within EU Member States. However, according to article 11 of the said Directive, any third-country incumbent who wishes to acquire assets downstream and have its TSO operate on the territory of a Member State must request prior certification for the TSO from the responsible national regulatory authority in the Member State. In this way, it is ensured that both the third-country incumbent and its TSO will respect the unbundling provisions implemented in the Member States, where the respective TSO is supposed to function. Hence, liberalization provisions designed for the regulated segment of the gas chain might exert an influence upon its nonregulated ends.

The interpretation of gas as a network industry brings additional complexity to the legal framework accompanying the natural gas market, by underlining its infrastructure aspect which, in the context of the single market project, has to respond to requirements of interconnection and interoperability of national grids, as well as network access demands.

---

22 *E.ON/MOL*, Case M.3696.

23 Cameron, 2007, p. 25.

2.2 AN HISTORICAL SKETCH OF FACTS AND LAW

Traditionally, each Member State's gas market has been dominated by national monopolies, which have been encouraged by their individual governments. As in the case of electricity, the perception of gas as a service submitted to public service obligations has been dominated by the belief that the influence of market forces should be kept minimal, through state-inflicted regulation.[24] The major advantage of such a monopolistic structure was the fact that it facilitated the investments required to satisfy the European energy demand and to develop the European gas grid. National governments retained an important share in or owned all transmission companies, which in return exercised monopoly on the wholesale of gas and on the operation of transmission systems, as well as monopsony on the buying of gas from third-country producers. Long-term supply contracts between third-country producers and a limited number of EU-based national companies generated and maintained market concentration, by securing the domination of traditional incumbents on their national markets, the partitioning of national markets and the lack of any gas-to-gas competition.[25]

By the time the Single European Act (SEA) was adopted, the European gas market had developed at both national and European level, despite the late breakthrough of European gas field discoveries.[26] Until that moment, natural gas consumption in Western Europe accounted for less than 1 per cent of the total energy consumption[27] and the European gas industry was dominated by manufactured gas. Even after the major discovery, the use of gas throughout the European Union had been limited to high-value applications and was extended to power generation only as late as the end of the 1980s. The motivation arguably lay in the determination of Member States' governments to keep stable the employment numbers in the coal and nuclear industry, as well as to avoid any possible dependency on Russian gas.[28] A further reason for such limitation derived from the fact that gas was perceived as a "premium fuel" owing to its higher energetic value in comparison to other hydrocarbons and the relative limitation of its reserves within Europe. That special

---

[24] Soriano, 2001.

[25] This was achieved through linking residential gas prices to the price of oil.

[26] http://www.geoexpro.com/history/groningen/

[27] Estrada *et al.*, 1988, p. 9.

[28] Spanjer, 2008, pp. 1–11.

status imposed further constraints upon the use of gas for power generation.[29]

The late introduction of the use of natural gas in Europe explains to a certain extent the rigidity and scarcity of the pipeline grid throughout the continent, which determined the fact that most Member States served as both destination and transmission hubs for the gas flux initiated on production sites. At the time as SEA entered into force, the European gas market was segmented by three separate regional grids: a Benelux grid that reached France and Germany, an Italian grid traversing the Southern part of the Union's territory and a Central and Eastern European one.[30] The segmentation was further nurtured by the horizontal integration between incumbents in a certain region of the market, correlated with the vertical integration of production and supply within the same undertaking. The latter case allowed the emergence of entities able to dominate the national production of gas, as well as retain important downstream ownership positions in pipelines and storage operators.[31]

Towards the beginning of the 1980s, pressure started to be exerted upon the traditional patterns of organizing gas markets. That trend appeared in a context where anxieties surrounding the security of gas supply had decreased considerably, mostly owing to an increase in the volume of available gas. As a consequence, the support of producers for the traditional model of government-dominated markets started to collapse. Although not part of the initial Single Market project, energy has always played a significant role in European integration. It came thus as little surprise that, in the aftermath of SEA ratification, the European Commission attempted to quickly bring to the table the matter of opening up gas and electricity markets to competition.[32]

In addition to the former's efforts to enact sector-specific regulations, the European Courts have developed their own case-law on energy issues, and especially those resulting from the specific structure of network industries. In a series of cases originating in the 1990s,[33] the Court of

---

[29] Stevens, 2010, pp. 4–5.

[30] Keyaerts *et al.*, 2008, p. 6.

[31] For example, Shell and ExxonMobil were in charge of the production of gas in the Netherlands and the North Sea. At the same time they retained an important share of Gasunie, and subsequently of a German pipeline and storage operator.

[32] Commission of the European Communities, 1988: The Internal Energy Market, /*COM/1988/238 final*/.

[33] *Commission v France*, Case C-159/94; *Commission v Italy*, Case C-158/94; *Commission v Netherlands*, Case C-157/94.

Justice of the European Union denounced the monopolies over the import and export of gas and electricity, justifying its findings on the basis of article 37 TFEU (former article 37 TEC). This required Member States to adjust progressively any state monopolies of a commercial character, so as to prevent discrimination between Member States' nationals regarding the conditions under which goods were produced and marketed. Notwithstanding the Court's jurisprudence, Talus[34] notes that the institution that has advocated most fervently the idea of an energy internal market has been the European Commission. In its turn, the Court has at times omitted to follow in the steps of the Commission or else has been inclined towards a more cautious approach to the matter. This had been specifically obvious in cases where the Court was asked to assess whether certain Member States respected the proportionality constraints in article 106(2) TFEU (ex-article 86(2) TEC). In those case-laws, the Court expressly stated that it eluded its attributions to undertake an assessment of the means which a Member State could have adopted "to ensure the supply of electricity on national territory on the basis of costs and prices capable of guaranteeing the balanced economic development of the country".[35] In fact, the Court maintained that it was incumbent upon the Commission to prove its allegations that a Member State had not fulfilled its obligations and to provide the Court with the required information, which could enable it to form an opinion whether such an obligation had or had not been fulfilled.[36] In the absence of proofs brought by the Commission, the Court could not be determined to rule against the questioned Member States.

Certainly the 1990s represented the beginning of the European Court's jurisprudence in the energy field and the decisions of the time reflected an attempt by the Court to maintain a balance between the Commission's novel ideas and Member States' sovereign rights upon national energy matters. The cautious approach of the Court was justified by the absence of Treaty provisions guaranteeing supranational competencies in this area, as well as the inability to rely on sector specific secondary law until towards the end of the decade. The following chapters will delve further into the role played by the European Courts in bringing about an integrated gas market.

---

[34] Talus, 2007a, pp. 435–448.

[35] *Commission v Italy*, supra note 33, para. 59. See also, *Commission v Netherlands*, *supra* note 33, para. 63; *Commission v France*, *supra* note 33, para. 106.

[36] *Commission v Italy*, *supra* note 33, para. 55.

2.2.1 Substantive Changes to Gas Market Governance Brought by the Single European Act

It could be argued in neofunctionalist terms that integration in the energy field has been stirred by economic integration in several other sectors, such as agriculture and environment, to which energy is functionally linked.[37] In particular, it appears that around 1990 environmental policy exerted significant pressure towards change within EU energy policy in a direction encompassing a high level of protection of health, environment and consumer interests.[38]

However, it would be inaccurate to limit the explanation accompanying the integration of European energy markets to a mere functional spill-over. Instead, it should be emphasized that SEA brought substantive changes to the European Union's functioning, at the same time as traditional national market structures started to be challenged from within. One of the major developments under SEA, with essential relevance for the establishment of a single energy market, has been the introduction of qualified majority voting in the Council regarding an array of sectors previously reserved to unanimity voting.[39] In this sense, SEA allowed for majority voting when taking decisions regarding the establishment of the internal market.[40] Such a provision facilitated the adoption of harmonization measures across Member States despite the obstacles contained in article 94 TEC (now article 115 TFEU). Following this development the Commission was provided with room for manoeuvre into areas formerly protected by state sovereignty, on the condition that it found enough support for its anti-monopolistic energy initiatives within a majority of Member States.

In addition to the above, SEA brought two other important substantive changes.[41] The first regarded the progressive establishment of an internal market, defined as an area without internal frontiers, in which the free movement of goods, persons and services was guaranteed. The internal market definition seems to be inclusive enough for the case of gas, both in its understanding as a commodity traded on national and European markets and as a service provided to European citizens. The second change regarded the enlargement of EU competences to areas that did not previously enjoy an express Treaty basis. In this respect, the extension of

---

[37] Gehring, 1996, pp. 225–253.
[38] Andersen, 2000.
[39] Craig and de Búrca, 2007, p. 13.
[40] Article 114 TFEU (ex-article 95 TEC).
[41] Craig and de Búrca, 2007, p. 13.

Union competences to research and technological development, and to environmental policy, bestowed added legitimacy upon the European Commission's efforts to establish an internal gas market.

2.2.2 Internal Changes in National Governance Patterns

It should be noted that the adoption of SEA coincided with major transformations in the modes of governance. These occurred as a result of the passage from the positive to the regulatory state. In the new order, "territorial boundaries of states no longer [coincided] with the extent or the limits of political authority over economy and society".[42] As a consequence, a shift of power from public to private actors took place, challenging the known forms of state-democracy. That has determined a global change in styles of governance, resulting in a limitation of the role of the interventionist/positive state and an enhancement of that of the regulatory state.[43] Under the latter, various types of public-sector failure have been identified and publicly owned companies have been brought under scrutiny for being unable to fulfil their social and economic tasks.[44] Against this background, supranational institutional models emphasizing the market organization of energy sectors registered growing support, which facilitated the inclusion of gas on the Single Market agenda by 1988.[45]

In between the 1988 European Commission's Communication on the Internal Energy Market and the 2009 Third Energy Package, which entered into force in March 2011, the European gas market and its afferent legislation underwent a process of continuous transformation. This was blocked at times by the economical and institutional differences between the energy sectors of different Member States.[46] Although the Commission's first liberalization steps regarding the transparency of energy prices[47] and transit[48] did not encounter major opposition from Member States, its initial proposal encompassing a strategy for the

42 Strange, 1996, p. 124.

43 Majone, 1997.

44 Majone, 1997, p. 142.

45 Andersen and Sitter, 2007, pp. 7–8.

46 Correljé, 1997, p. 49.

47 Council Directive 90/377/EEC advocated the idea that transparency of energy prices could contribute to the creation and smooth operation of the internal market.

48 Council Directive 91/296/EEC regarded transit rights for grid operators for the transit of natural gas across the European high-pressure transportation networks.

deregulation and liberalization of national gas markets faced important obstacles raised by both Member States' governments and by the industry.[49] Such response from the gas sector was justified by the residual continuation of a traditional approach to energy supply and distribution, which saw the former as an attribute of the state and of regional monopolies, while binding the latter to the dynamics of transmission networks, essentially controlled by producers.[50]

The 1998 compromise regarding gas market liberalization between the national and the supranational level was forged in the aftermath of the 1996 agreement on electricity liberalization. An important role was played by the Commission's slight change of discourse in a direction that focused on market functionality, security of supply and public service obligations, rather than on deregulation.[51] Following the enactment of the first Directive aimed at liberalizing the traditionally monopolistic gas market, two subsequent legislative packages have been adopted by the Commission in view of achieving a real integrated gas market. As further shown, the three legislative packages bear numerous resemblances to each other. However, with each regulatory step taken, the Commission has gained more and more clarity in defining the rules governing the common gas market. That has translated into stricter unbundling; third-party access and storage requirements; an increased application of competition law to gas long-term contracts and gas mergers; and enforced protection of energy consumers.

2.3 THE EVOLUTION OF GAS REGULATION IN THE EU

Since the 1990s, the European Commission has begun a process of reforming the European energy sectors through the adoption of sector-specific legislation aimed at dismantling long-established monopoly structures in gas and electricity and gradually replacing the state as the sole regulator of energy network industries. Roggenkamp *et al.*[52] consider this task unique among the other examples of EU market integration. Such an observation is only partially true. Simultaneously with the development of the single energy market, parallel market integration processes have been initiated with regard to other network industries and

---

49 Eikeland, 2004, pp. 4–6.
50 Padgett, 1992, pp. 53–76.
51 Andersen and Sitter, 2007, p. 12.
52 Roggenkamp *et al.*, 2007, p. 250.

the financial sector. Traditionally, those have been also considered as regulatory domains of the sovereign state. However, among the above examples, the energy sector seems to have had the slowest advancement towards a liberalized and integrated European market. A potential explanation of this fact could come from a simple anthropological observation. The more developed a society is, the greater its dependency on a stable and continuous supply of energy. In such an environment, energy is not only one of the sources of life, but also the heart of the mechanism sustaining life at a certain level and pushing it towards further progress. Consequently, the transfer of sovereignty over energy matters from the hands of Member States to the EU needed to be tackled cautiously from the very beginning of the single energy market process. At the same time, energy resources across Member States have been characterized by asymmetric patterns of availability and consumption. Such elements have brought additional difficulties for the harmonization of national energy regulatory regimes among themselves and with the EU level.

Until the present day, EU's efforts in creating an internal energy market have been concentrated under three consecutive energy packages. Each package has seen a gradual increase in the opening to competition of the EU energy market, an expanse of the regulatory oversight over market players, a bolder application of competition tools to market-harming behaviour, a reinforcement of the security of supply, a greater protection of consumers and ultimately deeper energy market integration. Before the adoption of the Third Energy Package, an Energy Sector Inquiry[53] (2005–2007) performed by Directorate-General (DG) Competition found that, despite the positive outcomes of the first two energy regulatory reforms, market failures continued to persist. In the aftermath of the Sector Inquiry, a new legislative package was adopted. The main directions taken within this package will be discussed in the following.

2.3.1 The EU Gas Market Under the Third Energy Package

Gas sector regulation in the EU is ensured by means of a Directive and three Regulations.[54] These legal instruments represent the latest stage of EU's gas policy. The new set of legislative measures constitutes the outcome of a series of legal, geopolitical and economic factors that have directly influenced its final shape.

---

[53] DG Competition Report on Energy Sector Inquiry, /*SEC/2006/1724/.

[54] Directive 2009/73/EC; Regulation (EC) No. 715/2009; Regulation (EU) No. 994/2010; Regulation (EC) No. 713/2009.

From a legal point of view, some of the major aspects that contributed to the adoption of the TEP have been

- the 2004 and 2007 EU enlargements, which almost doubled the number of EU Member States and increased the pressure of regulatory convergence among national gas systems, in order to achieve an internal EU gas market;
- the anticipation of the adoption of the Lisbon Treaty, which has been expected to create a concrete legal basis for EU action on the energy market;
- the reinforcement of consumer protection at an EU level;[55]
- global environmental challenges, which required a faster speed to achieve energy efficiency;[56] and
- the strengthening of the use of competition law on the EU gas market, as a result of the adoption of a "modernized" legal framework for competition law at EU level,[57] which allowed for gas investigations carried out through the Commission's 2007 Energy Sector Inquiry.[58]

Among the elements presented above, the recent EU enlargement has produced not only legal, but also geopolitical effects for the TEP. Ten of the 12 new Member States coming from Central and South-Eastern Europe are dependent on Russian gas supply. The degree of dependency ranges from very high levels to absolute.[59] This element – along with the accelerated decline of the North Sea gas production and the UK becoming a net importer of gas – has contributed to the increase in the total dependency of EU on gas imports. The group of new Member States is homogenous with respect to their capacity for influencing EU policy individually, which is very reduced or nonexistent. This places them in a

---

[55] Reflected in the 2008 Commission's Green Paper on Consumers Collective Redress /*COM/2008/794 final*/ and the 2008 Commission's White Paper on Damages Actions for Breach of the EC Antitrust Rules /*COM/2008/165 final/.

[56] Green Paper /*COM/2007/140 final*/.

[57] Mainly this refers to the adoption of Regulation (EC) No. 1/2003 of 16 December 2002 on the implementation of the rules on competition laid down in articles 81 (now 101 TFEU) and 82 (now 102 TFEU) of the Treaty [2003] OJ L 1/1, as amended by Regulation (EC) No. 411/2004.

[58] Article 17 of Regulation (EC) No. 1/2003 allows the Commission to engage in reviews of sectors that appear to resist competition.

[59] Kubcik and Bielecki, 2012.

category of so-called small or minor states.[60] Small or minor states have been found to be more receptive to adopting common European policies, as a result of their inability to achieve certain national interests alone and owing to the direct benefit to them of being an integrated part of supranational institutional arrangements.[61] As a result, the presence of the new Member States in the European Council has tilted the decision-making balance towards a more favourable approach to supranational authority on the EU gas market. Furthermore, in between the entering into force of the Second Energy Package and the adoption of the TEP, EU security of gas supply has been twice severely challenged by Russia. The two gas crises between Russia and Ukraine in 2006 and 2009 underlined the weakness of EU's emergency preparedness and response mechanisms to unforeseen supply disruptions.[62] The immediate EU reaction has been to initiate a policy of reducing gas supply dependency on one supplier by diversifying both energy sources and supply routes.

The economic component of the context leading to the adoption of the TEP is two-fold. First, the recent empowerment of EU consumers has put additional pressure on the traditionally monopolistic gas market to work more efficiently under free competition conditions. This should mean the guarantee of sufficient and stable gas supply at reasonable, but also cost-reflective prices. Secondly, Europe has been facing increasing competition from other regions of the world that are dependent on gas and oil.[63] As such, in order to ensure the security of its gas imports, the EU requires an integrated energy policy, which in time should be completed with an integrated external representation.[64]

In comparison to its predecessors, the new gas legislative package brings forth several novelties, aimed at establishing a real integrated market in gas. In the aftermath of the Energy Sector Inquiry, it became clear that several obstacles preventing the achievement of an internal and integrated gas market were still in place. A considerable number of such obstacles resulted from the lack of harmonious application of the existing

---

60 For more details on the small states theory see Hanf and Soetendorp, 1998, p. 4; Tonra, 2001, p. 47.

61 For more detailed reading see Maass, 2007; Hanf and Soetendorp, 1998, p. 4; Tonra, 2001, p. 47.

62 Commission Staff Working Document: The January 2009 Gas Supply Disruption to the EU: An Assessment, /*SEC/2009/977 final*/.

63 For example, China, India and other Asian markets.

64 Haghighi, 2007, pp. 9–37.

liberalization rules within Member States.[65] Specifically, Member States' divergent approaches could be noted in regard to, among others:

- the level of market opening;
- the degree of independence of national regulatory authorities;
- the position towards traditional models of gas supply and market balancing regimes – exerting a direct influence upon network access availability;
- the protection granted to consumers;
- and the degree of the attachment to services of general economic interest.

The new legislation brings forth both substantive and institutional changes. Regarding the former, it envisages the correction of problems identified during the Energy Sector Inquiry, as well as the increase of competition in electricity and gas markets. Regarding the latter, the legislation modifies the agency structure by strengthening the power of national regulators, on the one hand, and by establishing new regulatory bodies on the other. The goal of market liberalization having been achieved to a certain extent,[66] harmonization and cooperation become key concepts in the new legislation, together with the strengthening of regulatory oversight, transparency and consumer protection.

2.3.2 Selected Topics

The scope of the third gas legislative package extends over common rules for transmission, distribution, supply and storage, network access and security of supply, as well as the organization and the functioning of the gas sector. The developments brought by this package in comparison to

---

[65] For example, see ERGEG's response to the European Commission's Communication "An Energy Policy for Europe", Ref. C06-BM-09-05, 6 February 2007, http://www.ergeg.org/portal/page/portal/ERGEG_HOME/ERGEG_DOCS/ERGEG_DOCUMENTS_NEW/Energy%20documents/C06-BM-09-05_RespToSER_2007-02-06.pdf

[66] As will be further discussed later in the book, despite the complete opening of the energy market, the positive effects of liberalization are hampered by a continuation of anticompetitive behaviour of traditional incumbents, barriers to third-party access, unequal national regulatory authorities, differing or contradictory regulatory regimes across Member States, regulatory lag and poor market interconnection.

its predecessors are three-fold and regard structural, functional and social aspects, inherent to achieving an internal gas market.

A specificity of the TEP consists of the open pursuit of an *integrated EU gas market*. The first two energy packages focused on opening up the European gas markets to competition and referred to market integration as solely a subsidiary achievement deriving from the accomplishment of a liberalized market. The Energy Sector Inquiry concluded, however, that the completion of the liberalization process in the European gas markets had not resulted in market integration.

Under the new legislation, the objective of an integrated gas market is clearly articulated. In this sense, Directive 2009/73/EC requires Member States to pursue regional integration of their national markets and cooperation of system operators at EU and regional level only as an intermediary step preceding a fully integrated common energy market. An integrated gas market is expected to stimulate competition between producers, contribute to allocative efficiency, market size enlargement and expansion of firms' production, and stimulate innovation through dissemination of technologies across Member States.[67] Furthermore, such a market should provide a better response to security of gas supply crises[68] and should ensure the gradual absorption of "isolated systems" forming "gas islands" in the Union.[69]

In order to attain an integrated market, the TEP provides legal instruments to achieve harmonization of cross-border regulatory rules. These are EU-wide network codes, made legally binding by the Commission through the comitology procedure, and nonbinding framework guidelines, developed by the Agency for the Cooperation of Energy Regulators (ACER). Additionally, regional cooperation and regional market development are reinforced through the imposition of cooperation obligations on Member States, NRAs and other network operators. Despite such positive developments, at the moment it is still too early to anticipate the effect of these measures upon the progress of market integration. Specifically, while a closer look at the European gas market reveals residual divergent approaches of Member States and major national stakeholders to releasing control over energy assets. This case is

---

[67] Geradin, 2004.

[68] Article 6 of Directive 2009/73/EC and article 194 TFEU provide Member States with guidelines regarding energy market integration and reaction ahead of security of supply threats. So far, the effectiveness of the new provisions has not been yet tested and it is difficult to anticipate how the Member States will react when faced with a new security of gas supply crisis.

[69] Article 7 of Directive 2009/73/EC.

illustrated by the coexistence of three unbundling models under the new regulatory regime.

One of the most important features of the new package regards the introduction of *effective unbundling* in the EU gas market.[70] This represents a step forwards from the legal and functional unbundling fostered by the Second Energy Package,[71] which proved to be inefficient in opening up the EU gas market to competition.[72] Unbundling refers to the separation of competitive activities linked to supply and production from the more monopolistic segments of the gas chain – the transmission and distribution network, storage and LNG facilities. Its necessity is justified by the following scenario. When market players control transmission and distribution systems, there is a tendency of incumbents to prevent third-party access to transmission networks, or to allow such access in exchange for excessive tariffs paid for network use.[73] Effective unbundling ensures that the profit gained in the latter case is not reinvested in the competitive side of the business, thus generating an unjust advantage, positioning incumbents above their competitors.

Following comprehensive deliberations at the level of market participants and stakeholders[74] and political discussions in eight Member States on the original proposal of the European Commission, Directive 2009/73/EC foresees three options for unbundling.[75,76] Member States have a choice between the most radical separation of network activities affecting the structure of ownership in vertically integrated undertakings and less expropriating measures, entailing in return higher levels of regulatory supervision.[77]

The new rules applicable to unbundling are detailed in articles 9–23 of Directive 2009/73/EC. According to article 9(1)(b), the same person is not entitled to exercise direct control over an undertaking performing any

---

70 Recital 6 of Directive 2009/73/EC.

71 Recitals 3 and 10 and article 9 of Directive 2003/55/EC.

72 European Commission, 2006: Energy Sector Inquiry Preliminary Report, p. 5, http://ec.europa.eu/competition/sectors/energy/inquiry/preliminary_report_1.pdf

73 See case COMP/39.316, *GDF gas foreclosure* and case *E.ON gas foreclosure*, COMP/39.317 regarding third-party access refusal.

74 France, Germany, Austria, Bulgaria, Greece, Latvia, Luxembourg and Slovakia.

75 Kroes, 2007.

76 Proposal for a Directive of the European Parliament and of the Council amending Directive 2003/55/EC […] {SEC(2007) 1179} {SEC (2007 1180) /*COM/2007/0529 final –2007/0196/COD */.

77 Article 9(8) of Directive 2009/73/EC.

of the functions of production or supply and at the same time exercise any right over a transmission system operator or a transmission system. The incompatibility of ownership rights thus works both ways. This provision represents the essence of the ownership unbundling model and the trigger of most academic controversies around it, which question its legitimacy in the light of property rights.[78] Despite such controversies, this continues to be the most endorsed type of unbundling at EU level, owing to its anticipated impact on third-party access to the gas network, on investments facilitating market interconnection and on the enhancement of social benefits – for example, consumer protection and security of gas supply.[79] Currently, this is already the practice in UK, Romania, Denmark, the Netherlands, Spain and Sweden. According to recital 18 of Directive 2009/73/EC, Member States have a right to opt for full ownership unbundling, which ensures that no undertaking operating on their territory may set up an independent system operator or an independent transmission operator. At the same time, all undertakings concerned must respect the provisions of article 9(2)(b). These provisions have an added significance for third-country suppliers, which despite being subjected to EU competition rules while supplying EU consumers, still maintain a monopolistic structure at their domestic level and might be tempted to replicate such a model within the EU gas infrastructure.

As an alternative to ownership unbundling, the Commission suggested in its original proposal the establishment of an independent system operator (ISO). It is expected that the ISO model will also prove to be effective in removing any conflicts of interest between producers, suppliers and transmission system operators to create incentives for necessary investments and to guarantee the access of new market entrants. According to article 14 of Directive 2009/73/EC, such an operator must be independent from supply and production interests. This does not contradict the property rights of the vertically integrated undertaking, which may maintain its ownership of network assets. Article 14 presents the ISO as an undertaking or entity, entirely separated from the vertically integrated company and expected to perform all of the functions of a network operator. To guarantee its independence, detailed regulation and extensive regulatory control mechanisms are put in place through article 14(6). The regulatory tasks are to be exerted by both national competition authorities and national regulatory authorities, together.

---

[78] See, e.g., McHarg *et al.*, 2010; Baur, Pritzsche and Pooschke, 2008.

[79] For example, European Parliament resolution of 10 July 2007 [...], OJ C 175 E, 2008 7 10, p. 206; Commission Staff Working Document Accompanying the legislative package on the market for electricity and gas, /*SEC/2007/1179/.

The third unbundling model refers to the setting-up of an independent transmission operator (ITO). Resulting from article 17 of Directive 2009/73/EC, the features of ITO are similar to those of ISO. However, this option does not involve either ownership separation or an independent system operator. Instead, according to article 20 of the Directive, the transmission system operator is equipped with a Supervisory Body in charge of taking decisions with significant impact on the value of assets of shareholders. Such decisions regard specifically the approval of the annual and longer-term financial plans, the level of indebtedness of TSOs and the amount of dividends distributed to shareholders. Such decisions will not include day-to-day activities of TSOs, the management of the network and the activities necessary for the preparation of the 10-year network development plan. In order to ensure the independence of ITO,[80] additional regulatory arrangements are required, including a pillar of organizational measures relating to the governance of TSOs. Such measures regard investment, connecting new production capacities to the network and market integration through regional cooperation. The regulatory supervision of ITO is stricter than the supervision that occurs within the previously described models. Accordingly, the responsible NRA closely monitors its activities, especially in those cases where such activities include financial and commercial relations with the vertically integrated undertaking. Additionally, the ITOs are supposed to submit any financial agreement for approval and notify the NRAs of their financial resources.

According to the Commission's impact assessment of the new gas legislation, the implementation of ownership unbundling requires a lower level of regulation than that of the ISO or ITO models.[81] Both ISO and ITO options entail higher regulatory burdens. In the former case, the designation of ISO by national regulatory authorities requires prior approval from the Commission. Prior approval is required also for an ISO's decisions on financing investments in transmission capacities. Such decisions are binding upon network owners, but they have to comply with a 10-year network investment plan proposed by the national regulator. The ITO case is made all the more cumbersome by additional regulatory arrangements accompanying market operations and stricter and more comprehensive supervision.

---

80 Articles 18 and 19 of Directive 2009/73/EC.

81 Commission Staff Working Document [...] /*SEC/2007/1179/, *supra* note 79.

The new unbundling rules regard not only the transmission system as seen above, but also the distribution system and storage. Unlike its predecessors, Directive 2009/73/EC makes a clear difference between the unbundling obligations imposed on transmission system operators (TSO) and those that apply to distribution system operators (DSO). Stricter unbundling conditions are imposed on TSOs, while it is considered that these operators are the ones most likely to be influenced by production interests and congestion issues. Therefore, third-party access discrimination has more chances to occur within a TSO than within a DSO. Under the new Directive functional and legal unbundling rules imposed on DSOs are considered adequate to ensure effective unbundling. However, having regard to Member States' discretion in choosing an unbundling regime and in granting exemptions in accordance to article 26(3) of Directive 2009/73/EC, the legislator has linked the effectiveness of such unbundling to proper implementation, close monitoring and a more clear definition.[82]

A further regulatory novelty brought by the Third Package regards the clarification of provisions regarding access to storage and ancillary services, such as LNG storage, as means of both transmission support and line pack. Storage plays an important role in the guarantee of security of gas supply, specifically in Member States with high levels of dependency on natural gas imports.[83] Regulation (EC) 715/2009 aims among other things at setting nondiscriminatory rules for access conditions to LNG and storage facilities.[84] In view of enhancing the enforcement of storage provisions, the 2009 gas Directive requires Member States to define and publish a new and nondiscriminatory clear framework that determines the appropriate regulatory regime applicable to storage facilities.[85] Under article 33 of the Directive, access to storage may occur either in a regulated or a negotiated manner. For the first time, storage system operators are required to comply with legal unbundling provisions,[86] in view of their increased independence. Furthermore, the access of storage operators to valuable market information is enhanced, in direct connection with the Commission's demand for more transparency of market operations.[87]

---

82 Article 26 of Directive 2009/73/EC.
83 Regulation (EU) 994/2010.
84 Article 1(b) of Regulation (EC) 715/2009.
85 Article 33(1) of Directive 2009/73/EC.
86 Article 15 of Directive 2009/73/EC.
87 Recital 24 of Directive 2009/73/EC.

Moving away from unbundling, the second structural development brought about by the TEP refers to the reinforcement of *market interconnection*. Of relevance here are the dispositions of Directive 2009/73/EC regarding investments in new infrastructure, those of Regulation (EC) 715/2009 regarding network access and specifically the establishment of network codes, and those of Regulation (EU) 994/2010 regarding cross-border interconnection. Directive 2009/73/EC and Regulation (EU) 994/2010 link market interconnection to the development of a true internal market in natural gas and a stable gas supply.[88] Article 3(7) of the Directive provides that, where appropriate, Member States may offer adequate economic incentives to ensure the maintenance and construction of necessary interconnection capacity. Recital 18 of Regulation (EC) 715/2009 requires the inclusion of necessary regional interconnection in the 10-year Union-wide network development plan drawn up by the European network of transmission system operators (ENTSO for gas).[89] The necessity of such interconnection is judged according to its commercial and security of supply relevance.

Such provisions should be read in conjunction with article 36 of the Directive, which refers to exemptions in case of investments in new infrastructure. Under the new text, exemptions provided for in Directive 2003/55/EC will continue to apply until the scheduled expiry date indicated in the exemption decision. However, additional comments are made on the exemption regime, in order to avoid biased decisions. National energy regulators are able to grant only partial or temporary derogations, in particular with respect to new pipelines transporting gas from third-country suppliers into the EU. Furthermore, where the new infrastructure project represents an interconnector between two or more Member States, the Agency for the Cooperation of Energy Regulators established through Regulation (EC) 713/2009 becomes the last-resort institution handling the exemption request. This provision allows more legal certainty with regard to market interconnection projects, by adding a supranational level to the decision-making process. ACER is expected not only to adopt impartial decisions,[90] but also to forge more harmonization between regulatory regimes across Member States[91] and thus prepare the ideological background required for more structural market integration.

---

[88] Recitals 40 and 57 of Directive 2009/73/EC.

[89] Article 5 of Regulation 715/2009.

[90] Articles 1–15 and 18 of Regulation (EC) 713/2009.

[91] Through establishing a cooperation framework for national regulatory authorities. Article 7(3) of Regulation (EC) 713/2009.

The tendency towards an integrated approach to market interconnection is confirmed in Regulation (EU) 994/2010. According to recital 18 of the Regulation, where new cross-border interconnections are needed or existing ones must be extended, the Member States concerned, the Competent Authorities and the national regulatory authorities are expected to engage in close cooperation from an early stage. In order to foster market interconnection, new infrastructure should be designed with the aim of ensuring bi-directional flow at cross-border points.[92]

Transmission system operators have always played an important role in market interconnection. Under the TEP, however, this role is expanded and modified, as a direct consequence of new unbundling rules and the reinforcement of market interconnection requirements. TSOs thus become responsible for enabling permanent bi-directional capacity on all cross-border interconnections between Member States which are not leading to production, LNG or distribution facilities and which are not exempted from such an obligation.[93]

Of particular interest for the present subsection is the duty of gas TSOs to cooperate at EU and regional level within the ENTSO for gas.[94] Additionally, NRAs and Member States are required to facilitate the cooperation of TSOs at regional level and foster the consistency of their legal, regulatory and technical frameworks.[95] *Increased regional cooperation* – fostering deeper market integration – between homologous bodies in the Member States and at EU level represents a key feature of the functional changes brought by the TEP. The areas of cooperation between TSOs regard the drafting and adoption of network codes, monitoring the implementation of network codes, coordination of network exploitation and joint investment planning. The role of ENTSO for gas encompasses the elaboration of network codes regarding network security and reliability rules; network connection rules; third-party access rules; data exchange and settlement rules; operational procedures in an emergency; capacity-allocation and congestion-management rules; rules for trading related to technical and operational provision of network access services and system balancing; transparency rules; balancing rules; rules for the harmonization of transmission tariff structures; and rules regarding the energy efficiency of gas networks.[96] Furthermore, according to article 8(3) of the 2009 gas Regulation, the major tasks with

---

[92] Article 6(5)(b) of Regulation (EU) 994/2010.
[93] Article 5(5) of Regulation (EU) 994/2010.
[94] Article 12 of Regulation (EC) 715/2009.
[95] Article 7(2) of Directive 2009/73/EC.
[96] Article 8(6) of Regulation (EC) 715/2009.

which ENTSO is entrusted regard the adoption of common network operation tools for the coordination of network operations in normal and emergency conditions; a nonbinding EU-wide 10-year network development plan; and recommendations relating to the coordination of technical cooperation between Union and third-country TSOs. Through the elaboration of network codes, ENTSO plays an important role in the resolution of cross-border network issues and in establishing an integrated gas market.

In similarity to the cooperation of TSOs within ENTSO, NRAs are expected to cooperate on cross-border issues within ACER. Resulting from article 1(2) of Regulation (EC) 713/2009, ACER has been established with the purpose of assisting the regulatory authorities at EU level in exercising their regulatory tasks performed in their respective Member States and, where necessary, to coordinate their actions. Article 7(3) of the Regulation provides that the Agency should offer a framework in which NRAs may collaborate. The tasks of the Agency regard the cooperation of TSOs, cooperation with and assistance given to NRAs, the setting of terms and conditions for access to and operational security of cross-border infrastructure, monitoring and reporting on energy sectors, and other tasks – for example, decisions on exemptions from third-party access to new cross-border infrastructure.[97]

Along with the establishment of the Agency, NRAs see their competencies increased. In the aftermath of the Energy Sector Inquiry it has become clear that the minimal requests for NRAs powers, independence and discretion were insufficient. The effectiveness of regulation was frequently hampered owing to the inability of energy regulators to take decisions in relation to all regulatory issues. Article 39(4) and (5) of Directive 2009/73/EC provides a series of guarantees for the independence of NRAs from their national governments. These include:

- legal and functional independence of NRAs;
- independence of their staff and managerial body from any market interest or any instructions from a private or public body, including the government;
- possibility for NRAs to take autonomous decisions independently from any influence;
- separation of annual budget allocations and autonomy in the implementation of the allocated budget and own human and financial resources;

[97] Articles 5–9 and 11 of Regulation (EC) 713/2009.

- fixed appointment of five to seven years of the members of NRAs' board and of the top management, which should function on an appropriate rotation scheme.

The role of NRAs is among others to ensure that balancing tariffs are nondiscriminatory and cost-reflective, in order to allow market access for all market players, including new entrants. They should be able to fix and approve transmission and distribution tariffs or at least methodologies for calculating these tariffs, keeping in mind the nondiscrimination requirements mentioned above.[98] Under the new legislative package, NRAs should have power to impose binding decisions on gas undertakings, to sanction or impose penalties on those undertakings, or to propose that a competent court apply such penalties on the said undertakings.[99] Moreover, the 2009 Directive provides that energy regulators can request information from natural gas undertakings, settle disputes and make appropriate and sufficient investigations. Such provisions curtail the discretion of Member States in allocating regulatory powers to NRAs. In this respect, under the new Package all NRAs are expected to be granted the powers listed above as a minimum basis for regulatory harmonization across the EU.[100] NRAs are also expected to decide on appropriate measures ensuring customer benefit through the promotion of effective competition.[101] Regulators should be granted powers to ensure high standards of public service in compliance with market opening, as well as full effectiveness of consumer protection measures,[102] including protection of vulnerable customers.[103]

Such a protection remains at the heart of EU energy policy and is enhanced under the TEP. Before 2008, the only sector-specific rules addressing energy consumer rights were those stemming from the electricity Directive 2003/54/EC and the gas Directive 2003/55/EC. Those, however, lacked clear guidelines on the enforcement of consumers' rights and access to justice. Article 3 of Directive 2003/55/EC stated that Member States should ensure high levels of consumer protection regarding, among others, dispute settlement mechanisms. Nevertheless, it remained vague as to what those mechanisms were. In a

---

98 Article 41(6) of Directive 2009/73/EC.

99 Article 41(4) of Directive 2009/73/EC.

100 Article 41(4) of Directive 2009/73/EC.

101 Article 40(g) of Directive 2009/73/EC.

102 Article 41(1)(o) of Directive 2009/73/EC.

103 The specificity of this category of consumers will be further detailed in Chapter 5.

similar manner, the provisions of Annex A on measures on *consumer protection* mentioned only that consumers had a right to a gas supply contract, which indicated the method used to initiate procedures for settlement of disputes, without going into further details. Certainly, the above quoted Directives could be read in connection to general consumer law principles[104] – in order to gain more insight on consumer redress procedures – as well as Treaty provisions, requiring that consumer protection should be taken into account when defining and implementing other Union policies and activities.[105] However, the lack of cross-reference between these multiple strands of legislation has favoured the maintenance of an opaque access of energy consumers to the legal tools protecting their rights.

Cseres observes that, in a liberalized market where market failures persist, consumers are very likely to face overdue or inadequate legislation aimed at protecting and assisting them, with questionable effective enforcement, information asymmetries, unfair trade practices, unfair standard contract terms, high switching costs and imperfect decision-making processes.[106] The TEP attempts to change this pattern, by introducing several novelties regarding the subject matter. According to recitals 1 and 3 of Directive 2009/73/EC, all consumers benefit from equal protection and the advantages of a competitive market. Furthermore, supplier switching procedures are rendered more transparent and efficient,[107] the quality of the provided gas service becomes a central responsibility of natural gas undertakings[108] and information asymmetries are reduced.

Of a particular relevance in this sense is the obligation of Member States to ensure transparency of supply contracts signed between consumers and their gas providers.[109] The increase in contractual transparency presupposes that consumers have access to their consumption data and associated prices and to all necessary information regarding

---

[104] For example, those contained in Council Directive 93/13/EEC and Directive 97/7/EC of the European Parliament and of the Council.

[105] Article 12 TFEU.

[106] Cseres, 2008, p. 77.

[107] Article 3(6)(a) of Directive 2009/73/EC provides that the switch should be performed by the responsible operator in a deadline of three weeks while Annex I to the Directive stipulates that a final statement of the switching should be given to the consumer within six weeks after the actual switch has taken place.

[108] Recital 48 of Directive 2009/73/EC.

[109] Article 3(3) of Directive 2009/73/EC.

their rights, the current legislation and the dispute settlement means and mechanisms.[110] Hence, article 3(9) of the 2009 Directive requires Member States to ensure the provision of single points of contact, designed to provide consumers with all necessary information regarding their rights, current legislation and means of dispute settlement. Furthermore, article 3(12) foresees the adoption of a clear and concise checklist of energy consumer rights, which should be made publicly available by the Commission and relevant stakeholders.[111] Despite the obvious legal limitations of such a checklist, its added value consists nevertheless of providing consumers with a structured representation of what market liberalization benefits they can take advantage of.

In close relationship to the reinforcement of consumer protection, the new legislative package has adopted a comprehensive approach to the *security of gas supply*. Regulation (EU) 994/2010 establishes rules for ensuring a secure gas supply to customers that should be protected.[112] In particular, these rules refer to consumers that find themselves under severe conditions.[113] The premise for reinforcing security of supply provisions draws on the importance of gas in the EU energy mix, the increase of gas consumption in the Union, the existence of so-called gas islands[114] and the lack of a crisis response mechanism among Member States.[115]

One of the paths of ensuring a sufficient level of security of gas supply consists of the real integration of Member States' isolated markets within the EU gas market. The argument is sustained by the real connection that exists between the security of energy supply, the efficient functioning of the internal market and the integration of said isolated markets. The existence of such a connection is evidenced by the requirements of

110 Recital 47 and article 3(9) of Directive 2009/73/EC.

111 This latter provision seems to derive from the Commission's proposal of setting up a Charter on the Rights of Energy consumers, aimed at encouraging effective enforcement of consumer rights in the Member States.

112 Such customers are defined in article 1 of Regulation (EU) 994/2010 and include households and, depending on the role of gas in the national energy mix, district heating installations, small and medium-sized enterprises and essential social services.

113 For example, seven days of temperature peak, 30 days of exceptionally high gas demand or disruption of the single largest gas infrastructure under average winter conditions. Article 8(1) of Regulation (EU) 994/2010.

114 Member States dependent on the supply of gas from a single external source and generally lacking the necessary infrastructure links with other Member States.

115 Preamble of Regulation (EU) 994/2010.

2009 Directive, which demands that Member States ensure that national markets are integrated at one or more regional levels and that gas islands in isolated regions – such as the Baltic countries – are not left aside.[116]

Closely related to the regional integration of Member States' gas markets is their obligation to cooperate for the promotion of regional and bilateral solidarity. Such a cooperation is necessary in order to safeguard a secure gas supply throughout the internal market.[117] It is furthermore needed in more exceptional cases of severe supply disruption, which affect one or more Member States. In the latter case, there is a demand for the coordination of national emergency measures, as well as for the identification, development and upgrading of natural gas interconnections. Complementary to this, an obligation for mutual assistance is formally put in place, in order to avoid the vulnerability of a segmented EU gas market against the political volatility of its external suppliers. In this respect and matching the solidarity requirements of the Lisbon Treaty, Directive 2009/73/EC requires Member States to establish a mechanism for monitoring the security of supply, in view of avoiding the negative effects of a potential gas crisis.[118]

The responsibility for ensuring the security of gas supply is shared between Member States, gas undertakings and the Commission.[119] As a measure to prevent emergency situations from occurring or from producing negative effects upon the gas supply to the EU, article 3(2) of Regulation (EU) 994/2010 provides for the establishment of national preventive action and emergency plans, based on a detailed assessment of the risks presented by the situation. Such an assessment takes into consideration all relevant national and regional circumstances, various scenarios of exceptionally high gas demand and supply disruptions, interaction and correlation of risks with other Member States and the maximal interconnection capacity of each border entry and exit point. It is important that national emergency plans do not endanger the security of supply of other Member States or of the EU.[120]

In case of a supply crisis, three levels have been identified: early warning, alert and emergency.[121] Member States are expected to collaborate with each other at any of these levels. Under the new rules, the Commission may declare a Union-wide or a regional emergency, if one

---

[116] Article 7(1) of Directive 2009/73/EC.

[117] Article 6 of Directive 2009/73/EC.

[118] Article 5 of Directive 2009/73/EC.

[119] Article 3(1) of Regulation (EU) 994/2010.

[120] Article 10(7)(b) of Regulation (EU) 994/2010.

[121] Article 10 of Regulation (EU) 994/2010.

of the competent authorities of a Member State has declared a national gas supply emergency. The possibility becomes an obligation when two or more Member States find themselves in an emergency situation.[122]

2.4 THE LEGAL BASIS FOR EU ACTION ON THE GAS MARKET

The Union's competences over matters formerly making the object of Member States' sole interest and sovereignty – and including energy policy – have expanded over the years. This happened as a result of the gradual migration of such matters from the national to the EU level, the influence of the European Court's case-law and the increasing necessity of EU's actions to achieve Treaty objectives.

Before TFEU entered in force, the Union was able to act in relation to the EU gas market only by making use of its competences in areas touching upon energy matters[123] or by adopting specific sector legislation, on the basis of its competence in the internal market (Title I TFEU) and the environment (articles 191–193 TFEU). Starting with the Maastricht Treaty, certain provisions hinted at measures in the field of energy on behalf of the Union.[124] However, the sole mention of energy matters in the article could not constitute a basis for EU energy competences.[125]

Unlike its predecessors, TFEU contains an entire title dedicated to energy, which provides the legal basis for Union action in this field.[126] Moreover, the Treaty recognizes explicitly the international legal personality of the Union,[127] as well as the capacity of the Union to conclude international agreements on different matters,[128] including issues related to the gas market. Such an observation confirms the strengthening of

---

[122] Article 11 of Regulation (EU) 994/2010.

[123] Deriving from rules on the free movement of goods (Title II TFEU), the right of establishment (articles 49–55 TFEU), free movement of capital and services (articles 55–66 TFEU), rules on competition (articles 101, 102 and 106 TFEU) and rules on state aid (article 107 TFEU).

[124] Article 3(1)(u) TEC.

[125] Kahl, 2009.

[126] Title XXI TFEU.

[127] Article 47 TFEU.

[128] Article 216 TFEU.

EU's competences on energy matters, in close relation to the specifics of the geopolitical context in which they have been adopted.[129]

Presently, the EU's direct competences on the gas market are explicitly stated in article 4(i) TFEU on shared competencies between the EU and the Member States and in article 194 TFEU on EU energy policy. Article 194 TFEU, specifically, provides that Union policy regarding energy is four-fold and regards:

- the ensuring of the functioning of the energy market;
- the ensuring of the security of energy supply in the Union;
- the promotion of energy and energy saving, including the development of new and renewable forms of energy;
- and the promotion of the interconnection of energy networks.

All of these objectives are to be pursued by the Union in a spirit of solidarity between Member States. However, according to article 194(2) TFEU, Member States preserve their sovereign right to autonomously establish their national energy mixes. This provision hints at a vertical division of competences between the Union and the Member States in the field of energy. As such, the action of the Union on the European gas market must occur in accordance with the principles of conferral, subsidiarity and proportionality. The conflict between the articulation of EU competences in article 5 TEU and the reality of the institutional practice are not the subject of this book. Certain aspects have, however, been thought useful for the following discussion and hence will be briefly discussed.

According to article 5(1) and (2) TEU, the Union is deemed able to act only within the limits of competences conferred to it by the Member States. The principle of conferral is applicable both to internal and external legal acts of the Union[130] (the latter regarding relations with third countries and international organizations). In the interpretation of the Court of Justice, the Union's competences are considered specific and cannot be extended beyond the general framework of the Treaty.[131] This means that, in the absence of a clear attribution of competences upon the

---

[129] For example, the three major gas crises generated by Russia, as the biggest gas supplier of the EU, in response to transit problems through Ukraine and Belarus.

[130] Delvaux and Guimaraes-Purokoski, 2008.

[131] *Accession by the Communities to the Convention for the Protection of Human Rights and Fundamental Freedoms*, Opinion 2/94, paras 25 and 30.

Union, these remain with the Member States. However, each modification of the Treaty that saw an enlargement of EU's competences did not happen overnight. Rather, as in the case of energy, the Union has made use of its competences in adjacent areas to increase its margin for manoeuvre with regard to particular issues.

Articles 114 TFEU (former article 95 TEC) and 352 TFEU (former article 308 TEC) allow for "creeping competences". Article 114 TFEU regards the adoption of harmonization measures in relation to the establishment and functioning of the internal market in gas. This is a general Treaty basis, considered also as a residual provision, which only applies save where otherwise provided in the Treaty. Of high relevance for the understanding of the use of article 114 TFEU is the *Spain v Council* case-law.[132] In this case-law the Court maintained that the harmonization power of the internal market could be used to prevent any potential future obstacles to free movement or undistorted competition.[133] The critique brought to the use of article 114 TFEU, however, was that any national diversity of laws could be regarded as an obstacle to internal market principles. As such, the competence of EU could be virtually extended to any subject.[134] That did not mean, however, that the Union could automatically assume competence over new matters without providing a precise legal basis. The imperious requirement of the latter was established by the Court of Justice of the European Union (CJEU) in its *Commission v. Council* judgment, which maintained that neither the parties concerned nor the Court should be "left uncertain as to the precise legal basis" for Union action.[135] The result of the Court's jurisprudence[136] is that the provisions of article 114 TFEU are of strict interpretation and should not be regarded as creating a general power for the Union to regulate the internal market. Moreover, the Court ruled that a measure adopted on the basis of article 114 TFEU should genuinely aim at improving the conditions for the establishment of the internal market.[137]

Resulting from *Commission v. Germany* and more recently from *Kadi*,[138] the strict approach to the interpretation of Union competences

---

[132] *Spain v Council*, Case C-350/92.

[133] *Ibid.*, para. 35.

[134] Weatherill, 2011.

[135] *Commission v Council*, Case 45/86, para. 9.

[136] Case C-376/98 *Germany v Parliament and Council.*

[137] Case C-376/98, *supra* note 136, paras 83–84.

[138] *Commission v Germany*, Case C-103/01; *Kadi and al Barakaat International Foundation v Council*, joined cases C-402/05 and C-414/05.

has, however, not been constant in the Court's case-law. Specifically in the latter case, the Court held that the Union had competence to impose sanctions on individuals, despite the *lex specialis* of 75 and 301 TFEU (ex-articles 60 and 301 TEC). In fact, the application of the general law in this particular jurisprudence resulted in the modification of the provisions of article 75 TFEU, in a manner reflecting the position adopted by CJEU in its decision.

In addition to article 114 TFEU, article 352 TFEU has also been used extensively by the Council and CJEU to extend the Union's competences beyond Treaty provisions. According to this article, if action from the Union is necessary within the framework provided by the policies defined in the Treaties, in the absence of specific powers mentioned in EU legal texts, the Council, acting unanimously at a proposal from the Commission and after consulting the Parliament, shall adopt the appropriate measures. The article is regarded as a loop-hole provision, which allows the Union to adopt legal measures in order to attain Treaty objectives in those cases in which it does not benefit from a Treaty basis to do so. The broad generality of Treaty objectives with regard to the common market and the "ever-close Union" provided the grounds for the almost limitless extension of the latter's competencies in matters reserved to state sovereignty.[139] Article 352 TFEU refers to neither internal nor common market. Instead it makes reference to the policies defined in the Treaties. This represents a further extension of the scope of application of free movement and competition principles. Until 2002, around 700 legislative measures had been adopted based on article 352 TFEU.[140] An important limitation to the continuation of the use of article 352 TFEU as a legal basis for the Union action on the European gas market derives from the jurisprudence of the Court of Justice. The approach of the latter has been that, in areas regulated by specific sector legislation, article 352 TFEU may not be used.[141] As a result of the long-standing jurisprudence of the Court, article 352(3) TFEU has included a new provision according to which the use of the article cannot lead to harmonization of Member States' laws or regulations in cases where the Treaties exclude such a harmonization. In the case of the EU gas market, the harmonization of national laws is certainly not excluded by the Treaty. However,

---

139 Conway, 2010, p. 969.

140 Schütze, 2009, p. 134.

141 The application of *lex specialis* precludes the application of *lex generalis*. *Hauptzollamt Bremerhaven v Massey-Ferguson*, Case C 8/73; *Spain v. Council*, *supra* note 132.

the TEP provides enough solid bases for regulatory harmonization to render the use of article 352 TFEU on the gas market somewhat obsolete.

With respect to the application of subsidiarity to EU competences on the gas market, article 5(3) TEU provides that, in areas which do not fall in its exclusive competence, the Union shall act only in so far as the objectives of the proposed action cannot be sufficiently achieved at the level of Member States and require Union intervention. The article contains a double condition justifying the legitimacy of the Union's intervention. The proper attainment of certain objectives by the sole action at Member States level must be impossible. Complementary to this, Union action should prove a superior efficacy in reaching the said objectives. The principle has attracted vivid critiques, based on the reasoning that, when compared with the purposes of a common market without barriers, Union intervention will be always considered necessary.[142] Such an opinion is supported by the approach taken by the Court of Justice in *Commission v. Germany*.[143] The case underlined that the coexistence of different national provisions governing the same legal matter could potentially harm the common market, hence favouring the emergence of harmonizing Union action. The subsidiarity principle has exerted a particular significance for the development of an EU gas single market. On the one hand, at the beginning of the common energy policy, it allowed the Union to make use of implicit competences – deriving from the inability of Member States' action to properly achieve Union objectives – where it lacked explicit ones. On the other hand, it also limited its scope to impose strict regulatory measures on the harmonization of Member States' gas markets.[144] Nevertheless, it is important to notice that the subsidiarity principle has been generally perceived as a restrictive clause for Member States' action. In this sense, once the Union has exercised its powers under the Treaty to regulate a specific matter in a certain way, Member States are not allowed to enact measures conflicting with Union regulation.[145] Hence, even the early sector specific regulation, which allowed for a great level of Member State discretion regarding the national application of European gas market rules, could foresee limitations on such a discretion, deriving from Treaty provisions.

---

[142] Conway, 2010, p. 969.

[143] *Commission v Germany*, *supra* note 138.

[144] Commission: Amended Proposals for a European Parliament and Council Directive on common rules for the internal market in gas, /*COM/93/643 final – 1993/0385/COD*/.

[145] Craig and de Búrca, 2007, p. 102.

The principle of proportionality as resulting from article 5(4) TEU requires that the content and form of Union action shall not exceed what is necessary to achieve the objectives of the Treaties. The relevance of the application of this principle to the utilities market regards not only the action of EU to achieve Treaty objectives, but also the necessity and scope of complementary measures of Member States.

In between the period immediately preceding the adoption of the Single European Act and that prior to the entry into force of the TEP, the approach of the CJEU towards the justification of state intervention for reasons of security of supply has changed. In *Campus Oil*,[146] the Court concluded that petroleum products represented a fundamental energy source for a country's development. The case regarded a contestation by Irish oil companies of an order of the Irish government demanding them to purchase at fixed prices about 35 per cent of their supplies from the only state-owned Irish oil refinery. Among its findings, the CJEU deemed the objective of security of energy supply an integral part of the "public security" concept, covered by article 36 TFEU (ex-article 30 TEC). This allowed Member States to adopt complementary national measures required to guarantee public security, despite the application of EU rules to the same matter. In later cases though, the Court seems to have changed its opinion on the necessity of additional national measures. In *Commission v. Spain*,[147] the Court concluded that, although national governments may act to protect a superior public interest, the process of approving or rejecting a merger with Union dimension on reasons of public security must be coordinated with the corresponding European process.

According to article 4(i) TFEU, the EU and its Member States share competences over energy, as well as the internal market, environment, transport and trans-European networks and research and technological development. This is essential because, although the Union benefits from a direct legal basis for regulating the EU gas market under TFEU, it continues to enjoy an implicit scope of action deriving from its competences in the above-named areas. The division of competences between national and supranational levels over these fields has an impact on the final outcome, which is the EU gas policy.

As seen above, despite certain limitations, article 114 TFEU regarding the internal market continues to offer the general basis for the EU

---

146 *Campus Oil Limited and others v Minister for Industry and Energy and others*, Case C-72/83.

147 *Commission v Spain*, Case C-196/07.

creating a single European gas market. Articles 170–172 TFEU on trans-European networks and energy infrastructure provide the basis for promoting the interconnection of gas transmission networks and third-party access to gas infrastructure. At the same time, they codify the special care that should be granted to gas islands or landlocked, peripheral territories, in the sense of article 49 of gas Directive 2009/73/EC. Article 171(3) TFEU provides the ground for Union-wide cross-border infrastructure projects with third-country suppliers aimed at ensuring the security of gas supply towards Member States. Furthermore, articles 191–193 TFEU regarding the environment might be seen as a direct basis for the pursuit of energy sustainability by the EU. Of particular relevance for the EU gas market is the formulation of article 191(1) TFEU, which provides that the Union's policy on the environment shall contribute to the prudent and rational utilization of natural resources. This is an important provision in the context of the accentuated depletion of natural gas in the EU and worldwide, as well as the raising dependency of the EU on external gas supply. Article 181 TFEU includes research and technological development on the EU gas market in order to render it more efficient. Additionally, article 122(1) TFEU contains a specific reference to energy solidarity, which allows the Council, on a proposal from the Commission, to decide in a spirit of solidarity between Member States on measures appropriate to difficult economic situations in the area of energy. This could potentially increase the EU's margin of reaction in case of a supply crisis or other emergency affecting security on the gas market.

3. Setting of relationships with natural gas producers

Despite visible integration efforts, the European gas market continues to present itself as a two-dimensional market, encompassing a national and a European level. Although the Member States' gas markets have been opened to competition, they continue to remain different and separated from the neighbouring gas markets.[1] According to Eurostat's most recent report on EU energy and transport in figures, 60.3 per cent of the EU gas consumption in 2008 was ensured via imports, out of which Russian gas exports accounted for 34 per cent of the total quantity imported.[2] Eurostat figures and the International Energy Agency's reference scenario show that Europe is heavily dependent on gas imports, which are estimated to grow from 51 per cent in 2000 to 81 per cent in 2030. Fortunately enough, the EU is surrounded by over 60 per cent of world's gas resources, to which it has access through pipelines or LNG tanks.[3]

The EU's relationship with external gas producers is highly influenced by the geographical distribution of natural gas resources and the specific gas dependencies of the Member States. A quarter of all the energy consumed in the EU is represented by natural gas and almost a fifth of this quantity corresponds to Germany's inland consumption.[4] Some of the Member States possess natural gas resources and others have the potential to develop unconventional gas.[5] The largest gas fields in the EU are situated in the UK and the Netherlands, which in the last 20 years have seen their production declining along with the depletion of the

---

[1] Harris, 2008.

[2] European Commission, Eurostat, 2010: EU energy and transport in figures, http://ec.europa.eu/energy/publications/statistics/doc/2010_energy_transport_figures.pdf

[3] Cligendael International Energy Programme, 2008.

[4] European Commission, Eurostat, 2010: Consumption of Energy, http://epp.eurostat.ec.europa.eu/statistics_explained/index.php/Consumption_of_energy

[5] Nevertheless, the production and supply of unconventional gas do not currently benefit from a specific regulatory framework; instead they are covered by adjacent EU legislation. See Philippe and Partners, 2011.

North Sea reserves. The UK transitioned from a net exporter to a net importer in the winter of 2005–2006,[6] while 75 per cent of the Netherlands gas fields are in decline or in the tail-end production phase.[7] Such a development has rendered the EU gas consumption more prone to import dependency.

In the following, market arrangements will be analysed by looking at EU relationships with natural gas producers and their effect on the regulatory developments occurring on the European gas market.

3.1 LEGAL INSTRUMENTS APPLICABLE TO THE EU'S GAS TRADE RELATIONSHIPS WITH THIRD-COUNTRY SUPPLIERS

Russia, Norway and the Mediterranean region have been traditional energy partners of the EU. Of the three, Russia remains the uncontested largest provider of foreign gas into the Union, followed by Norway.[8] The Mediterranean region, representing the third gas supplier to the EU,[9] also serves as a transit area for energy resources from neighbouring regions.[10] Additionally, in recent years, several European pipeline projects have been developed in an attempt to route gas directly from formerly unexplored partners such as the Caspian–(Central Asia and South Caucasus) and the Gulf region.[11]

---

[6] Energy Charter Secretariat, 2009: Fostering LNG Trade: Developments in LNG Trade and Pricing, p. 8.

[7] EBN, 2010: Focus on Dutch Gas, http://www.ebn.nl/files/ebn_focus_on_dutch_gas_2010.pdf

[8] Almost all Norwegian gas is exported to the EU via the Langeled pipeline, which has five receiving terminals in the Union.

[9] See Commission Staff Working Document /*SEC/2008/2871*/.

[10] For example, the Persian Gulf and the Caucasus region.

[11] Recent developments concern the failure of the Nabucco pipeline project in June 2013 and the signing in 2012 of an Inter Government Agreement and a Host Government Agreement to implement the Trans-Anatolia Gas Pipeline (TANAP) between Turkey and Azerbaijan. The Nabucco pipeline project has been regarded as the EU's most important attempt to diversify its dependency on Russian gas imports. In a nutshell, the project aimed at connecting Europe with the world's largest natural gas deposit in the Caspian/Middle East region. In June 2013, following geopolitical and business pressure, correlated with a nonconsistent response from EU Member States involved, the project collapsed. In September 2013, Turkey appeared to attempt a revival of the project. See also http://www.naturalgaseurope.com/category/pipelines/nabucco-pipeline. The

Relationships between the EU and its external suppliers are complex and entail economic, political and legal aspects. From an economic point of view, external gas supplies contribute to the general welfare and development of EU Member States, while representing at the same time a source of governmental income for the producing countries. This is specifically the case for Russia, for which the revenue from hydrocarbons represents 8 per cent of the country's gross domestic product (GDP).[12] Concerning the political element, this is reflected in the control that producing states are exerting upon their main incumbents and the deeming of the energy sector as a strategic one. Such tendencies can be observed in the case of all EU external suppliers.

The legal aspect is closely related to the political one. A favourable approach to the EU energy market model by non-EU partners allows the development of specific frameworks aimed at institutionalizing market liberalization practices at intergovernmental level. However in its gas trade relations the EU is faced with gas suppliers and producers that emerge from a variety of legal, political and economic systems. Such systems are not necessarily similar to or compatible with the general principles that ensure the functioning of the Union. In certain situations, such ideological differences are accompanied by a volatile political environment, which could act as a deterrent for potential investment and which could put a strain on infrastructure and international industrial cooperation. An example of this is provided by the lack of an EU common strategy for energy cooperation with the countries around the Persian Gulf, despite the large reserves of gas that this region holds.[13]

Apart from Norway, which through its membership in the European Economic Area agreement is obliged to apply most of the EU energy rules, the rest of the main gas-producing countries and their incumbents are not under the direct influence of EU law and have no perspective of future integration in the Union. Nevertheless, a certain segment of

---

TANAP agreement foresees the construction of a gas pipeline from the eastern Turkish border to the western one. The pipeline is supposed to be linked to the expanded Southern Caucasus pipeline and to further transport gas into the EU. TANAP represents the first direct gas corridor between the EU and the Caspian Sea.

[12] If it were to take into consideration only the net contribution of Gazprom. In Grätz, 2009.

[13] The Middle East as a whole is regarded as possessing the largest gas reserves to production ratio in over 200 years, four times bigger than those of Europe and Eurasia put together. See Spanjer, 2008, p. 37.

smaller non-EU energy partners accepts the energy *acquis communautaire*. In this sense, non-EU countries that are members of the Energy Community Treaty[14] demonstrate an integrated commitment towards embracing EU energy policy principles.[15] Such a situation is, however, not unique. The Euro-Mediterranean partnership established in 1995 and re-launched in 2007[16] has provided a framework for the import of European institutional energy market structures into the Mediterranean region. The Mediterranean Transmission System Operator (Med-TSO) is based on the model of EU's European Network of Transmission System Operators (ENTSO) for electricity, and like its forebear it is aimed at and expected to facilitate the integration of electricity systems within its area of operation, through the implementation of common criteria and harmonized, nondiscriminatory and transparent rules of access to grids.[17]

Several legal instruments may be applied to EU's gas trade relationships with its external suppliers. First, energy activities at both global and regional levels are governed by public international law.[18] In this sense, inter-regional trade is the object of several bilateral investment treaties, concluded among states and specifically regarding infrastructure development, stable investment rules and dispute settlement mechanisms. Second, trade relationships with the EU condition third-country suppliers to abide European competition rules. The applicability of such rules to gas trade relations has been underlined in the GFU joint gas sales cases and other cases involving restrictive clauses in long-term contracts between Gazprom and European incumbents.[19] Third, Directive 2009/73/EC requires the certification of all TSOs operating in the EU. In the case of foreign TSOs, such a certification is subject to conformity with the

---

[14] The Treaty entered in force in July 2006 between the EU on the one hand and Albania, Bosnia and Herzegovina, Kosovo, FYR of Macedonia, Moldova, Montenegro, Serbia and Ukraine on the other.

[15] Roggenkamp *et al.*, 2007, p. 211.

[16] European Neighbourhood and Partnership Instrument (ENPI) Regional Strategy Paper (2007–2013) and Regional Indicative Programme (2007–2010) for the Euro-Mediterranean Partnership.

[17] Med-TSO – Mediterranean Transmission System Operators Multilateral cooperation for the integration of the Mediterranean electricity systems, Rome, 4 May 2012. Intervention by Mr Fabrizio Barbaso, http://ec.europa.eu/energy/international/doc/20120504_med-tso_conference.pdf

[18] Roggenkamp *et al.*, 2007, p. 15.

[19] *GFU*, Case COMP/36.072; *Territorial restrictions 1) Algeria gas transport contracts, 2) Expansion of TAG pipeline*, Case COMP/37.811; *Territorial restrictions – Austria (Gazprom/OMV)*, Case COMP/38.085; *Territorial restrictions in Germany (Ruhrgas/Gazprom)*, Case COMP/38.307.

unbundling regime adopted by the Member State, where the said TSOs intend to carry out their operations. Recital 21 of the 2009 Directive demands in a similar line that fully effective separation of network activities from supply and production is reciprocally applied to both Union- and non-Union-based undertakings.

3.1.1 International Law and International Agreements

International law is responsible for regulating inter-state activities – including energy trade – at a global and regional level. The EU – whose legal personality has been expressly recognized by the Lisbon Treaty[20] – is a direct subject of international law and of the rights and obligations that might arise under international agreements to which the Union is a party.[21] In this sense, article 216(2) TFEU provides that international agreements concluded by the Union are binding upon its institutions and the Member States.[22] Furthermore, according to a well-established jurisprudence, international law principles and rules of customary international law are binding upon the EU's institutions and form an integral part of the EU's internal legal order.[23]

The following subsection will study the effect of applying international law rules to EU's gas trade relationships with third-country suppliers. It will do so by looking at rules stemming from the agreements forming the World Trade Organization and the Energy Charter Treaty (ECT). Due attention will be granted to conflicts that might arise from accommodating the autonomy of EU's legal order with the Union's participation in agreements or organizations setting up bodies whose powers might conflict with those of the EU's institutions.

---

[20] Article 47 TEU. Until the Lisbon Treaty the legal personality of the Union was referred to rather implicitly, while only the Community enjoyed explicit legal personality (article 281 TEC).

[21] Any entity possessing separate international legal personality may be a subject of international law and international agreements. See UN, General Assembly, International Law Commission Fifty-fifth session, 2003.

[22] This reaffirms a long-established jurisprudence stating that, once an international agreement concluded by the Union enters into force, its provisions become integral part of the Union's internal legal order. See *Haegeman*, Case 181/73, para. 5.

[23] *Anklagemyndigheden v. Poulsen and Diva Navigation*, Case C-286/90, para. 9; *Racke GmbH & Co. v Hauptzollamt Mainz*, Case C-162/96 5, para 46; *Commission v. Germany*, Case C-61/94.

3.1.1.1 Status of natural gas under WTO law

The World Trade Organization (WTO) is an international organization that deals with the rules and commitments of general trade between nations. Its purpose is the foundation of an open, transparent and nondiscriminatory global trading system.[24] The core of the WTO law stems from a number of "covered agreements", signed and ratified by its Member States. Rules and obligations under the WTO are applicable only to countries that are members to it. Both the EU and its Member States are WTO members. At the same time, five of EU's major gas suppliers enjoy a similar status. These are the Russian Federation,[25] Norway, Nigeria, Qatar and Egypt.[26] Algeria, which is responsible for the largest supplies of LNG towards the Union, is not a part of the WTO.

The main WTO agreements of interest for the EU gas market are the General Agreement on Tariffs and Trade (GATT) 1994,[27] regarding rules on trade in goods; the General Agreement on Trade in Services (GATS), regarding rules on trade in services; and Trade-related Investment Measures (TRIMs), regarding trade-related aspects of investment.[28]

Neither of these agreements refers specifically to energy or even more specifically to gas. The reason for this derives possibly from the fact that original rules on the trade of goods did not provide any specific approach in this sense. Energy was regarded as a strategic resource of nation states, whose inclusion in the GATT provisions would have not benefited the interest of any of the founding members, which were not great energy producers.[29] The situation has significantly changed in recent decades owing to the increasing global awareness of the importance played by

---

[24] About the WTO – A statement by the Director – General http://www.wto.org/english/thewto_e/whatis_e/wto_dg_stat_e.htm.

[25] Since 22 August 2012.

[26] WTO Members and Observers, http://www.wto.org/english/thewto_e/whatis_e/tif_e/org6_e.htm

[27] GATT 1994 placed on a formal treaty footing the provisionally applied GATT 1947, which regarded the reduction and elimination of tariff and nontariff barriers to trade.

[28] There are also other "cover agreements" encompassed by the WTO such as the Agreement on Trade Related Aspects of Intellectual Property Rights (TRIPs), Technical Barriers to Trade (TBT) and Sanitary and Phytosanitary Measures (SPS).

[29] Van Hoorn, 2009, p. 62.

energy and its trade in general economic and political development and the modification of interests represented in the WTO.[30]

The first WTO jurisprudence regarding energy dates from as early as 1996.[31] Since 2002, several proposals have been formulated by different WTO members to include references regarding "pipeline and grids" and goods transported via fixed infrastructure among GATT's provisions regarding transit.[32] Despite such a development, explicit mentions of "energy", "energy resources" or "energy trade" continue to be lacking from the WTO agreements. Furthermore, no energy-specific sector agreement has been put in place. GATS sector-specific rules refer to the movement of natural persons, financial services, telecommunications and air transport services.

In the absence of particular rules regarding energy, natural gas is submitted to general rules regarding trade. As a commodity, which can be stored and traded in both gaseous and liquefied form across borders, the trade of gas is regulated by GATT rules. However, as seen in the above subsections, natural gas can be perceived as both a good and a service. As such, its transmission, distribution and related actions are covered by GATS rules. Concerning the application of TRIMs, this regards trade-related effects of investment measures. With respect to investments in the gas sector, the recourse to TRIMs has been rather controversial,[33] owing to TRIMs' exclusive application to the trade of goods and not to the trade in services.

Under GATT rules, international trade has to abide the principles of nondiscrimination (the most-favoured nation principle in article I GATT), national treatment (article III GATT) and general prohibition on the use of quantitative restrictions (article XI GATT). Of importance for natural gas and the security of its supply are the provisions of article V GATT regarding transit.[34] Accordingly, the freedom of transit is guaranteed through the territory of each contracting party, via the routes that are

---

[30] Already in 2003 countries representing 58 per cent of the total world petroleum production were members of the WTO and others – representing in total 27 per cent of the petroleum production – had begun the accession process.

[31] *United States – Standards for Reformulated and Conventional Gasoline* ..., WT/DS2/AB/R.

[32] Communication from the European Communities, WTO Trade Facilitation ... G/C/W/422; Communication from Armenia ..., 2006 TN/TF/W/133; Communication from Cuba ..., 2009, TN/TF/W/146/Rev. 2.

[33] Roggenkamp *et al.*, 2007, p. 137.

[34] As resulting from the following, all gas supply crises that affected gas security in the EU touched upon transit matters.

considered most convenient for international transit to or from the territory of other contracting parties.

As shown above, owing to the geographical positioning of main gas producers and suppliers, transit plays a central role in EU gas trade relations. As such, it is important to understand the legal framework that accompanies it. In this respect, it should be observed that article V GATT has attracted substantial critiques in various legal studies and even among WTO members, owing to the large margin of interpretation allowed by its wording.[35] Specifically, extensive discussions have been carried out on the interpretation of the "freedom in transit" principle. During the accession of Ukraine to the WTO, the argument fluctuated between regarding transit states as solely responsible for determining the "most convenient route" and granting both transit states and states requiring transit the right to choose which route is most convenient for them.[36] Shortly after, the WTO Panel brought additional clarifications on the matter, which seemed to indicate that the appreciation of the "convenience" concept should be left at the discretion of the transit state.[37]

Notwithstanding the polemics accompanying article V GATT, its relevance for energy trade and the EU gas market draws on the fact that it has served as an inspiration for article 7 of the ECT.[38] Nevertheless, its own applicability to natural gas transit is limited. In this sense, the convenience test in the case of gas transit is based on certain particularities deriving from the specific gas transmission infrastructure which are absent from the transit of other goods that do not require a grid. Owing to the scarcity of gas pipelines, any of them could be deemed convenient for transit, disregarding the geographical position of the producer/supplier vis-à-vis the consumer. Hence, it could be assumed that any country that is a WTO member is almost always obliged to grant transit for natural gas supply from another WTO member. A similar transit obligation does not occur in the case of other goods.

---

35 See Azaria, 2009, p. 566; Communication from Turkey and Georgia – Proposal on a Free and Transparent Transit Regime in Road Transportation – Revision, Negotiating Group on Trade Facilitation, TN/TF/W/146/Rev. 1, 10 March 2008, 1.

36 World Trade Organisation: Working Party on the Accession of Ukraine …, WT/ACC/UKR/152, 25 January 2008, para. 363.

37 *Colombia – Indicative Prices and Restrictions on Ports of Entry*, Case WT/DS R366. See also Azaria, 2009, p. 571.

38 Article 7(1) ECT contains a clear reference to the principle of freedom of transit circumscribed by article V GATT. This article represents an essential element for guaranteeing energy security to ECT members.

The above reasoning is only partly correct. As the European Commission and the WTO Panel have observed, the enforcement of transit obligations must not conflict with the provisions of GATS.[39] Under GATS, the obligation of a transit country to provide transport service through its territory depends on specific commitments that the said country has undertaken. This derives from the principle that national treatment commitments in sectors protected by GATS are applicable only to the extent that a WTO member has accepted such commitments in that specific sector.[40] Commitments are enshrined in annexes regarding sector-specific rules.[41] The fact that a country has assumed commitments under an annex does not entail their automatic application. Each country disposes of a national schedule to apply them.

At the moment, GATS does not contain any annex on special commitments in the case of natural gas. However, general rules regarding the cross-border supply of services, the consumption of a service abroad, commercial presence and the supply of service through the presence of natural persons,[42] capture the essence of gas supply into the EU gas market. Indeed, gas supply to the EU entails the cross-border transmission of gas from one WTO member to another.[43] The consumption of a service abroad relates to the situation in which a consumer from a WTO member consumes a certain service in the territory of another WTO member. While such a situation is perfectly compatible with the nature of the EU gas market, its relevance for the regulation of gas transmission has not been always recognized.[44] Commercial presence supposes that a natural gas supplier from a WTO member is present through a subsidiary in the territory of another WTO member, for the purposes of production, sale or supply of natural gas.[45] Furthermore, the supply of a service

---

[39] Communication from the European Communities, WTO Trade Facilitation …, G/C/W/422; WTO Analytical Index: General Agreement on Trade in Services, para. 16.

[40] Cottier *et al.*, 2011.

[41] GATS annexes regarding sector-specific rules provide for movement of natural persons, financial services, telecommunications and air transport services.

[42] These are the so-called four modes of GATS, which represent different ways of providing a service among WTO members.

[43] As seen in the above, both the EU and its major gas suppliers are WTO members.

[44] Energy Charter Secretariat, 2002: The Energy Charter Treaty A Reader's Guide, http://www.encharter.org/fileadmin/user_upload/document/ECT_Guide_ENG. pdf

[45] French GDF Suez is a leading producer in Belgium and Netherlands through Electrabel, German WINGAS is active in Germany, Belgium, France,

through the presence of natural persons is mostly relevant for natural gas investments, while it refers to a natural person of a WTO member being present in the territory of another WTO member with the purpose of providing a service.

Despite the absence of an annex on special commitments in the case of natural gas, energy-related products and services are indirectly mentioned in the services sectors' classification system used by GATS members.[46] In this sense, fuel transportation is included in the broader category of land transport services, while services incidental to energy distribution are dealt with as part of agricultural, mining and manufacturing services.[47] However, in the absence of express commitments regarding natural gas, it could be assumed that under GATS transit countries are not obliged to guarantee transport through their territory for gas originated in another WTO member.

Having regard to all of the above and bearing in mind that GATT and GATS are not inconsistent with each other, it is prudent to consider that the obligation under WTO of a transit country to guarantee natural gas transport through its territory must be assessed on a case-by-case basis.

3.1.1.2 The Energy Charter Treaty

The limited applicability of TRIMs to the trade of gas makes the existence of the ECT all the more necessary for the protection of investors and investments in energy. The ECT represents an important example of international rule-making applicable to EU gas market.[48] The treaty has a double function: that of codifying customary international rules regarding the trade of energy resources and that of steering the conduct of energy partners in a direction compatible with EU liberalization principles. Its relevance for the case study is presented in the following subsection, explaining the choice of its detailed treatment in the next paragraphs. Indeed, it could be argued that the ECT has been specifically envisaged as a legal framework addressing the energy supply relationships between the EU and Russia. The affirmation is supported by

---

the UK, Austria, the Czech Republic and Denmark through locally established subsidiaries, while Italian ENI is present in 20 EU Member States.

46 The so-called Services Sectoral Classification List was adopted during the Uruguay negotiation rounds and represents the reference document for WTO members to schedule commitments regarding different sectors. There are 11 sectors identified in this document, all of which are divided under different subsectors, entries and headings.

47 Provisional Central Product Classification of the United Nations.

48 Article 1(4) ECT and Annex EM Energy Materials and Products.

the fact that the ECT incorporates the standards of GATT regarding the most-favoured nation and national treatment.[49] Also in this sense, article 6 ECT requires the removal of energy monopolies from ECT markets, while article 22 ECT renders Member States responsible for the contractual misconduct of energy undertakings under their control. In general, external incumbents supplying gas to the EU are entirely or partially owned and controlled by their state. As a consequence, their resilience to comply with EU gas market legislation is directly influenced by national policies and politics. The two articles read together stir the EU's external suppliers towards a competitive advancement of their gas markets' arrangements.

The ECT represents the first binding multilateral agreement covering the promotion and protection of foreign investment and the first legal instrument applying nondiscriminatory transit rules, in accordance to the principle of freedom of transport.[50] The treaty has been conceived as a mixed agreement. This presumes a shared competence between the Member States and the Union, which are both parties to the agreement. It offers its signatory parties the legal directions for cooperation on energy matters, "based on complementarities and mutual benefits, in accordance with the objectives and principles of the Charter".[51] As a further step from WTO, which regulates trade only among signatory countries, the ECT is applicable also to local communities and public undertakings controlled by states. This is a particularity of the treaty, which is justified by the capacity of energy undertakings to engage into actions that could disturb the positive effects of the ECT.[52]

The ECT's framework addresses important aspects of the gas trade, including security of supply, transit, investment and cross-border flows. Such aspects are interrelated. Investment for the modernization and renewal of the energy transport system contributes to the fluidity of cross-border flows and increases the security of supply. The guarantee of nondiscriminatory transit from the producing state to the destination country, through the territory of a third signatory party, has become all the more significant in the context of the 2006 cuts of Russian gas supply to Ukraine. These have raised the EU's awareness on the frailty of its gas supply security.[53] Additionally, they have added to general energy supply concerns, stirred by climate change and afferent policies, energy

---

49 Sodupe and Benito, 2001, p. 170.
50 Energy Charter Secretariat, *supra* note 44.
51 Article 2 ECT.
52 Laffont and Simonet, 2005.
53 Stern, 2006.

resources scarcity, the possibility of terrorist attacks on energy infrastructure, technical breakdowns and insufficient investment. In the following, four particularities of the ECT will be briefly discussed in relation to their significance for third-country suppliers:

- the applicability scope of the treaty;
- the limitation clause;
- the rules on transit; and
- the dispute settlement under the ECT.

In general, according to article 18 of the Vienna Convention on the Law of Treaties (VCLT), a treaty becomes imposable on its signatory members upon its entry into force. However, article 25 VCLT allows the *provisional application* of a treaty, which refers to a signatory's commitment to apply entirely or partly the said treaty, before the accomplishment of the ratification process or its entry into force. According to article 45 ECT, the treaty is provisionally applicable to the signatory states regarding all the investments made upon and after its entry into force and it continues to protect these investments during the next 20 years following the withdrawal of any of its members. The provisional application of the ECT and the fact that it still produces legal effects on signatory members that have terminated their participation in the ECT have been affirmed in the *Yukos* case law.[54]

According to article 45(1), each signatory party to the ECT agrees to apply provisionally the treaty, until its entry into force. Such a clause derives from a well-established concept in international law. In this sense, article 25(1) VCLT maintains that "a treaty or a part of a treaty is applied provisionally pending its entry into force if: (a) the treaty itself so provides; or (b) the negotiating States have in some other manner so agreed". Article 45(3) ECT provides in its turn that the termination of the treaty by a signatory country has no retroactive effects and cannot attract the dissolution of the protection granted by the treaty to investments made in the said country prior to the ending of the provisional application. Such a protection will continue to be accorded for the next 20 years following the termination of the treaty.

The importance of article 45(3) ECT stems from the fact that it prevents Russia, and any other signatory member that would decide to withdraw suddenly from the ECT, from relying on article 25(2) VCLT,

[54] *Yukos Universal Limited v. Russian Federation*, PCA Case No. AA 227.

which allows the termination of the provisional application of a treaty upon a signatory party's decision to denounce the said treaty.

There are two important *limitations to the provisional application* of the Treaty. These stem from article 45(1) and (2) ECT. Article 45(2) ECT provides that a signatory state is exempted from the temporal application of the treaty's provisions as a result of its written opt-out declaration. Presently, five of the ECT's signatory members have not yet ratified the Treaty: Russia, Belarus, Australia, Norway and Iceland. Only the last three have expressly opted out from provisionally applying the Treaty. With Russia's withdrawal from the ECT in 2009, Belarus has remained the only ECT signatory party that applies the treaty provisionally. According to article 45(1) ECT, the provisional application of the treaty must be consistent with the constitution, laws or regulations of the signatory country that decides to apply the ECT provisionally. In *Kardassopoulos v. Georgia*,[55] the tribunal appointed under the ECT decided that a state could not invoke the inconsistency of the treaty's provisional application with its internal laws and hence upheld its jurisdiction to decide whether or not Georgia expropriated lawfully the claimant's interest in an exclusive concession to transport oil and gas via pipelines on Georgia's territory. In *Yukos*,[56] Russia invoked the application of the exception in article 45(1) ECT arguing that each temporally applied provision of the ECT should be tested for consistency with national laws, even if the country that applies the provision has willingly accepted the ECT's provisional application. Following the decision in *Kardassopoulos*,[57] the arbitrators in the *Yukos* case decided against the "piecemeal" approach to the ECT invoked by Russia. Instead, they reaffirmed the "all-or-nothing" approach, which meant that by signing the ECT any contracting party agreed to apply the treaty in its entirety, except where found conflicting with its internal laws, regulations or constitution.

For the members of the ECT who have ratified or apply the treaty provisionally, its rules on *transit* hold an important significance, owing to the variety of legal, political and economic conditions to which "energy in transit" is submitted. In order for the ECT rules of transit to apply, the transit state and either the supplying or the receiving state must be parties

55 *Ioannis Kardassopoulos v Georgia*, ICSID Case ARB/05/18.
56 *Yukos*, *supra* note 54.
57 Case ARB/05/18, *supra* note 55.

to the treaty.[58] According to article 7(1) ECT, contracting parties are expected to take all measures necessary to facilitate the transit of energy materials and products, notwithstanding their origin, destination or ownership. Freedom of transit should be guaranteed and no discriminatory prices, delays, restrictions or charges based on nationality criteria should be imposed. As seen above, the text of article 7(1) ECT sends to the provisions of article V GATT. However, the provisions presently under discussion are stricter in what concerns the obligation of the transit country to guarantee passage through its territory. Not only should the freedom of transit be respected, but also transit should be actively facilitated where necessary. Unlike article V GATT, its ECT correspondent neither limits the transit regime to "convenient routes" nor disregards the importance of the construction of new infrastructure for the purpose of transit.[59]

One of the most important aspects of energy transit provisions for the EU gas market are contained in article 7(6) ECT. This refers to the obligation of contracting parties not to interrupt transit for energy materials and products in the event of a dispute over any matter arising from that transit. Two exceptions are permitted. One refers to the conclusion of dispute resolution procedures set out in article 7(7) ECT. The second regards express provisions in contracts or agreements regarding the said transit or a decision allowing the interruption pronounced by a conciliator. Despite the existence of such exceptions, article 7(6) ECT offers additional guarantees to the security of energy supply through the territory of a third-country. An important limitation of article 7 ECT however, comes from the fact that the obligation of not interrupting transit during a dispute that concerns it falls entirely under the responsibility of the transit country. There is no mention of a similar obligation on the side of the supplier, should a conflict occur between the former and the transit country. Both the 2006 and the 2009 transit disputes between Russia and Ukraine arguably fell within the scope of the conciliation mechanism provided for in article 7(7) ECT. However, in neither case was the conciliation procedure invoked. This was due to the fact that the conflict could not be considered a transit dispute, since the interruption of energy flow was determined by Russia – the supplying country – and not Ukraine – the transit state. Hence, contrary to

[58] Article 7(10)(a)(i). Paragraph (ii) of the same article provides for the situation when the origin state is also the state of destination. However, such a situation is not relevant for the EU natural gas market.

[59] Article 7(4) and (5) ECT.

conclusions formulated by other authors,[60] the present study maintains that the ECT's conciliation mechanism regarding transit disputes is severely impaired by the unilateral obligation of the transit country to preserve an uninterrupted energy flow in times of conflict.

Concerning the application of the ECT to gas trade relationships within the Union, certain limitations may be observed. Thus, several divergent opinions[61] have been formulated in relation to *dispute settlement* mechanisms under the ECT and the competent jurisdictions for the delivery of judgments in the respective disputes. In the case *Aes-Tisza v Hungary*,[62] it has been maintained that article 26(6) ECT prevents EU law from being applicable to disputes arisen under within the ECT's framework. This goes in line with CJEU ruling in *Haegeman*,[63] according to which once an international agreement enters into force, its provisions become an integral part of the Union law. As a consequence, international agreements concluded by the EU under article 218 TFEU are binding on its institutions and Member States.[64] However, without contesting the hierarchy of laws in the EU legal system, it should be observed that article 344 TFEU states the exclusive jurisdiction of the CJEU over disputes concerning the interpretation or application of the EU Treaty.[65] This provision receives an added significance with the inclusion of an explicit EU competence over foreign direct investment in article 207(1) TFEU. In the *Mox Plant* case law,[66] the CJEU maintained its exclusive jurisdiction upon mixed agreements, therefore precluding Member States from referring a dispute to an international tribunal. In this sense, CJEU noted that "a breach [of article 344 TFEU] involved a manifest risk that the jurisdictional order laid down in the Treaties and consequently the autonomy of the Community legal system may be adversely affected".[67] From the cumulative reading of articles 7, 260 and 344 TFEU and the

---

[60] Azaria, 2009, p. 584.

[61] *Aes Summit Generation Limited and Aes-Tisza Erőmű Kft. v Hungary*, ICSID Case No. ARB/07/22. Expert opinion of Professor Piet Eeckhout, 2008: The Law Applicable to the Dispute under the Energy Charter Treaty 1994, http://italaw.com/documents/AESEeckhoutExpertOpinion.pdf; see also Grigoriadis, 2008.

[62] *Aes Summit Generation Limited and Aes-Tisza Erőmű Kft. v Hungary*, ICSID Case No. ARB/07/22.

[63] *Haegeman*, Case 181/73, para. 5.

[64] Craig and de Búrca, 2007, p. 202.

[65] Case C-459/03 *Commission v. Ireland* [2006] ECR I-4635.

[66] *European Commission v Ireland* ('MOX Plant'), Case C-459/03.

[67] *Ibid.*, para. 154.

joined *Frankovich* cases,[68] EU Member States that have ratified the ECT are more likely to be held liable for their energy breaches by EU instances rather than by ECT forums.[69]

The efficiency of the ECT in cases regarding disputes opposing Member States is brought under scrutiny by the lack of consensus among the latter regarding regulatory measures that should have been included in the Charter. Discussions on mandatory third-party access to export and transit pipelines and on access to foreign investors[70] have demonstrated the persistence of contradictory opinions across the EU regarding gas market liberalization and integration. Hence, it is unclear to what extent the ECT might provide an effective harmonization of the regulatory regimes of its signatory parties.

Resulting from the above, the applicability of WTO and ECT rules to the EU's relationships with third-country gas suppliers is submitted to important limitations, which derive principally from the still inaccurate treatment of energy matters under WTO agreements and the lack of obligation for natural gas suppliers under the ECT to maintain an uninterruptible energy flow in times of disputes regarding transit. In the following, the analysis will turn to legal tools that EU disposes of internally to regulate natural gas supply from third countries.

3.1.2 The Applicability of EU Competition Law to Foreign Grid Companies

Through their behaviour on the EU gas market, third-country producers and suppliers of natural gas may determine competition restrictions, higher prices and a decrease in the security of supply, which could ultimately lead to EU consumer harm. In the following, the scope of application of EU competition rules to activities of third-country suppliers is examined. The analysis is carried through by looking first at anticompetitive behaviour of third-country undertakings operating on the EU gas market, before moving into considerations upon the extra-territorial application of EU competition law to activities of foreign gas suppliers outside the territory of the EU.

---

[68] *Frankovich and Bonifaci and others v Italy*, joined Cases C-6/90 and C-9/90.

[69] Grigoriadis, 2008, pp. 45–64.

[70] Grätz, 2009.

3.1.2.1 Restraints in supply relationships

The application of EU competition law to activities performed by third-country gas undertakings on its territory is covered by articles 101 and 102 TFEU, regarding antitrust measures. EU's jurisdiction in this case derives from the territoriality principle and is reaffirmed in article 7 of Council Regulation (EC) No. 1/2003. According to the territoriality principle, EU may exercise its executive and judicial jurisdiction over offences committed on its territory, notwithstanding the nationality of the offender.[71] Article 7 of Regulation (EC) No. 1/2003 states that "the Commission, when finding that there is an infringement of [articles 101 and 102 TFEU], may impose on the relevant undertakings any behavioural or structural remedies which are proportionate to the infringement committed and necessary to bring the infringement effectively to an end".[72]

According to article 101(1) TFEU agreements between undertakings which may affect trade between Member States and distort competition within the internal market are prohibited. Owing to the structure of the EU gas market and the particularities of natural gas as an energy resource, trade agreements between EU-based undertakings and non-EU gas producers and suppliers represent a major part of the EU gas supply. As such, they play an important role in the development of the EU gas market and in the manner in which competitiveness requirements are applied. Relationships between EU and non-EU undertakings may occur both vertically (between undertakings situated at different levels in the market) and horizontally (between companies at the same level in the market).[73]

Competition concerns arising in vertical supply relationships are generally associated with territorial sale restrictions, profit-splitting mechanisms, long-term exclusive purchase obligations and use restrictions. Regarding the gas markets, such restrictive practices may occur both in upstream markets – between producers and large wholesalers – and downstream markets – between wholesalers, on one side, and regional distributors, commercial or industrial clients and private households, on the other. European Commission's principal focus on the regulation of upstream markets has led non-EU producers and suppliers to a direct confrontation with the liberalization process.[74] An example of this is provided by the Commission's investigations on the conformity of

71 *Grosfillex-Fillisdorf*, Commission Decision 64/233.

72 Council Regulation (EC) No. 1/2003.

73 Roggenkamp *et al.*, 2007, p. 256.

74 Nyssens and Schnichels, 2007, p. 1405.

including territorial restrictions clauses[75] into gas supply contracts with the provisions of article 101 TFEU.[76] As a result of the investigations, restriction clauses have been removed from the said gas supply contracts. In another investigation regarding Nigerian gas incumbent LNG Ltd, the Commission's intervention has given way to even more overarching outcomes. LNG Ltd not only agreed to remove restriction clauses from one of its contracts with an EU customer, but also committed itself not to introduce such clauses in future contracts.[77]

Horizontal agreements between gas undertakings may be concluded with the clear purpose of restricting competition or may only incidentally produce competition-restrictive effects. In the former case, such agreements generally consist of price-fixing or market-sharing activities and are by default considered incompatible with article 101 TFEU. In a recent decision regarding agreements and concerted practices in the gas insulated switchgear (GIS)[78] sector, the European Commission imposed 750,512,500 Euros fines on a series of French, German, Austrian, Swiss and Japanese undertakings.[79] The latter had conducted cartel activities, which lasted from April 1988 until May 2004, when it became obvious that the Commission's inspections had been set in motion. Such activities encompassed the coordination of the allocation of GIS projects worldwide, according to agreed rules that respected historic market shares and fixed price levels, while reserving some territories to certain producers. The decision of the Commission in the GIS case is also relevant for the natural gas sector, while it sanctions not only the conduct of EU undertakings, but also that of non-EU companies.[80] As in the case of GIS, natural gas is traded beyond national borders. The gas market, although not stretching worldwide, is a regional market gathering producers/suppliers and buyers from different states and continents. As a consequence, its functioning is affected by the behaviour of both EU and non-EU nationals. Although talks of a gas cartel have never been

---

[75] Such clauses prevent wholesalers from reselling gas outside the originally agreed supply area.

[76] *GFU*, Case COMP/36.072; *Territorial restrictions 1) Algeria gas transport contracts, 2) Expansion of TAG pipeline*, Case COMP/37.811; *Territorial restrictions – Austria (Gazprom/OMV)*, Case COMP/38.085; *Territorial restrictions in Germany (Ruhrgas/Gazprom)*, Case COMP/38.307.

[77] IP/02/1869 LNG Ltd of 12 December 2002.

[78] Gas insulated switchgear is used to control energy flow in electricity grids.

[79] *Gas Insulated Switchgear*, Case COMP/38.899.

[80] See extraterritorial application of EU competition law in the next subsection.

concretized among major gas producers,[81] it is reassuring to observe that the enforcement of EU competition law in cases of competition distortion caused by non-EU nationals is effective.

Natural gas horizontal agreements that incidentally affect competition on the EU gas market in a negative way are generically known as cooperation agreements.[82] An illustrative example of such agreements is given by the Gas Negotiation Committee (GFU) case on joint selling/marketing.[83] The Norwegian GFU was traditionally composed of Norway's largest two gas producers, Statoil and Norsk Hydro. The two incumbents entered a series of long-term supply agreements with European gas operators on behalf of all Norwegian natural gas suppliers. This was done through the joint selling scheme provided by the frame of GFU, which allowed for the fixing of prices and quantities sold of natural gas exported from Norway into the EU. As a result, the European gas markets suffered from significant rigidity and lack of liquidity. In 2001, the European Commission opened an investigation into the GFU selling scheme and demanded effective access to European pipelines for Norwegian gas. The investigation was grounded in articles 101 TFEU and 53 EEA, which ban cartels and other damaging concerted practices. In the aftermath of the investigation, the Commission requested that the Norwegian joint market regime for gas was discontinued. The reasoning was that 30 Norwegian natural gas suppliers, as opposed to one, would allow a better protection of EU consumers' interests. Furthermore, it was required that third-party access became effective, cost-reflective and nondiscriminatory.

Competition on the EU gas market might be distorted not only by agreements between undertakings or concerted practices, but also by abuse of a dominant position. Article 102 TFEU stipulates that any abuse of a dominant position by an undertaking within the internal market or in a substantial part of it shall be prohibited. Such an abuse may consist of imposing unfair trading conditions (including prices), applying dissimilar conditions to equivalent transactions, limiting production, markets or technical development and subjecting the conclusion of contracts to nonrelevant conditions. The exact scope and applicability of article 102 TFEU on the EU gas market will be discussed in detail in Chapter 5. For the purpose of the present discussion it is important to observe that, although article 102 TFEU has been predominantly used to correct the

---

[81] Orttung and Overland, 2011, pp. 53–66.
[82] Geradin, 2004.
[83] *GFU*, Case COMP/36.072.

reinforcement of the dominant position of EU incumbents in their respective markets,[84] it can also be used as a corrective measure for the anticompetitive behaviour of non-EU nationals. Sufrin notes that abuses involving trade with non-EU countries enter the scope of article 102 TFEU if they produce competition restrictive effects in the EU.[85]

In September 2012 the European Commission opened a formal investigation with regard to allegations that Gazprom, the main Russian gas producer and supplier, might be hindering competition in Central and Eastern European markets in breach of EU antitrust rules. Specifically, the investigation refers to article 102 TFEU owing to concerns that Gazprom might be abusing its dominant position in upstream gas supply markets. There are three types of potential competition harm that Gazprom might have caused, which the investigation is focused upon:

- a possible division of EU gas markets by hindering the free flow of gas across Member States;
- the prevention of security of supply diversification;
- the imposition of unfair prices on its consumers by linking the price of gas to oil prices.[86]

The investigation is still on-going at the moment, but should it be confirmed that Gazprom has abused its dominant position, the Commission can decide to impose a fine of 10 per cent of the undertaking's total turnover, order interim measures or accept a commitment to cease breaches of EU competition law.[87]

3.1.2.2 The extension of the territoriality principle

It has been observed that in general the EU may impose competition law-based sanctions only on foreign grid companies or the supplier that operates on the territory of the Union.[88] What happens if gas market competitiveness is affected by the behaviour of third-country nationals outside the EU? The mechanics of EU natural gas supply entail the movement of goods, services and capital across the borders of Member States and the Union itself. Major gas incumbents, as well as the largest amount of natural gas resources, are located outside the EU. At the same

---

[84] As resulting from the Commission's Energy Sector Inquiry. See also Subiotto *et al.*, 2010.

[85] Jones and Sufrin, 2004, p. 342.

[86] *Gazprom*, Case COMP/39.816.

[87] Articles 8, 9 and 23 of Council Regulation (EC) No. 1/2003.

[88] Cottier *et al.*, 2011.

time, natural gas activities presume the establishment of subsidiaries of non-EU undertakings on EU territory, mergers among internationally active incumbents and upstream and downstream acquisitions. Owing to the particular legislative framework of the internal gas market and the approach to competition of external gas suppliers, it is possible that activities conducted by a third-country gas undertaking may have an impact upon the conditions of competition within the territory of the EU, even if no Member State or its territory is involved. Hence, the Union's jurisdiction under EU competition law should be able to reach anti-competitive behaviour that occurs outside its territory.

As a result of the territoriality principle introduced in the preceding section, a state cannot take measures on the territory of another state by way of enforcement of its national laws without the consent of the latter. Hence, jurisdiction to impose penalties, demand information and exercise investigation powers is confined to a national territory, or in the case of the EU to the Union's territory. There are, however, exceptions from the principle. According to the "effects doctrine",[89] the jurisdiction of a state is extended upon activities committed by foreigners outside its territory if they produce effects within its territory. The same reasoning applies to the EU. The controversies surrounding the application of the "effects doctrine" in antitrust legislation in EU and US law have become the subject of extensive academic discussions[90] and preclude the scope of the present subsection. However, it should be briefly observed that part of this controversy at a European level stems from the reluctance of the Court of Justice of the European Union to expressly recognize the "effects doctrine",[91] despite referring to it in a number of case laws.[92]

As a result of a well-established jurisprudence, the Court has tended to extend the territorial jurisdiction of the EU based on the "single economic entity doctrine"[93] or on the "implementation doctrine".[94] According to the former, parents and subsidiaries are considered to form

---

89 Shaw, 2008, p. 612.

90 See Brownlie, 2008, pp. 309–311; Shaw, 2008, p. 612; Sato, 2010, pp. 33–49.

91 Floudas, 2007.

92 *Béguelin Import Co. v S.A.G.L. Import Export*, Case 22/71; *Gencor Ltd v Commission*, Case T-102/96, *Imperial Chemical Industries Ltd. v. Commission of the European Communities*, Case 48/69, paras 126–130.

93 *Imperial Chemical Industries, supra* note 92; *Viho Europe BV v Commission*, Case C-73/95 P.

94 *Ahlström Osakeyhtiö and others v Commission of the European Communities* (*Woodpulp*), Cases C-89/85, C-104/85, C-114/85, C-116/85, C-117/85 and C-125/85.

the same undertaking for the purpose of applying competition rules. Hence, responsibility for an infringement of competition law by the subsidiary may be imputed to the parent if it can be shown that the former does not decide independently upon its conduct, but takes instructions from the latter. In the *Dyestuffs*[95] case law, the CJEU acknowledged the competence of the Commission over actions undertaken by foreigners. The case regarded concerted practices between an undertaking established in the UK – which at the time was not a Member State – and its Belgian subsidiary. The European Commission denounced the agreement as contrary to article 101 TFEU and imposed a fine on the British company. The latter appealed against the Commission's decision before the Court of Justice, claiming that the Commission had no power to apply competition rules to non-EU undertakings. The Court, basing its judgment on the "single economic entity doctrine", concluded that the Commission had jurisdiction over the concerted practice and the British company. The reasoning of the Court was that the subsidiary did not enjoy real autonomy, but acted on the instruction of the parent company. Hence, it was correct for the Commission to fine the British undertaking as "unity of conduct on the market as between a parent company and its subsidiaries overrides the formal separation between those companies resulting from their separate legal personality".[96]

The "implementation doctrine", in its turn, maintains that the EU has jurisdiction over non-EU undertakings that participate in an anti-competitive agreement reached outside the Union. The condition is that such an agreement is implemented on the territory of the EU – for example, the said undertakings sell their products to buyers in the EU. In the *Woodpulp*[97] case law the Court of Justice of the European Union observed that article 85 EEC (now article 101 TFEU) is applicable to all agreements between undertakings and concerted practices that might affect the trade between Member States and that would result in disturbed competition in the Common Market.[98] The case regarded a fine applied by the Commission for price-fixing in the wood pulp industry. In a similar manner to the gas market, the global dimension of the wood pulp market entailed trade relationships between European purchasers – situated in Member States – and third-country suppliers. According to paragraph 12 of the decision, the direct sale of resources from producers established outside the Union to purchasers in the Common Market, as

---

95 *Imperial Chemical Industries*, *supra* note 92.

96 *Ibid.*

97 *Woodpulp*, *supra* note 94.

98 *Ibid.*

well as engagement in price competition in order to attract partners, were actions under the jurisdiction of EU competition principles. Paragraphs 13 and 14 concluded that the coordination of prices among producers had the effect of restricting competition in the territory of the Union. Therefore, competition rules enunciated in the Treaty were considered applicable to the undertakings under scrutiny, even if their offices were registered outside the Common Market. The judgment expressly recognized the Commission's jurisdiction to apply EU competition rules to anticompetitive conduct of third-country undertakings, where such conduct was "implemented" in the territory of the Union.

There are certainly important limitations regarding the three doctrines exposed above. The "effects doctrine" suffers from a significant ambiguity caused by the coexistence of a multitude of effects test and the absence of a real definition associated with the idea of "effects". From the jurisprudence of CJEU it follows that the effect of foreign anticompetitive conduct on competition in the common market must be *appreciable*.[99] However, the Court has not qualified this concept.[100] The "single economic entity doctrine" entails a certain connection with the territory of the EU, which is not required by the "effects doctrine". As an example, competition rules apply to the parent company situated outside the EU, based on the conduct of the subsidiary established in the territory of the EU. This is an important limitation of the "single economic entity doctrine", while its application does not seem to be justified in cases of anticompetitive conduct of subsidiaries and parent companies situated outside the EU which produce effects in the EU. With regard to the "implementation doctrine", doubts have been raised as to the justification of its application, in cases where no marketing organization of the trespassing company may be found in the EU (branch, agency or subsidiary).[101] Although the "implementation doctrine" does not require that the non-EU companies have subsidiaries or agents in the EU, the legitimacy of the doubts is given by the fact that it is hard to prove that a company not active in a market has led any activity in that market.[102] Hence, it appears that the "effects doctrine" can be relied upon in cases that escape the scope of the other two doctrines exposed above. Despite the ambiguity of the "effects doctrine" and the position of the CJEU,

---

99 *Béguelin, supra* note 92.

100 The Commission however has attempted to correct this omission in its guidelines on the effect on trade. Commission – Notice – Guidelines on the effect on trade concept contained in articles 81 and 82 of the Treaty.

101 Kiriazis, 2001.

102 Sato, 2010, pp. 33–49.

both the Commission and some Advocates General appear to have constantly been in favour of employing this particular doctrine to justify the extraterritorial application of competition law.[103] In the *GIS* case, the Commission maintained that it had jurisdiction to apply both articles 101 TFEU and article 53 of the EEA Agreement, owing to the fact that the cartel under investigation had an appreciable effect on trade between Member States.[104]

Both articles 101 and 102 TFEU refer directly or indirectly to the idea of external anticompetitive behaviour exerting effects on the common market. From the formulation of article 101 TFEU it ensues that the article applies to vertical restraints in supply and transport relationships and horizontal agreements between undertakings, traders or consumers. A restrictive agreement in the sense of the treaty is represented by any agreement between undertakings, decisions by associated undertakings, as well as concerted practices that could affect the trade between Member States and that have as their object or effect the prevention, restriction or distortion of competition within the internal market. It should be observed that no part of the article specifies that the anticompetitive agreements must occur in the territory of the internal market, or that the undertakings concerned should be based in the EU. The only condition considered is that such agreements should produce anticompetitive effects within the internal market. Article 102 does not employ the exact term of effect, but refers to a conduct which might affect the trade between Member States.

3.1.3 Obligations and Conditions Issued from the Specific Gas Sector Regulation

Under TEP, a TSO controlled by a third-country supplier and wishing to perform its functions on the territory of a Member State must receive certification prior to its establishment in the EU.[105] The certification procedure represents a validation of the fact that the transmission system operator complies with the requirements of article 9 of Directive 2009/73/EC regarding ownership unbundling of production, supply and network assets within the controlling undertaking.

According to article 11 of Directive 2009/73/EC, the certification procedure in relation to third countries takes place as follows. The NRA is obliged to notify the Commission, where certification is requested by a

---

[103] *Imperial Chemical Industries*, *supra* note 92; *Woodpulp*, *supra* note 94.

[104] Case COMP/F/38.899, para. 220.

[105] Article 11 of Directive 2009/73/EC.

third-country TSO or TSO owner. Further notification is required without delay if any circumstances occur that could result in a foreign control acquired over a transmission system or TSO. The latter is obliged to inform the responsible NRA of any such circumstances. Following the preliminary steps, the NRA is supposed to adopt a draft decision on the certification of the TSO within four months from the date of request and notify it to the European Commission. In adopting its decision, the NRA must ensure that the TSO complies with unbundling requirements and that such a certification will not put at risk the security of energy supply of the Member State and the Union. The burden of proof in this case lies with the respective TSO. The Commission is obliged to deliver an opinion on the NRA's decision within the two months following notification. While formulating the opinion, the Commission may request the views of ACER, the Member States concerned and the interested parties. This two-level decisional mechanism should ensure the correctness of the certification procedure and guarantee that NRAs will not be biased by the opinion of their national governments. However, despite the fact that NRAs are encouraged to take the "utmost account" of the Commission's opinion, they might still decide to go astray from it. The Member State concerned has the final say in the certification procedure. In this respect, it may decide to go against the decision of the NRA and refuse certification, where such an act would bring risk to its security of supply or to the security of supply of another Member State. The appreciation of the risk to the security of supply is left at the discretionary appreciation of the Member States. The directive remains silent as to the case where an NRA would refuse the granting of a certification and the concerned Member State would adopt a contrary decision.

The reasons behind the introduction of the certification procedure in TEP can be related to the proposal of several Member States asking for a "reciprocity clause" to be included in the new sector regulatory provisions. The use of reciprocity in international law has aimed primarily at creating similar rights and obligations between homologous partners.[106] For the purpose of the EU gas market, the initial clause was regarded as a measure to mitigate the effects of vertically integrated third-country companies entering the European transmission and retail market business. As such, it imagined the prevention of non-EU gas-producing undertakings from having decisive influence upon EU TSOs' assets, in the

[106] For example, the granting of licences for the establishment of foreign bank subsidiaries in the territory of another state or the granting of mutual landing rights as a manner of organizing air travel. Cottier et al., *supra* note 40.

absence of a bilateral agreement on mutual market access to transmission assets situated in the investor's country of origin.[107] In other words, this meant that foreign investments in EU infrastructure would be permitted only on the basis of an international agreement between the EU and the country of the investor. Such a clause was considered necessary to compensate for past acquisitions and ventures in the European downstream natural gas business by foreign transmission system operators. Quite soon it became visible that the clause expressly targeted Gazprom and what was deemed its massive buy-out of strategic EU energy assets.[108] The Member States backing the proposal feared that a unilateral split-up of EU companies would weaken their power of negotiation regarding Gazprom and would allow the latter to exert an even greater grip on European energy supply.[109]

Despite the advocated benefits of such a clause, Russia's argument against it has been supported by eight EU Member States, led by Germany, France and Italy, whose companies have been competing with each other to receive better gas import conditions.[110] As a result, the clause was erased from the Third Energy Package proposal at the Energy Council in October 2008. Instead article 11 on the certification of third-country TSOs was introduced in Directive 2009/73/EC.

Since both EU and its major gas suppliers are members of the WTO and its cover agreements, there has been a large discussion on the conformity of the reciprocity and certification requirements with WTO provisions and especially the most favoured nation and national treatment principles under GATS.[111] While, at the moment, the only reciprocity requirements contained in TEP are those in article 37 of Directive 2009/73/EC regarding the eligibility of customers under different national systems, the question remains as to the compatibility between article 11 of the 2009 gas Directive and WTO rules. It should be observed that the Directive contains rules regarding the certification procedure for both EU TSOs and non-EU TSOs. The difference between the two categories is that the former are not required to pass a security of supply test in the same way as the latter. Therefore, it is sensible to conclude that non-EU

---

107 Grätz, 2009, p. 77.

108 European Parliament, 2009: 3rd Energy Package gets final approval from MEPs, http://www.europarl.europa.eu/sides/getDoc.do?type=IM-PRESS&reference=20080616FCS31737&language=EN

109 Eikeland, 2004, pp. 29–30.

110 Molis, 2011.

111 See Van Hoorn, 2009 p. 62; Cottier *et al.*, 2011; Ćwiek-Karpowicz and Gawlikowska-Fyk, 2012.

TSOs are granted a less favourable treatment than EU ones. However, according to article XIV(c) GATS exemptions from the principles enunciated in the Agreement are allowed if they are necessary to secure compliance with laws or regulations that are not inconsistent with the provisions of the Agreement. In relation to the certification procedure, the article appears to set two conditions. The first one refers to the necessity of the certification requirements, while the second one refers to the consistency of EU gas liberalization regulation with GATS. The latter aspect is easily confirmed while looking at the purposes of GATS in the beginning of the legislative text of the Agreement.[112] With regard to the necessity requirement, the following reasoning could be maintained. In the absence of mutual cooperation in matters of competition control between the EU and a third-country supplier, there is a need for an instrument to ensure that the foreign operator will not engage in anticompetitive activities, for example, access denial or cross subsidization.[113] To this end, the certification procedure fulfils the function of ensuring that a foreign TSO is properly unbundled. Furthermore, the introduction of a supplementary condition in article 11 of Directive 2009/73/EC regarding security of supply is justified by article XIV bis GATS. In accordance to this article, any contracting party is allowed to take any action necessary for the protection of its essential security interests.

In the following, the effectiveness of the application of article 11 will be analysed by looking at its strengths and weakness. Regarding the former, the certification procedure is closely linked to the guarantee of security of supply in the EU. Recital 22 of the 2009 Directive provides that persons from third countries should be allowed to control a transmission system or a TSO only if they comply with requirements of effective separation that applies within the Union. Such a condition is justified at EU level by the belief that the security of gas supply towards the Union requires among other things the independence of network operations. As seen above, the certification of a foreign TSO must be refused if it is thought that it might put at risk the security of supply of the Member State, where the TSO demands certification, and the Union. Member States have the final right to refuse certification based on security of supply considerations. A further strength of article 11 comes

---

112 Annex 1B General Agreement on Trade in Services, http://docsonline.wto.org/GEN_searchResult.asp?RN=0&searchtype=browse&q1=%28@meta_Symbol+LT%FCUR%FCA-1B%FCS%FC1%29+%26+%28@meta_Types+Legal+text%29

113 Cottier *et al.*, 2011.

from the fact that, despite the coexistence of three different unbundling models at EU level, a foreign TSO attempting to enter the gas market of a Member State should obey effective unbundling rules. Even more so, when certification is required for an independent system operator controlled by a non-EU person, additional rules are applicable.[114] These derive from article 14(2) of Directive 2009/73/EC and provide that such an ISO might be certified only when:

- it has demonstrated that it disposes of the required financial, technical, physical and human resources to carry out its tasks regarding transmission, storage and/or LNG system operators;
- it has undertaken to comply with a 10-year network development plan monitored by the regulatory authority;
- the candidate operator has demonstrated its ability to cooperate with TSOs at European and regional level, as well as to comply with other obligations resulting from Regulation (EC) 715/2009.

Despite its strengths, article 11 also presents the following weaknesses. The first, most obvious one comes from the time-frame granted by the EU for the transposition of the provisions of Directive 2009/73/EC into national law. According to the schedule, the Directive entered into force in March 2011. Exceptionally, article 11 should have been applied from March 2013. This has allowed foreign producers to enjoy roughly two additional years of downstream acquisitions without having to conform to the certification procedure and the unbundling requirements therein.

At the end of the grace period, the gas directive demanded that external suppliers demonstrate that their subsidiaries comply with effective unbundling regulations. There are two important obstacles to the effectiveness of such a demand. The first stems from the territorial application of EU law. As previously shown, the EU's control on the effective separation of network and supply activities of a foreign supplier is limited exclusively to the part of the foreign undertaking operating within the territory of the Member States. It is only this segment of the foreign gas supplier – TSO or system operator – that is required to apply effective unbundling, when attempting to become established in a Member State. EU unbundling rules cannot be further imposed on the rest of the activities of the foreign undertaking or on its country of origin.

[114] Commission Staff Working Paper Interpretative Note on Directive 2009/72/EC Concerning Common Rules for the Internal Market in Electricity and Directive 2009/73/EC Concerning Common Rules for the Internal Market in Natural Gas The Unbundling Regime.

Nevertheless, in Member States opting for ownership unbundling, vertical integrated incumbents established in third-countries cannot acquire transmission operators.[115] This comes as a result of an extended application of the level playing field principle to TSOs controlled by a foreign person. According to this principle, an undertaking performing any of the functions of generation or supply in a Member State must comply with the rules on ownership unbundling when acquiring rights in another TSO established in a Member State having opted for ownership unbundling.[116] The sought-after effect of such a provision is the prevention of the EU gas market foreclosure through anticompetitive behaviour of third-country suppliers.

The second obstacle derives from the wording of article 11. According to this, foreign TSOs attempting to establish themselves in an emergent gas market or an isolated gas island are not expected to comply with unbundling requirements in article 9 of the Directive. This is an important limitation to the effectiveness of article 11. The Member States concerned by the above paragraph are also the most prone to having their markets foreclosed by the market presence of unbundled TSOs. This happens owing to their absolute or almost absolute dependency on a single external gas supplier.

Having regard to all of the above, it is difficult to anticipate at the moment to what extent the application of article 11 of the Directive will influence gas trade relationships between the Union and its third-country suppliers. It is an important step that the Directive requires the highest degree of unbundling of foreign TSOs, without offering less restrictive unbundling conditions for TSOs aiming to enter a market unbundled in accordance with the ISO or ITO model. Considering the regulatory burden of such unbundling models in the EU and the complexity of institutional mechanisms established to guarantee the independence of TSOs under the ISO and ITO system, it is questionable how a similar level of supervision would practically take place in the country of origin of the foreign TSO.

---

[115] Grätz, 2009, pp. 66–79.

[116] Commission Staff Working Paper Interpretative Note on Directive 2009/72/EC, *supra* note 114.

3.2 EU AND RUSSIAN GAS RELATIONSHIPS: A CASE STUDY

EU gas trade relationships with its suppliers are not identical. Different levels of dependency, contrasting legal backgrounds and specific geo-political conditions contribute to the diversity of such partnerships. Despite the heterogeneity of content, similar benchmarks can be applied when analysing EU's external gas relationships. These benchmarks will be further highlighted in a case study on Russian gas. As mentioned at the beginning of this chapter, Russia is an important player in the world energy market, responsible for the largest amount of EU's external gas supply.[117] As such, its role in ensuring the energy stability of the Union is essential and to a certain extent unmatched by that of other providers.

In spite of this, in the aftermath of the cuts of Russian gas supply in 2006, which affected both transit and destination countries, the EU has ignited a mobilization process for diversifying gas routes and resources reservoirs, as a way to decrease its gas import dependency on Russia. In contrast to EU's position, several scientists[118] and the industry have questioned the seriousness of the security of supply risks posed by the EU's dependency on Russian gas. This was done specifically while pondering upon the EU's alternative supply options, which, as mentioned before, with the exception of Norway, raise at times more severe questions in terms of political stability and adaptability of the respective legal systems to the EU model.

The following discussion on EU gas trade relationships with Russia is taking into consideration two aspects. These are the legislative framework encompassing the said relationships – including Russia's own approach to natural gas market liberalization – and the specifics of the traditional gas trade relationships between EU Member States and Russia. Furthermore, the case study explores the limitations of the existing legal framework for energy cooperation between the two sides.[119]

---

[117] According to the International Energy Agency, in 2011 Russia was the biggest producer of oil and gas, the largest exporter of gas and oil products and the second largest exporter of oil, while holding a key position in the production and exportation of coal and nuclear electricity to the rest of the world. International Energy Agency, 2011: Key World Energy Statistics, http://www.iea.org/textbase/nppdf/free/2011/key_world_energy_stats.pdf.

[118] Stern, 2006; Talus, 2007b, pp. 535–548.

[119] This study addresses only the legal framework encompassing the EU–Russia gas relationships. For a broader approach reflecting the political as

3.2.1 EU Level

EU–Russian gas supply contacts have crystallized in the aftermath of the Cold War, owing to their relative geographical proximity, the already existent pipeline grids and the promising political stability of the former Russian Federation. Presently, a quarter of the total EU gas consumption is ensured by Russian exports. At the same time, Gazprom obtains more than 60 per cent of its revenues from EU markets. This seemingly mutual dependency is more facetted in reality than it appears at a first glance. Russia's monopolistic gas giant Gazprom holds important shares in major energy incumbents of several Member States.[120] According to Kaderják, Central and Eastern Europe are dominated by joint ventures between Gazprom and German companies over ownership of gas assets and gas supplies.[121] Gazprom's presence in the EU and particularly in Central and Eastern Europe is owed to a certain extent to the two first waves of liberalizing the European gas sector, which have opened up the downstream and retail market to external acquisitions. A similar, but reversed situation does not hold true since Gazprom continues to be owned almost in absolute majority by the Russian state. The only EU incumbent having shares in Gazprom is German E.ON, whose ownerships over the latter's shares is insignificant.[122]

The background of EU–Russia energy relationships includes the creation of Common Spaces,[123] EU–Russia bilateral summits,[124] the signing

well as development cooperation policies between the two, please refer to Haghighi, 2007, pp. 341–381.

[120] In 2007 Gazprom held 37.2 per cent from the shares of Estonia's Eesti Gas, 48 per cent of the Polish Europolgaz, approximately 50 per cent of Wingas, the energy subsidiary of BASF (Germany's chemical and energy group), as well as various amounts of shares in the Latvian, Lithuanian, British and Finnish TSOs. See Gazprom in questions and answers, book 2011, http://eng.gazpromquestions.ru/index.php?id=4

[121] Kaderják *et al.*, 2007.

[122] http://eng.gazpromquestions.ru/index.php?id=4

[123] At a summit in St Petersburg in May 2003, four common spaces representing areas of common policy interest for Russia and the EU were established. These are the Common Economic Space, the Common Space of Freedom, Security and Justice, the Common Space of External Security and the Common Space of Research and Education, http://eeas.europa.eu/russia/common_spaces/index_en.htm

[124] These take place twice a year.

of a Partnership and Cooperation Agreement in 1997,[125] Russia's tentative tango with the Energy Charter Treaty between 1994 and 2009, the establishment of the Energy Dialogue, and most recently Russia's accession to WTO. Among these instruments, the last four are of particular relevance for EU–Russia energy relations and hence will be dealt with in more detail in the following.

3.2.1.1 The general framework encompassing EU–Russia gas relations

The Partnership and Cooperation Agreement (PCA) has been designed as a legal framework for the general cooperation relations between the two parties.[126] The key idea behind the partnership has been to create a free trade area between the two partners and to help Russia's accession to WTO. Nevertheless, until 2007, when the 10-year-long agreement expired, neither of the two objectives had been achieved.

Regarding energy, the main success of the Partnership has been the introduction of a specific article in the agreement, aimed at improving energy relations between the parties and at integrating the Russian energy market. Article 65 PCA has served as the legal basis for the establishment of the Energy Dialogue and as one of the main legal frameworks for the development of EU–Russia energy relationships, along with the ECT. Under PCA, the definition of energy cooperation between EU and Russia has been very broad. The partners have expressed the need for an EU energy policy targeting the Russian energy sector, in view of increasing trade and investment in the field.[127] However, no clear methods for reaching the objectives have been identified. Article 65(1) envisioned the development of EU–Russia energy cooperation in accordance with market economy principles and the ECT's provisions. Paragraph 2 of the same article underlined the necessity of formulating an energy policy addressing the objectives of both parties and of introducing a range of institutional, fiscal, legal and other conditions necessary to encourage increased energy trade and investment. In 2011 the EU–Russia Energy Roadmap identified persistent regulatory, commercial and technological barriers to gas relationships between the two parties and urged

---

[125] Agreement signed between Russia and the EU in 1997 with the aim of creating a legal basis for the EU–Russia relationships. See Council and Commission Decision of 30 October 1997 on the conclusion of the Partnership and Cooperation Agreement between the European Communities and their Member States, of the one part, and the Russian Federation, of the other part.

[126] Article 55 PCA.

[127] Haghighi, 2007, p. 344.

their removal. Presently, negotiations are underway to re-launch the Partnership, in the frame of the Energy Dialogue. However, no actual deadline has been set up so far.

Until 2009, the most efficient legal instrument addressing EU–Russia energy could have been considered the ECT. In 1994, Russia signed the document, but failed purposefully to ratify it. This could be regarded as a direct outcome of national stakeholders' opposition in both private and public spheres. Critics maintained that the purpose of the ECT ran in contradiction to Russia's economic interests, "allowing other countries easier access to Russia's natural resources and perpetuating the country's reliance on its raw materials".[128] The critique was adopted by Gazprom, which opposed ratification while arguing that the freedom of transit principle guaranteed by the ECT would mean third-party access for Central-Asia producers and Russia's consequent loss of control over the flow of energy.[129] The Russian government, in its turn, did not envisage a complete rejection of the ECT's principles from the very beginning. This is demonstrated by the fact that Russia accepted to become a signatory party to the ECT. However, while the failure to ratify the Treaty might have been influenced by industry pressure groups, it also reflected a shift in government position regarding the strategic importance of national energy resources and an increasingly sceptical approach to foreign investors. In 2009, through its decision to withdraw from the ECT, Russia escaped the only legal framework able to supervise its contracting parties in forming and executing their investments in the energy sector. Despite Russia's withdrawal, the ECT's provisions are still applicable to investments concluded during the period of time in which the country was a signatory member. The affirmation is sustained by the combined reading of paragraphs (1) and (3) of article 45 ECT. Nevertheless, there are no guarantees for new investors that further competition and a secured provision of energy supply will be ensured.

As previously noted, the ECT's creation was originally partly envisioned to substitute for Russia's lack of membership in the WTO. In an interesting turn of events, three years after its withdrawal from the ECT, Russia became the 156th member of WTO.[130] As a member of WTO, Russia has to abide by the rules of its core agreements regarding the most-favoured nation, national treatment, market access and domestic regulation. As seen above, the effectiveness of such provisions in

128 Johnson, 2005, p. 273.
129 Belyi, 2009.
130 On 22 August 2012, 19 years after the beginning of its accession process.

regulating EU–Russian natural gas relations has important limitations. Hence it would appear that the guarantees offered by the ECT's provisions – for example, regarding cross-border energy trade – are stronger and more specific than those of WTO, which omit to refer expressly to energy. The ECT is a sector-specific treaty, with a focus on energy cooperation between its parties, in order to ensure energy security. Such an objective is not present in any of the WTO agreements. Thus, while the most vulnerable part of EU–Russian gas relationships remains the security of energy supply, it is questionable whether Russia's WTO membership will be able to substitute for the country's lack of participation in the ECT.

3.2.1.2 EU–Russia Energy Dialogue

Apart from the above presented general framework, EU–Russia bilateral gas relationships are currently regulated by the Energy Dialogue, established during the Paris summit on 30 October 2000. The present structure of the Dialogue comprises four thematic groups – the Energy Markets and Strategies Group, the Group on Electricity, the Energy Efficiency and Innovations Group and the Group on Nuclear Energy. Additionally it features a Gas Advisory Council. The Dialogue's leadership is equally divided between the so-called Delegated Coordinators – the Director General of DG Energy of the European Commission and the Deputy Minister of the Russian Ministry for Energy. At the 2000 summit, Russia and EU confirmed their allegiance to strengthen their strategic partnership, based on democracy, the respect of human rights, the rule of law and the market economy.[131] Furthermore, in the first progress report on the Energy Dialogue, the parties underlined common concerns for ensuring stable energy markets, reliable and growing imports and exports, and a pressing need to modernize the Russian energy sector, to improve energy efficiency and to reduce greenhouse emissions from energy production.

3.2.1.2.1 Potential and limitations The Energy Dialogue aimed to push the simple producer–consumer relations to a political partnership in energy, focusing on energy security.[132] By comparison, the PCA provided a too broad economic framework to tackle in sufficient detail Russia–EU energy relations. Nevertheless, during the initial years of its existence, the

---

[131] EU–Russia Summit Paris, 30 October 2000 Joint Declaration.

[132] EU–Russia Energy Dialogue Synthesis Report presented by Russian Vice-Prime Minister Victor Khristenko and European Commission Director-General François Lamoureux.

Dialogue developed in the shadow of the PCA, which represented the real legal basis for EU–Russia relationships. The Energy Dialogue was not intended to produce legal effects in either of the two parties. Rather, it was conceived as a platform for highlighting common interests,[133] recognizing cooperation areas of mutual benefit in the field of energy, and ensuring the development of the gas markets on both sides, as well as the security of supply and mutual trust between the partners. In this sense, the first progress report dating from 2001 provided that, in order for the mutual commitments of the parties to become legally binding, they had to be approved within the PCA framework. Despite the lack of legally binding effects, several concrete results have been achieved in the 12 years following the Dialogue's initiation. Between December 2002 and October 2003, various round tables on gas, energy strategies and electricity matters allowed Russia and the EU to discuss developments on their respective energy markets, yearly energy strategies and recent evolutions in each sector.[134] The round tables allowed the two parties to have a correct overview of the state of advancement of their respective energy sectors. Regarding the integration of the two markets, it was thought that such a step would be mutually beneficial for both EU and Russian industries regarding the security of supply, business opportunities and competition.

However, such market integration has still not been attained and is far from becoming a reality, owing to several legal, political and economic reasons. From a legal point of view both sides remain attached to contrasting models of energy market liberalization. At a national level, the most recent reformation of Russian competition law demonstrates an evolution in line with its European counterpart.[135] Such a development derives from the main objective of EU–Russia Roadmap for a Common Economic Space.[136] This underlines the need to accomplish "harmonised and compatible standards, regulations and conformity assessment procedures, where appropriate, including through enhanced regulatory dialogues

---

[133] For example, introduction of cooperation on energy saving, rationalization of production and transport infrastructures, investment environment and possibilities, and energy relations between producer and consumer countries.

[134] Haghighi, 2007, p. 346.

[135] See Federal Law No. 135-FZ of 26 July 2006 on the Protection of Competition.

[136] Roadmap for the Common Economic Space – Building Blocks for Sustained Economic Growth.

and cooperation between responsible institutions and a reinforcement of the institutional capacities".[137]

Notwithstanding such a common objective, the application of legal harmonization to the natural gas sector appears to have remained limited. Although Russian gas regulation includes third-party access provisions since 1997, the entire pipeline network continues to belong to Gazprom, which also controls most of the gas fields with immediate commercial significance.[138] The structure of tariff and access regulation by Russian authorities poses further difficulties to the effective use of competition in this sector. Recent modification brought to Russian natural gas legislation have consolidated the positions of majority state-owned companies, and especially that of Gazprom, which exerts a monopoly on all gas exports and the legal right to be granted certain exploration licences, not auctioned for competition.[139] Russia does not have an independent energy regulator in the sense of EU's liberalization Directives. Instead, the Russian Federal Antimonopoly Service is responsible for ensuring the respect of third-party access and for handling discrimination and lack of transparency, in relation to available capacity in networks.[140] Additional obstacles to free competition derive from the influence exerted by Gazprom in tariff formation and the repeated decisions of Gazprom's TSOs to block access to its gas networks.[141] It could be concluded that Russian gas reforms seem to have limited themselves to acknowledging third-party access rights, while failing to establish an independent body with regulatory oversight. Furthermore, the regulatory framework embodying the Russian natural gas industry continues to allow the dominant player to preserve its position on the market and hence depends on its cooperation for ensuring access to the networks.[142]

From a political and economic point of view, the two parties share different visions on what regards the perpetuation of the current arrangements satisfying EU's energy needs. Without contesting the important role played by Russia in the satisfaction of European energy demands, in

---

137 *Ibid.*

138 Belyi, 2011.

139 Federal Law No. 57 – FZ of the Russian Federation.

140 EBRD: Russia Country Profile, http://www.ebrd.com/downloads/legal/irc/countries/russia.pdf

141 However the production capacities of independent gas producers have increased during the last decade and Gazprom saw its share in the total national gas production decreased by approximately 20 per cent in the same period of time. See Belyi, 2009.

142 Stern, 2005, p. 195.

the aftermath of the 2006 Russia–Ukraine gas transit crisis the EU has started to look for alternative gas resources and transit routes curtailing its main gas partner. Such an initiative has been met with reluctance and opposition on the Russian side.[143] On the Russian side the stakes are given by the interest to preserve and possibly increase the stable rent of gas issued from the trade with EU Member States. There are several ways in which Russia could ensure the maintenance of its revenues from gas sales. Among these the most important are the acquisition of new supply markets and the maintenance and reinforcement of its shares in traditional markets. Arguably, diversifying towards new markets might prove to be the least productive option between the two, owing to geographical constraints and higher transaction costs (e.g. necessity of large capital deployments to develop new pipelines, lack of harmonization between legal systems). On the contrary, preserving the revenues from trading with old partners could guarantee the required stable contribution of gas rents to Russian GDP, despite sparking further security of supply debates at the EU level.[144]

The Caspian region represents an example of the EU's attempt to diversify its foreign energy policy. The European Neighbourhood and Partnership Instrument encompassing the former Technical Assistance to the Commonwealth of Independent States (TACIS) programme, as well as the Transport Corridor European–Caucasus–Central Asia (TRACECA), the Black Sea Regional Energy Centre (BSREC), the Black Sea Environmental Program (BSEP) and the Interstate Oil and Gas Transport to Europe (INOGATE), provide the general lines of cooperation between the EU and the Caspian region. However, none of these initiatives has been able to secure a steady export of gas towards the Union. There are two explanations for this situation. The first draws on the lack of a clear cooperation framework between EU and the Caspian region, in the sense of an international cooperation or association agreement, regulating the terms of energy supply towards the Union. The second explanation may come from the fact that the latter is geographically and politically situated within Russia's immediate proximity. Russia's leverage upon the Caspian region is two-fold. On the one hand, the region almost entirely depends on Russia's infrastructure to export and transport its oil and gas production. On the other hand, Russia remains one of the main energy trading partners of South Caucasus and

---

[143] For a brief description of Russian-initiated pipeline projects competing with EU's energy diversification policy see Ratner *et al.*, 2013.

[144] Paillard, 2010, pp. 65–84.

Central Asia. As demonstrated by the limited reach of the INOGATE programme, Russia's disagreement with the Union's involvement in the Caspian region seems to have put a negative strain on the level of openness of the region towards energy cooperation with the EU.

3.2.1.2.2 Main achievements in the framework of the Energy Dialogue In spite of the above explained obstacles to an integrated EU–Russia energy market, the positive effects of setting up a common purpose to achieve energy market integration could be observed in the initiation of energy infrastructure projects of common interest.[145] Furthermore, awareness about the importance of sustainable environment was raised in discussions held between the two parties within the Energy Dialogue's framework.[146] As a result, Russia ratified the Kyoto Protocol in October 2004, which came into force in February 2005. Additionally, the Dialogue offered the premises for the creation of the EU–Russia Technology Centre in 2002, which was meant to reinforce energy research and technology cooperation between the two parties and to provide a forum for the exchange of ideas and information. The leadership of the centre has been divided from the very beginning between a representative of the European Commission and one of the Russian Federation. The functions of the centre have been further specified with each Progress Report. In the fourth one, it was emphasized that the role of the centre included the identification, evaluation, facilitation and assistance in the promotion of concrete, large-scale, technology cooperation projects in energy sectors. It appeared that in the long-run the activities of the Centre were aimed at attracting mutually beneficial energy investments for the parties. As shown in the fifth progress report, the Energy Technology Centre has been mostly successful in promoting energy efficiency in Russia, while remaining quite ambiguous about its contribution to the improvement of the investment climate between the EU and Russia.[147]

As a further result of the Energy Dialogue, progress has been made towards the removal of "destination clauses" from long-term supply

---

[145] For example, the North European Gas Pipeline project. See EU–Russia Energy Dialogue Fourth Progress Report, Moscow/Brussels 2003.

[146] EU–Russia Energy Dialogue Fifth Progress Report, Moscow/Brussels 2004.

[147] In EU–Russia Energy Dialogue Fifth Progress Report the only mention of the role of the Energy Technology Centre in the improvement of the investment climate is that the latter together with the energy industry "steering group" "will contribute to the continued improvement of the investment climate between the EU and Russia, and identify more clearly the issues to be tackled".

contracts.[148] Such clauses aimed at curtailing supply competition in territories where the producer was also the main supplier. From the very beginning of the Energy Dialogue, Russia insisted on the importance of long-term energy contracts for energy security and a steady supply to the European market.[149] However, in order to mitigate the potential negative effects of such contracts upon the energy market opening, the European Commission adopted a cautious approach to specific clauses featured in such contracts quite early on in the liberalization process. Although the Commission stated clearly that take-or-pay clauses in long-term gas contracts are indispensable, it tried to reach a mutually agreeable solution with Russia for the amendment or deletion of "destination clauses" in both existent and future long-term contracts. In 2002 already, two years after the creation of the Energy Dialogue, both parties appeared to be satisfied by the progress in the matter.[150] In the 2003 report, the tone of the European Commission seems to have taken a turn towards a stricter interpretation of what a mutually agreeable solution in the case of "destination clauses" removal would amount to. Thus were mentioned Russia's interests, as well as EU competition law and other legitimate interests of the Union.[151] Such a change of tone occurred probably owing to the EU's own reinforced energy policy and Gazprom's continued practice to include or maintain destination clauses in contracts with European undertakings. A series of 2003–2007 Commission investigations into the gas sector revealed the inefficiency of the Energy Dialogue in the matter of removing "destination clauses" from Gazprom contracts.[152] To a certain extent such a development could have been anticipated taking into consideration that the Commission itself acknowledged the limits of the Energy Dialogue in replacing private companies in matters regarding the trade of energy supplies.[153]

---

148 Talus, 2010.

149 EU–Russia Energy Dialogue Synthesis Report, Brussels/Moscow, 2001.

150 EU–Russia Energy Dialogue, The Second Progress Report, 2002, http://ec.europa.eu/energy/international/bilateral_cooperation/russia/doc/reports/progress2_en.pdf

151 EU–Russia Energy Dialogue, Third Progress Report. Brussels/Moscow, 2002.

152 *GFU*, Case COMP/36.072; *Territorial restrictions 1) Algeria gas transport contracts, 2) Expansion of TAG pipeline*, Case COMP/37.811; *Territorial restrictions – Austria (Gazprom/OMV)*, Case COMP/38.085; *Territorial restrictions in Germany (Ruhrgas/Gazprom)*, Case COMP/38.307.

153 Communication from the Commission: The Energy Dialogue /*COM/2004/777 final*/.

More recent advancements connected to the Energy Dialogue have been the adoption of the Early Warning Mechanism in 2009 and the establishment of a Gas Advisory Council in 2011. The first instrument is a direct result of the 2009 gas crisis, which accentuated the energy security concerns over gas supply from Russia. The Early Warning Mechanism (EWM) was adopted to facilitate an early evaluation of potential risks and problems related to energy supply as well as a rapid reaction in case of an emergency situation. From Section 4 of the memorandum accompanying EWM it follows that the mechanism applies in cases of an emergency situation in the energy supply from Russia to the EU. This qualifies and limits the interpretation of the envisaged energy emergencies. Sections 5 and 6 of the memorandum specify obligations for notification and exchange of information between the parties to the Dialogue in the shortest possible time. The direct link of the instrument to the Energy Dialogue framework is demonstrated by the inclusion of members of the Dialogue's four thematic groups in an Expert Working Group, established within the instrument. The main task of the Expert Working Group, which brings together EU and Russian representatives, is to facilitate consultations that should lead to the solving of energy supply emergency situations. EWM has also a Monitoring Group, which is called upon to examine the circumstances of an emergency situation to keep an objective recording of them. The Group reunites representatives of both sides, energy companies, international energy organizations and independent experts. In February 2011, an updated version of the EU–Russia EWM was signed by the Commission and the Russian Minister of Energy. The updated instrument provides for joint actions, as measures to overcome an emergency situation, comb its consequences and prevent a future re-occurrence.[154] At the moment, it is still too early to predict how efficient EWM will prove to be in the eventuality of a new energy supply crisis involving Russia and the EU. It is, however, important to note that the updated version of the mechanism provides that the Coordinators may decide to address their respective decision-making bodies with a proposal for additional action at bilateral or multilateral level, in order to increase the efficiency of the joint actions. This remark appears to provide a legally enforceable backup to a system based otherwise on consultation, monitoring and recommendation.

The latest development of the EU–Russia Energy Dialogue is the establishment of a Gas Advisory Council consisting of EU and Russian

---

[154] Early Warning Mechanism: Memorandum.

representatives of the gas industry and gas market research experts.[155] The Council is intended to assess the development of the gas markets and to provide recommendations for future energy cooperation. It may not take any legally binding decisions, but it may issue recommendations on mid- and long-term perspectives of the EU–Russia Energy Dialogue, as well as the EU–Russia roadmap until 2050. Moreover, it can issue recommendations for further analyses on specific topics making the object of interest of the four Thematic Work Groups or other experts.[156]

Despite the results achieved in the framework of the EU–Russia Energy Dialogue, the efficiency of the latter remains questionable. An important argument here is that the scope of the Dialogue is limited to issues that touch only the surface of EU–Russian energy relationships. The more profound topics of the EU's attachment to liberalizing the gas and electricity markets and Russia's encouragement of its national gas champion are left aside.[157] The fact that such matters are not dealt with by what allegedly should be one of the most effective international frameworks addressing EU–Russian gas trade generates additional instability in the relationship between the two partners. The Dialogue brings together divergent Russian and European interests. Whereas the former is focused on investments to export infrastructure, the latter envisages primarily issues of access to Russian markets and security of gas supply.[158] The strategies pursued by the two parties in the Energy Dialogue are a reflection of their respective energy policies and a starting point for further legislative developments accentuating their position. According to the Russian energy strategy, natural resources situated in subsurface areas with federal status are considered strategic. Gas is part of this category of resources. The creation of a round table of industrialists within the framework of the Energy Dialogue was thought to facilitate the emergence of a stable energy investment environment between EU and Russia, through annual recommendations regarding trade and regulatory barriers between the two parties. However, despite the establishment of the round table, investments in Russian undertakings, deemed strategic for the exploration and production of gas, continue to be subjected to strict rules emanating from Russian investment law,

---

155 Conclusions of the first meeting of the EU–Russia Gas Advisory Council, Vienna, 17 October 2011, http://ec.europa.eu/energy/international/russia/dialogue/doc/20111017_gac_conclusions.pdf

156 EU–Russia Gas Advisory Council Rules of Procedure.

157 Roggenkamp *et al.*, 2007, p. 216.

158 Talus, 2010.

which impose statutory restrictions upon foreign investors.[159] In its turn, the "reciprocity" clause introduced in the proposal for a third European energy package was a projection of the EU's policy of securing the supply of gas by entering the Russian upstream markets. The proponents of the clause bestowed a two-fold purpose upon it. On the one hand the clause was designed as a balancing measure against Gazprom's acquisition race in the European downstream gas markets.[160] On the other hand, the clause was aimed at counteracting market foreclosure strategies that accompany long-term contracts between the Russian gas incumbent and main European gas undertakings. As a result of Russian investment laws and EU's demands for reciprocity in asset acquisition practices, the collaboration between the two partners under the Energy Dialogue was challenged.

The present regulatory environment in the Russian gas sector and the lack of a specific legislative instrument encompassing the multilateral relations between Russia and the EU in the energy field burden the access of independent producers to the transport network, reduce the incentives of foreign investment and project legal uncertainty upon the EU–Russian gas trade relationships. The third European energy package attempted to reduce this legal uncertainty by establishing mutual standards of ownership unbundling for Member State TSOs and external TSOs looking to acquire shares in the European natural gas transmission network.[161] Nevertheless, Member States' disunity upon the minimum standards of unbundling that should be adopted throughout the EU allows the maintenance of specific bilateral relations between Russian gas exporters and the different EU Member States.

3.2.2 National Specificities

Under the TEP, Member States continue to play an important role in gas market opening and integration, as well as in cross-border energy exchanges. As part of achieving EU's energy objectives, Member States are responsible, among other things, for establishing national regulatory agencies and guaranteeing their independence,[162] for setting up TSOs in

---

159 Federal Law No. 57 – FZ of the Russian Federation, *supra* note 139.

160 For example, the establishment of Promgas in Italy and Wingas in Germany allowed Gazprom the possibility to distribute gas directly to its European customers, which put it in a more advantageous situation compared with its European counterparts.

161 Article 11 of Directive 2009/73/EC.

162 Article 39 of Directive 2009/73/EC.

compliance with Directive 2009/73/EC[163] and for establishing a supplier of last resort to ensure a stable supply of energy to vulnerable customers.[164] They are supposed to ensure that TSOs harmonize their balancing regimes and streamline charges and levels of balancing charges.[165] At the same time, they are responsible for laying down rules on penalties applicable in case of infringement of the provisions of the Third Energy Package.[166] Member States enjoy a large discretion with regard to measures ensuring security of supply,[167] consumer protection,[168] powers granted to national regulatory authorities[169] and public service obligations.[170]

In light of the above, the particularities of Member States' relationships with main gas suppliers receive a distinctive importance. As a completion of the EU–Russia energy cooperation framework, Member States may and do forge and maintain bilateral energy relationships with the latter. The EU gas market is internally divided. The result of this is the high short- and mid-term vulnerability of the fragmented Member States' gas markets as exemplified by the cut-off in 2009, which bestowed a pronounced political sensitivity upon gas issues.[171] The approximately three weeks of no Russian gas being exported through Ukraine underlined the Union's external dependency on Russian gas and the vulnerability of gas cross-border trade transiting intermediary countries.[172] Moreover, it emphasized the importance of having emergency plans and measures in place at a national level and the need to establish a coherent EU approach encompassing the Union's external energy security. The lack of a real pan-European pipeline network has contributed to the difficulties of forging an integrated EU natural gas market and has ensured a low level of cohesiveness among Member States. As a

---

163 Article 1 of Regulation (EC) No. 715/2009.
164 Article 3(3) of Directive 2009/73/EC.
165 Article 21 of Regulation (EC) No. 715/2009.
166 Article 27 of Regulation (EC) No. 715/2009.
167 Regulation (EU) No. 994/2010.
168 Article 3(3) of Directive 2009/73/EC.
169 Article 39 of Directive 2009/73/EC.
170 Article 3(3) of Directive 2009/73/EC.
171 Grätz, 2009, pp. 66–80.
172 For example, 80 per cent of the total gas traded from Russia to the EU transits Ukraine, which has been involved as a dispute party in both 2006 and 2009 energy cut-offs to the EU. See http://europa.eu/rapid/pressReleasesAction.do?reference=SPEECH/09/137&format=HTML&aged=0&language=EN&gui Language =en

consequence, the latter has decreased the leverage of the EU's bargaining power in its cross-border relations with its main supplier.

The relationships of both the Union and its Member States with gas producers vary considerably according to different aspects ranging from the volume of traded gas to the exporters' acceptance and replication of European market liberalization rules. As explained in the following, traditional gas supply relationships between EU Member States and Russia can be divided into four categories.

3.2.2.1 Traditional gas supply relationships between EU Member States and Russia

A group of countries formed by the old Member States – Germany, Finland, Italy and France – nurtures mutually positive gas trade relationships with Russia and Gazprom. Historical links and geopolitics are essential factors influencing the said relationships. Owing to their long common border, Russia is one of the natural trading partners of Finland. The country is 100 per cent dependent on Russia for its gas supplies, but unlike the Baltic States, which are in the same situation and resent this dependency for political reasons, Finland appears to be confident in the possibility of securing its vital interests through dialogue with Moscow.[173]

Germany and Italy are the recipients of the largest share of Russian gas supply to the EU and probably the ones who benefit the most from long-term contracts and the stability they offer in terms of price and capacity delivered.[174] The most recent partnership signed between ENI and Gazprom extended Italy's contracts for Russian gas supply up to 2035 and ensured ENI's participation in South Stream to the detriment of the Nabucco pipeline project. Russia is Germany's most important energy supplier, while Germany is one of Russia's leading investors in internal energy infrastructure. Furthermore, the strategic partnership between them links the two countries' mutual support in energy matters to a larger area of common interests including disarmament, arms control, the management of international crises and global challenges.[175]

---

[173] Whist, 2008.

[174] In Germany the main Russian gas incumbent, Gazprom, has been present for many years through joint ventures with E.ON Ruhrgas, Wingas, RWE and Bayernwerk. In contrast, in Italy Gazprom entered its gas sector for the first time in 2008, through the notified merger between Gazprom Germania GmbH and A2A Spa. See *Gazprom/A2A/JV*, Case M.5740.

[175] Federal Foreign Office: Strategic Partnership with Russia, http://www.auswaertiges-amt.de/EN/Aussenpolitik/RegionaleSchwerpunkte/Russland/Russland_node.html

France is the fifth largest gas consumer in the EU, with no obvious dependency on Russian gas. The country's main gas providers are Norway and the Netherlands. Despite the lack of a striking dependency, the French–Russian energy trade has developed in a beneficial manner for both parties since 1975.[176] A fine example of this is given by the extension in 2006 of the cooperation agreement between Gazprom and GDF Suez granting the latter access to Russian gas supplies until 2030.

The mutually positive gas relations between Russia and the above enumerated importing countries are reflected in partnership agreements initiated between their main incumbents. Such agreements generally envisage a specific goal, which might refer to acquiring knowledge of local laws and practices and reinforcing solid relations with hosting governments.

A second category of Member States depends either entirely or in the majority on natural gas imports from Russia. In general, these are Central and Eastern European countries, which have been vulnerable to the reinforcement of the Russian gas monopoly in their distribution markets during recent years.[177] Indeed, most markets in Central and Eastern Europe are tied to Russian supplies through one-way relationships with Gazprom, which ensures the coverage of their gas needs and the necessary investment in domestic markets.[178] Generally, such relationships take the shape of long-term contracts. The strengthening of the Russian presence in the territories of the new Member States draws upon economic reasoning. While most of Russia's money comes from long-term contracts with traditional partners, the best deals represented by assets acquirement have emerged in Eastern and Central European countries. In 2006, eight of the new Member States imported 77 per cent of their natural gas needs from Russia.[179] From those eight countries, the Baltic States and Slovakia[180] import all their gas from the former Russian Federation and their markets are generally dominated by joint ventures involving the participation of Gazprom. This reliance on Russian gas has increased with the accession of Romania and Bulgaria to the EU. The Bulgarian gas sector is fully dependent on Russian gas imports and the

---

176 Gazprom export: France http://www.gazpromexport.ru/en/partners/france/

177 Roggenkamp *et al.*, 2007, p. 200.

178 Dreyer *et al.*, 2010.

179 Kaderják *et al.*, 2007.

180 That is, the Slovakian gas market is dependent in totality on imports of Russian gas and it appears to be dominated by a joint venture between the Slovak government and Slovak Gas Holding, which is reputed to have links with Gazprom.

country's market intermediary is a subsidiary of Gazprom. Despite this absolute dependency, Bulgaria is one of the few examples in Central and Eastern Europe where concerns do not run extremely high about the risk that Russian investment may bring with it unwanted political influence.[181] By way of contrast, the Baltic States and Poland illustrate Member States that prefer to see the EU as a protective force for their energy policy. This vision is not shared by some of the old Member States, which would rather keep their energy security matters within national borders.

The UK and the Netherlands represent a third category of Member States, according to their position regarding gas trade relationships with Russia. These have recently developed closer relations to the latter, as a result of a change in their national energy resources. Shortly after becoming a net importer of gas, UK signed a deal with Russia to prospect the building of a pipeline under the sea, bypassing the existing European infrastructure. The plan has been designed to allow the UK to buy gas directly from Russia, without the risks and delays implied by transit countries.[182] At the moment, however, it is not clear whether such a pipeline will be built, owing to the continuing change in the perception of Russia as a reliable supplier. To date, there are no long-term supply contracts between the two sides, partly owing to the UK's self-reliance on gas resources until recently and the country's limited storage capacity and largely spot-based markets in both electricity and gas.[183] Together with the Netherlands, the countries are two of the main shareholders in the contested Nord Stream pipeline.

Spain and Portugal form a fourth category of Member States, whose dependency on Russian gas is absent or close to nil. With regards to Spain, its main source of imported gas is Algeria, which delivers the supplies through a pipeline and by LNG delivery. Additionally, Spanish natural gas imports come from Nigeria.[184] A similar situation occurs in Portugal.[185]

---

[181] CMS, 2008, p. 7.

[182] Roggenkamp *et al.*, 2007, p. 1172.

[183] Brinkley and McIlven, 2010.

[184] Spain – Energy mix fact sheet, http://ec.europa.eu/energy/energy_policy/doc/factsheets/mix/mix_es_en.pdf

[185] Energy Delta Institute: Portugal, http://www.energydelta.org/mainmenu/edi-intelligence-2/our-services/Country-gas-profiles/country-gas-profile-portugal

3.2.2.2 Elements contributing to the fragmentation of the EU gas market

The above listed categories are not exhaustive and the placement of a certain Member State in any of them is not definitive. However, they are useful in portraying the EU's gas market fragmentation. This stems from a convergence of various levels of gas imports dependency, Member States' attitudes towards market integration and market reform and the influence of stakeholders upon the position adopted by national governments.

Indeed, different gas supply dependency patterns correlated with specific energy mixes characterizing each Member State legitimizes asking if and how does the Russian energy policy contribute to the segregation of the EU gas markets? A first answer comes from the political deliberations surrounding the "reciprocity" clause. The leading countries demanding the scrapping of the said clause from the original Third Energy Package belong to the group of Member States that favours mutually positive economic relations with Russia. At the same time, they represent three of the four members of the Council of the European Union with the largest amount of votes. Among their opponents, Poland played a major role in advocating the need of the reciprocity clause.

A second relevant example stems from the discussions around the energy solidarity pact. This was proposed by Poland in the aftermath of the 2006 gas transit crisis. Poland based its proposal on an analogy with NATO's article 5 on mutual security[186] and it referred to the instalment of a rule of mutual help in emergency situations where one Member State's energy security had been violated.[187] From a structural point of view, the Polish vision included the construction of joint energy storage facilities and joint energy networks between Member States. From an ideological point of view, it entailed the consideration of any threat to the energy security of a Member State as a threat to EU's energy security. The structural argument coincided with Commission's position, which considered that "choices made by one Member State inevitably have an impact on the energy security of its neighbours and of the Community as a whole".[188]

---

[186] It should, however, be noted that, while referring to the provisions of article 5 NATO, Poland did not address any legal hierarchical relationship between EU and NATO on this matter. Specifically, its proposal was not intended to suggest that EU law was there to enforce NATO law. Instead, the reference to article 5 was maintained as a parallel.

[187] Geden *et al.*, 2006.

[188] Green Paper /*COM/2006/105 final*/.

The ideological argument could be considered as decisive for the initial failure of the solidarity proposal. Older Member States, like Germany and France rejected the solidarity idea for fear of hindering their political and economic ties to Russia, while newer Member States welcomed such a plan, while they expected it to decrease their energy supply vulnerability. The result of the opposition of several Member States has been the adoption of a fuzzy solidarity concept, comprising various intrinsic limitations. Specifically, article 6 of Directive 73/2009/EC conditions the solidarity action of Member States to a supply disruption that is deemed severe.[189] It could be deduced from such a formulation that the obligation of solidarity cannot be engaged automatically as a result of any security of supply threat. On the contrary, only severe disruptions oblige the Member States to act in solidarity.

The Lisbon Treaty retains and employs the concept of solidarity without really defining it. Article 122(1) TFEU provides that the Council, following a proposal from the Commission, may decide, in a spirit of solidarity between Member States, upon the measures that should be taken in case of severe difficulties in the area of energy. Not only the Council enjoys a large margin of appreciation in what concerns the course of action, but also the Member States. Article 194 TFEU underlines the need for cooperation between Member States. However, no consequences are mentioned for a Member State not respecting the solidarity impetus. Hence, it is questionable whether the solidarity mentioned in the Treaty represents an effective limitation to Member States unilateral actions regarding energy matters – for example, with regard to their relationships to external suppliers. Consequently, it could be inferred that the compromise agreed by the Member States on the solidarity clause has certainly guaranteed the introduction of the concept in the Treaty and the sector legislation, but at the same time might have seriously impaired it.

Traditional bilateral contracts continue to be the major tool regulating the gas trade between Russian and EU Member States at the expense of the energy security of some of the new Member States. This is an interesting observation when arguing that Russia's energy policy favours the preservation of gas market segmentation in the EU. When lacking a European perspective, bilateral agreements regarding the construction or enhancement of pipelines contribute to the poor interconnection of the

---

[189] This contrasts with the provisions of article 5 NATO, which do not request the qualification of an emergency situation as severe to attract mutual assistance.

European gas transmission grid. In this sense, the purpose of the Nord Stream project[190] between Germany and Russia has been the building of a pipeline under the Baltic Sea, linking the two countries and bypassing Poland, the Baltic States and Ukraine. Articles 60 and 79 of the 1982 UN Convention on the Law of the Sea[191] allow the construction of the pipeline according to the original plans and prevent Member States negatively affected by exclusion decisions from having any legal say in the matter.[192] However the applicability of the Convention has not obstructed the Baltic States and Poland from formulating security of supply objections to the Nord Stream project, owing to uncertainties related to the security of their own energy supply.[193]

The increased importance of natural gas for the energy cycles of industrialized economies, correlated with the growth of the volume of gas traded on international markets, has legitimately determined a shift in the focus of the large consuming countries towards the security and availability of their gas supply. An example of pro-market integration advocacy is provided by UK's intervention at the 10th Madrid Forum. This was aimed at influencing the discussions on the harmonization of gas qualities throughout Europe in order to ensure the interoperability of European transmission systems. The UK appears to be leading the way in gas quality standards harmonization, albeit steering the direction towards its own views on the matter.[194] Germany and France are counting themselves among the sceptics. Since the beginning of the liberalization process, Germany has manifested restraint in accepting EU energy policies that threaten its coal industry and could potentially cause high unemployment and social turmoil. Arguably, at that time the country was still at the beginning of its reunification programme. Regarding France, its tradition of state-planned industry policies and its attachment to

---

[190] The Nord Stream pipeline is a natural gas pipeline project initiated by Russia, aimed at bringing natural gas from Russia to the EU through the Baltic Sea and bypassing Poland and Ukraine. The project has been greatly encouraged by Germany, followed by the UK.

[191] Article 60 indicates who has the right to construct or supervise the construction of pipelines in an Exclusive Economic Zone. Article 79 provides for the right of all states to lay submarine pipelines and cables on the continental shelf of other states. Of additional interest for the EU–Russian gas relationships is article 124 of the Convention, which regards the right of access of land-locked states to and from the sea and the freedom of transit.

[192] Whist, 2008.

[193] SIPRI Yearbook 2007: Armaments, Disarmament and International Security.

[194] Haase, 2008, pp. 18–19.

nuclear energy have been maintained at times in contradiction to EU energy policies.[195]

In what concerns the influence of market stakeholders upon the market integration approach taken by the governments, it is suggested to look at Bulgaria and Slovakia. Both countries are interested in multiplying their market suppliers in order to conform to the liberalization packages, be able to guarantee the security of supply to their consumers and fulfil their public service obligations. However, the actors in the supply chain share powerful links with Gazprom, which determine the lack of their incentives to diversify imports or to invest in significant storage capacity.

3.3 REFLECTIONS ON THE SPECIFICS OF THE EU GAS MARKET

To a large extent, the perpetuation of the pre-liberalization *status quo* has been due to the weak unbundling between national incumbents and the transmission system operators. This has emphasized the structural disadvantages of new entrants on the market derived from the rigidity of the pipeline grid, and hence has allowed the maintenance of traditional gas trade relationships, which concentrate on the long-term pipeline capacity in the hands of the national incumbents. As a result, cross-border gas trade has been mostly translated into European mergers and acquisitions, allowing major energy utilities to strengthen their position on the gas market.[196] At this time, the effectiveness of the new unbundling measures in TEP to ensure a level playing field are still to be tested both among EU Member States and gas undertakings and at the level of external suppliers.

In what concerns the European dimension of the natural gas market, the EU has yet to develop an effective energy policy translated both in a more clear articulation of its energy objectives and in its ability to speak with one voice with regard to external energy matters. Article 17(1) TFEU, which provides the Commission with ensuring the Union's external representation in all fields of EU interest – except the common foreign and security policy – might constitute the future grounds for a real EU external energy policy. However, at the moment, this possibility is curtailed by the sharing of energy competences between the EU and the Member States and by the provisions of article 194(2) TFEU, which

195 Eikeland, 2004, pp. 26–27.

196 Haar and Marinescu, 2011.

could be interpreted as allowing the Member States to preserve their right of conducting bilateral energy relationships with non-EU suppliers as they see fit.[197] The solidarity spirit between the Member States, recommended when aiming at the establishment and functioning of the internal market,[198] has visible limitations owing to its rather vague formulation and the lack of accompanying sanctions, should the Member States decide to move in different directions. Taking a step further from the Treaty provisions, the 2010 Gas Regulation underlines the preventive aspect of the required solidarity and coordination among Member States in response to supply crises.[199] Certainly a positive reaction of the Member States' to the demand of adopting a unified stance in the face of security of supply threats would benefit the shaping-up of an integrated gas market. However, as seen in the following chapter, such reaction may be severely impaired by current energy exemptions and derogations,[200] which allow the national markets that most necessitate a boost of investments and connection to the rest of the EU to avoid improved competition conditions and thus continue to exist as isolated systems.

---

197 Braun, 2011.
198 Article 194 TFEU.
199 Regulation (EU) No. 994/2010
200 Articles 36 and 49 of Directive 2009/73/EC.

4. EU gas market structure

Market structure is understood as the sum of institutional and organizational arrangements of gas supply in the EU, resulting from the impact of gas market actors' behaviours upon gas market opening. For the purpose of analysing the conditional relationship between these two elements, the chapter discusses the levels of gas market concentration and gas market interconnection. Both aspects are essential when assessing the current stage of gas market liberalization. At the same time they provide the opportunity to observe how the instruments at hand can be better adapted to provide an integrated market structure.

The level of market concentration provides accurate signals about the barriers that still exist in the market, the number of actors who enjoy enough market power and the parts of the market that suffer the biggest foreclosure. In return, the level of market interconnection shows to what extent the EU gas market is still segmented into national constituencies. There is a direct link between the persistence of market foreclosure strategies and poor market interconnection. Thus, traditional market actors, who preserve an incumbent position on their respective national markets, only rarely decide to enter other markets as competitors. This favours the maintenance of a fragmentation of the EU gas market along national lines.

4.1 THE LEVEL OF CONCENTRATION IN THE EU GAS MARKET

One of the main findings of the 2007 Energy Sector Inquiry regards concentration in the European gas markets. According to the report, these have remained national in scope, as a result of main incumbents maintaining their dominant position in their traditional markets.[1] Taking into consideration the Energy Inquiry, several aspects of gas market structure are simultaneously considered. These include market strategies

[1] DG Competition Report on Energy Sector Inquiry, /*SEC/2006/1724/, p. 7.

of gas producers and main internal EU suppliers, the persistence of barriers to market entry and the typology of supply arrangements. Two of these elements regard the impact exerted by interactions between market stakeholders upon the level of market concentration. The third element refers to specific supply arrangements, inherent to rigid, grid-based industries. In discussing these aspects, the goal is to observe to what extent the combined application of competition tools and regulatory measures has determined a change of the old order, and how such a change – if any – falls in line with the provisions of the Third Energy Package.

4.1.1 The Market Players

The notion of gas market players is employed to describe industry stakeholders, which are active on the EU gas market and whose actions influence the organization of the market structure. In general, these are natural gas producers and exporters, gas importers and traders active at the wholesale level of the gas business, retail suppliers and providers of network and supporting services, large industrial consumers, local DSOs, small consumers and businesses, regulators, legal and competition authorities, financial service providers and the government.[2] The focus of this subsection is mostly centred on natural gas producers and European gas suppliers and on the strategies used by each of these categories to maintain their position in the market. The focus of the analysis was not chosen arbitrarily. Natural gas producers and European transmission system operators form one of the most influential interest groups with access to the formulation of EU energy objectives by EU institutions and Member States' governments.[3] In addition to the two categories identified above, the section discusses the emergence of transmission system operators as independent market players.

4.1.1.1 Gas producers and suppliers

The main gas producers continue to hold a monopoly over export in their respective countries owing to their governments' preference for strong national champions and the lack of effective sector liberalization.[4] Their

---

[2] See Keyaerts *et al.*, 2008.

[3] Eikeland, 2004, p. 20.

[4] This is no longer the case for Norway owing to its submission to European liberalization rules, but continues to be the case for the rest of the major European gas suppliers.

limited number has led in the past to the formation of an oligopolistic-like structure of their side of the market and has driven recent discussions on the legitimacy of emergent gas cartelization fears.[5] It is very interesting to observe that in general no real rivalry exists between third-country gas producers with regard to sharing the EU gas market. According to EU energy mix factsheets, in 2007, Italy, Germany, Belgium and France represented the only EU Member States featuring three or more non-EU gas suppliers.[6] In general most Member States import gas from only one external supplier, which in the majority of the cases is Russia. The latter remains by far the largest gas producer, whose competitive approach towards gas related projects is centred on launching pipeline construction plans as counter-acts to EU's diversification plans. Such actions are not necessarily directed against other external gas producers. Instead they are meant to satisfy Russia's own interest in maintaining its level of revenues from gas sales.

In the aftermath of EU gas market liberalization, major gas producers have adopted a dualistic approach. On the one hand, their policy has been to curb the outreaching effects of the European liberalization process on their national gas markets. On the other hand, third-country gas producers have started to take advantage of downstream liberalization, by reinforcing their policy of acquiring foreign assets.

Natural gas producers generate high levels of market concentration through their relationships with suppliers situated in the EU. Such relationships often have a commercial, an institutional and a legal component. Each component operates with different actors, outcomes, methods and dynamics of cooperation.[7] The commercial aspect of the relationship between producers and suppliers has traditionally taken the form of long-term contracts locking in important market shares or network capacity for a very long time. Such contracts have been encouraged by both sides owing to their ability to settle delivery and purchase for years and to make prices stable. In several cases, however, their side-effect has been the creation of significant barriers to entry into the gas market.

---

[5] Gorst and Crooks, 2008.

[6] Belgium – Energy mix fact sheet, http://ec.europa.eu/energy/energy_policy/doc/factsheets/mix/mix_be_en.pdf; France – Energy mix fact sheet, http://ec.europa.eu/energy/energy_policy/doc/factsheets/mix/mix_fr_en.pdf; Germany – Energy mix fact sheet, http://ec.europa.eu/energy/energy_policy/doc/factsheets/mix/mix_de_en.pdf; Italy – Energy mix fact sheet, http://ec.europa.eu/energy/energy_policy/doc/factsheets/mix/mix_it_en.pdf

[7] Talus, 2010.

The institutional part of the relationship occurs in a multilateral or bilateral pattern of cooperation between the national governments of the producing and supplying undertakings. Closely related to this, the legal component concerns the law applicable to the economic relationship between the two sides. Gas trade relations that take place in the territory of the EU are submitted to EU legislation and rules. Of specific importance for the current regulation of the position of EU gas suppliers in the EU gas market are the provisions of article 11 of the 2009 Gas Directive on the certification of third-country TSOs. In its essence, the certification procedure constitutes a compromise replacement of a proposal of assets' acquirement reciprocity. The removal of such a clause from the original proposal for a Third Energy Package cannot be directly assimilated to a strategic movement on the producers' side. Rather it underlines the influence of cooperation patterns between producing and importing countries upon the legal component of gas trade relationships. At the same time, it demonstrates the perfect symbiosis between producing countries and their gas incumbents with regard to downstream strategies.

Distortions to competition in the gas market have been regarded as an advantage by major European gas incumbents, which could only profit from a concentrated market in which their dominant position was ensured. Prior to the liberalization wave such incumbents formed an oligopoly of purchasers.[8] At the time, almost all Member States had one transmission company that was owned or partially owned by the national government and that monopolized wholesale gas. The situation has gradually changed through the successive gas sector regulatory reforms. Nevertheless, in 2007 the Energy Sector Inquiry revealed that the majority of European national gas markets were still highly concentrated, with just one or two significant players and some marginal competition within the competitive segments of the market. This happened while national governments have continued to back up the bargaining power of European gas incumbents. As an example of such a behaviour, Spain has both orchestrated the creation of a national giant in the utilities sector through the merger of Endesa and Gas Natural,[9] and further protected the position of its gas incumbent in the market by preventing foreign undertakings from take-over auctions.[10] Similarly, the rigidity and lack of

8 Spanjer, 2008, pp. 3–8.

9 *Gas Natural + Endesa*, COMP/37.542.

10 *Commission v Spain*, Case C-196/07; *E.ON/ENDESA*, Case M.4110.

liquidity in the European gas markets as a result of the GFU joint selling schemes occurred with the support of the Norwegian government.[11]

Under liberalization, market concentration occurs as a result of traditional players' strategies to avoid competition from potential new entrants in order to maintain their prominent position in the market and avoid being taken over by their rivals. Such strategies include the setting of end-user prices by state intervention and traditional players, which attempt to maintain their dominant position in the market. End-user price fixation interferes with the good functioning of the market. Such prices do not reflect the market's real needs and prevent free competition and market integration. As a result, consumers are deprived of their right to choose the best service in the market.[12] On the EU gas market, limited new entry of competitors to the retail markets has led to a reduced consumer choice in terms of supplier, an ineffective competition pressure and high levels of concentration similar to the pre-liberalization period. Case COMP/B-1/39.316 – *Gaz de France* is an illustration of this. The case refers to long-term bookings of almost all capacities on an essential facility. As a result of this, GDF ensured the preservation of its dominant position in the downstream markets. Owing to the important hindering effects that such a long-term capacity reservation would exert on third-party access and market competitiveness, the decision in the case deemed GDF actions as abusive, notwithstanding its real need to use the reserved capacities for its own transmission purposes.

Further strategies leading to market concentration have seen certain incumbents making use of their control over transmission networks to make market entry and expansion of competitors in the supply markets difficult. Both the regulatory provisions and the employment of competition law tools have allowed major incumbents to anticipate the loss of their market shares as a result of liberalization. The industry's response to this threat has been a renegotiation of former market arrangements

---

[11] For example, in 2001 the Norwegian government claimed that European competition law could not be applied to the GFU case, while the selling of Norwegian gas through the GFU scheme has been ordered by the government itself.

[12] For example, the Polish energy law requires energy enterprises to specify the end-user prices for gas and obtain prior approval from the NRA. Such approval must be obtained also by new entrants before having access to the market. The measure does not sufficiently differentiate between different consumer groups and its scope is too broad. In Romania, regulated end-user gas prices are still applicable to all consumers. See European Energy Review, 07/04/2011: Commission urges Italy, Poland and Romania to ensure freedom of choice for consumers.

resulting in cross-border and cross-sectoral mergers and acquisitions and assets swap between different segments of the gas network.

Four types of alliance and acquisition strategies meant to reinforce the position of certain incumbents in the market have been observed.[13] The first refers to the expansion of the incumbents' geographic range of activity in order to gain valuable leverage on the European level before the complete opening of the market.[14] The incumbents' reasoning has been to compensate the loss of advantages provided by their former monopoly through an increase of their customer base. This has been achieved through acquiring assets in other EU Member States and outside the EU. The second strategy has been practised by downstream local utilities. Their reasoning has been to protect themselves against foreign hostile takeover through the creation of stronger players "with sufficient bargaining power to agree favourable terms with strong European suppliers and competitors".[15] This has been achieved through national or regional mergers between previously less significant undertakings active on the downstream market. Thirdly, major gas incumbents started to see the potential of acquiring assets in other sectors in order to maintain their high profits. As a result of cross-sectoral mergers and acquisitions, costs generated by network activities could be reduced through the integration of different steps of the supply chain and national energy champions could be created. The fourth identified strategy observes newly created pan-European energy groups, which attempt to secure their market shares and supplies through vertical integration in both directions of the gas stream. These new types of vertical integration, which include cross-sectoral elements, may be interpreted as the incumbents' counter-action to growing uncertainties caused by the gas market liberalization.

It is interesting to observe that the commercial strategies of gas incumbents to maintain their dominating position in the market differ significantly from those of undertakings on other utility markets. In this sense, the preferred market foreclosure method of telecommunication incumbents has been the use of the margin squeeze.[16] According to Koch *et al.*, "a margin squeeze may occur if a vertically integrated firm which is dominant on an upstream market charges a price for a downstream product or service that prevents even equally efficient competitors from

---

[13] Cronshaw *et al.*, 2008, p. 13.

[14] Verde, 2008, pp. 1125–1133.

[15] *Ibid.*

[16] The procedure refers to a reduction by the dominant operator of the margin between the wholesale or retail price by either increasing the former, lowering the latter or a combination of the two.

achieving a margin which allows them to compete effectively on the downstream market".[17] The immediate effect of a margin squeeze is rendering the market entry more difficult for third-party access or facilitating the market exit.[18] On the gas market, examples of margin squeeze have been scarce. In the *RWE gas foreclosure*[19] commitment decision, despite allegations that the German undertaking established asymmetric conditions to network access and applied an uncompetitive margin squeeze between its gas transmission tariffs and its downstream gas supply tariffs, no breach was finally found.

The adoption of the TEP has challenged market foreclosure arrangements through the changing of the unbundling regime applicable to European gas market undertakings. The new regime follows the logic of preventing vertically integrated incumbents from using their monopolies over transmission networks in order to disturb competition in supply.[20] The envisioned unbundling has a triple action upon market concentration. At the level of vertically integrated undertakings, effective unbundling attempts to break the vassalage link between monopolistic and competitive segments, either through a division of ownership or through the creation of multiple accountability forums. At the level of the market, the immediate expected effect is the elimination of barriers regarding new entry. At the level of new entrants, effective unbundling targets the solidification of their position on the gas market by balancing competition demands and incentives for new investment.

Reduced incentives for investing in the network can be explained from both an incumbent and a new entrant point of view. It has been observed that, in general, market concentration strategies are accompanied by insufficient investment in the transmission networks.[21] This is also called strategic under-investment because the attempted effect is the weakening of supply competitors in the gas market, where the incumbent is the largest producer, importer or supplier. Owing to their long history in the market and favourable relationships with the government, such undertakings have the largest share of investing in supplementary transmission capacity, which could bring additional competition to their supply. This is why, in an inefficiently unbundled market, the lack of incumbents' incentives to invest in the network helps perpetuating market foreclosure. However, in a more efficiently unbundled gas market, the interest of new

---

17 Koch *et al.*, 2009.
18 Geradin and O'Donoghue, 2005.
19 *RWE gas foreclosure*, Case COMP/39.402.
20 Lowe *et al.*, 2007, pp. 23–34.
21 *Eni SpA v Commission*, Case T-558/08.

entrants to invest might be somewhat reduced by the uncertainties surrounding such an act, the increased competition in supply and the lengthy process of returns.

Under the TEP, such attitudes are counteracted by investment protection measures. Such measures regard the entrusting of transmission system operators with the annual production of a 10 year network development plan including all decided investments, as well as newly identified ones.[22] The plans have to be submitted to the regulatory authorities, which are responsible for their approval.[23] These provisions suggest that decision-making regarding new investments does not belong entirely to the management of a gas undertaking, but that under Directive 2009/73/EC it becomes the product of a collaborative process between different energy bodies. Article 36 of the Directive provides an additional safety net for investments in new infrastructure, by allowing their temporary exemption – under certain circumstances – from unbundling of transmission systems and transmission system operators, third-party access, access to storage and access to the upstream pipeline network. Additionally, recital 14 and article 7 of Regulation (EC) 715/2009, read together, require that network access tariffs or the methodologies of their calculation should provide incentives for investment – for example, by including appropriate returns. Such provisions are expected to increase the appeal of new investments in the transmission network.

4.1.1.2 Transmission system operators

As seen in Chapter 2, under the TEP, the role of TSOs with regards to ensuring effective unbundling has been increased. This has led to the appearance of a new type of gas market actor. Of course, transmission system operators existed also before the third liberalization package, but their attributions and powers were governed and monitored by vertically integrated gas undertakings. As part of a vertically integrated undertaking, TSOs have been able to restrict the access of competitors to information about the transmission network. At the same time, when allocating transportation capacity or access to storage facilities, they have shown a tendency to grant preferential treatment to the competitive segments of their own system. It should be observed that none of the unbundling regimes under the first and second liberalization packages has been able to ensure the independence of TSOs from the central command of vertically integrated undertakings. Under the current unbundling

---

[22] Article 22(2)(b) of Directive 2009/73/EC.

[23] Article 41(1)(g) and (3)(c) of Directive 2009/73/EC.

regime, three different levels of interdependency between TSOs and the production and supply of natural gas, as well as varied decision-making powers of the system operators coexist.[24] At the moment, only a limited number of Member States have opted for the ownership unbundling of their gas undertakings, which provides for the greatest degree of independence between TSOs and the production or supply ends of the chain.[25] Hence, legitimate questions may be formulated concerning the extent to which main incumbents are still able to influence decisions taken by their TSOs.

On the European gas market, the role of the TSOs is to facilitate access to the gas network through regulating balancing regimes. Such regulation is most frequently achieved through locking considerable amounts of flexible gas in long-term contracts or otherwise allocating long-term capacity rights to suppliers. These rights are essential for network users, because they allow them to be sure that their gas will be transported through the pipeline grid within the limits of the contracted capacity.[26] In this context, the important leverage that TSOs exert upon capacity allocation increases the risk of network access manipulation to the detriment of new suppliers.

It could be observed that, in general, incentives for TSOs' inefficient behaviour derive either from their close connection to the monopolistic segments of the market or from the incorrect reflection of balancing costs in imbalance charges.[27] The latter case is addressed by the new balancing provisions advanced by ACER, which demand the publication of transparent methodologies for the calculation of imbalance charges. Concerning the former case, this is supposed to be regulated through the enforcement of effective unbundling. In an attempt to overcome the difficulties raised by Member States' divergent unbundling preferences, Regulation (EC) 715/2009 provides in its article 4 for the cooperation of all TSOs at the level of the Union through the platform provided by the European Network for Transmission System Operators for GAS (ENTSO for gas). Such cooperation envisages the promotion of the completion and functioning of the internal market and cross-border trade. In fulfilling their tasks, TSOs are expected to increase the transparency and accessibility of relevant gas market information (e.g. with regards to available capacity, applicable tariffs, balancing rules etc.). This constitutes one of

---

[24] Geldhof and Vandendriessche, 2008, pp. 41–42.

[25] Article 9(1)(b) of Directive 2009/73/EC.

[26] Ascari, 2011.

[27] ACER Framework Guidelines on Gas Balancing in Transmission Systems.

the essential focuses of the new energy package, owing to the fact that unequal access to information regarding the physical status and efficiency of the system may create important unbalances among market participants. As a result of the above, traditional incumbents benefit from sufficient capital and market connections to ensure unabridged access to information regarding demand and supply, network capacity, balancing and availability of storage. This information is necessary for the effective competition and efficient functioning of the market in the sense of article 16 of Directive 2009/73/EC. The prevention of new market entrants from accessing such information may lead to the maintenance and proliferation of market foreclosure.

Summing up, current market arrangements concern horizontal and vertical mergers and acquisitions, which occur either within the borders of the same Member States or between incumbents belonging to different EU or non-EU states. These are circumscribed either solely to the gas sector or involve cross-sector alliances, specifically involving power incumbents. The new gas legislation foresees TSOs' certification as a method of reducing the external producers' expansion in the downstream and effective unbundling as a path of fighting market foreclosure caused by downstream players. Both mechanisms have yet to demonstrate their efficiency. Within this context, the consumers' demand for gas remains strong. Consequently, as long as switching procedures will continue to entail considerable transaction costs (information availability, cancellation and start-up fees) the market threatens to remain locked in a scenario in which major incumbents will preserve their dominance on the production and wholesale side. The maintenance of the *status-quo* is favoured by the industry's attachment to long-term gas supply contracts, which cover an essential part of the end-consumers, as well as the production from almost all existing gas fields and the majority of the pipeline capacity available in the downstream markets.

4.1.2 Long-term Contracts

The majority of gas-related European case law contains reference to long-term contracts, even if this specific feature of the gas trade has not always been brought under scrutiny for its potential harmful effects on market liberalization. In fact, opinions on the consequences of continuing to employ this type of contract on the EU gas market are divided at the level of regulators and industry.[28] Concerning the latter, traditional

[28] Talus, 2010.

incumbents are in favour of maintaining this type of contracting, whereas new entrants demonstrate reluctance. For the former category, long-term contracts represent a guarantee in terms of supplies and market shares. For the latter, they are harbingers of market foreclosure and barriers to entry.

However the industry appears to be unified in claiming that long-term gas contracts provide the least risk for long-term development and for the security of supply in Europe.[29] Additionally, such contracts are expected to reduce risks associated with large investment costs in gas infrastructure and to decrease adjacent costs, such as those derived from transportation.[30] The various aspects related to long-term gas contracts have been the object of analysis of many studies.[31] In general such studies have focused on the accommodation of long-term contracts with the application of European competition law and policy in the market. With a few notable exceptions,[32] the studies have not detailed the impact of sector regulation upon long-term gas contracts. Even in those cases hinting at the relationship between the two, reference has been limited to reflections on the manner in which sector regulation provisions are applied to certain contractual clauses. The present subsection will attempt to further the insight on the interaction between sector regulation and long-term gas contracts after introducing the specific features of the latter, evidenced through relevant case law.

4.1.2.1 The typology of long-term gas contracts in the EU

The predilection for using long-term contracts as the preferred method of trading and contracting gas has a two-fold explanation. First, such contracts represent a form of cooperation between external producers and internal suppliers, which guarantees both security of supply and security of demand. In their ideal configuration, long-term contracts ensure a fair sharing of risks between contracting parties regarding the stability of prices and the guarantee of provision and payment of contracted volumes. At the same time they provide valuable incentives for investment owing to the solid bilateral relations that they help forge between traditional gas suppliers and main European importers.[33] As a result,

---

[29] Eikeland, 2004, p. 31.

[30] Soligo and Jaffe, 2004.

[31] See Peter D. Cameron, Kim Talus, Anne Neumann and Christian von Hirschhausen.

[32] Neumann and von Hirschhausen, 2004, pp. 175–182.

[33] However, such investment incentives are generally targeting already established market incumbents, rather than new market entrants. This happens

long-term contracts offer a significant appeal for undertakings on the supply and demand side, which are motivated by the promise of stable revenues and a stable share of satisfied consumers. This affirmation is verifiable when looking at the above-presented supply arrangements between major incumbents.

Second, the use of long-term contracts is favoured by the specific structure of the gas sector. Low energy content per unit, infrastructure dependency, specificity of assets, capital expenditure and the need to secure a market at the early stages of the project have ensured that long-term contracts have remained for decades the backbone of natural gas supply.[34]

Most frequently, long-term contracts are signed for up to 30 years, a period that is considered sufficient to pay off infrastructure and upstream production investments. In general, they contain specific clauses that organize the behaviour of both sellers and buyers. The most frequent clauses are the take-or-pay clause, the market-value principle, the netback pricing and the destination clause. The logic of employing such clauses in long-term contracts regards the decrease or elimination of risks relating to traded volumes and prices of gas.

Despite their obvious benefits, these clauses have raised serious concerns regarding their impact upon market competitiveness.[35] The Energy Sector Inquiry has underlined that restrictive formulations in long-term contracts signed between traditional suppliers and distribution companies and industrial and commercial users can substantially hamper the opportunities of new entrants in the market.[36] In several Member States, an important segment of consumers are involved in gas supply contracts, which do not provide any incentives to switch to alternative suppliers. At the same time, suppliers themselves are sometimes locked in restrictive practices, which prevent them from choosing their delivery points or which foreclose their competitors from building their market share.[37] Such barriers result in a limitation of market competition and a weakening of consumers' rights protection. In the Distrigaz case, the

because the former are traditional partners in long-term contracts with third-country gas suppliers, unlike the latter.

[34] Talus, 2010.

[35] Clingendael international energy programme, 2008.

[36] European Commission, 2006: Energy Sector Inquiry, Preliminary Report. Nonconfidential version.

[37] Cameron, 2007, p. 325.

European Commission delivered a breakthrough decision on the elimination of market foreclosing restrictions from long-term contracts.[38] The case studied the impact of multiple long-term contracts signed by Distrigas with large industrial customers upon the opening of the Belgian gas market to competition. In its commitments following the Commission's inquiry, Distrigas proposed to remove any use, resale or destination restriction from existing contracts and to abstain from further introducing such restrictions in new supply contracts. Owing to the significant disadvantages brought by former Distrigas contracts to possible competitors, it was also established that all tacit renewal clauses would be removed and that no long-term contract of more than two years would be concluded with resellers.

Not all clauses historically introduced in long-term contracts have raised the same objections and fears among regulators. As an example, oil price indexation is allowed as long as its terms are clear and capable of ensuring that the price that is owed matches the relevant market price of the provided service and that the transaction is not liable to generate hidden profit.[39]

Furthermore, take-or-pay clauses are considered mandatory on the European gas market. Their role is to avoid capacity congestion and discriminatory third-party access to the transmission system[40] and to ensure security of supply and the fulfilment of public service obligations. Such clauses provide for the sharing of risks between producers and consumers regarding prices and volumes of traded gas. The buyer promises to pay for a certain amount of gas regardless of the actual consumption and the producer promises to maintain the stability of prices. Owing to their ability to simultaneously satisfy all parties involved, take-or-pay clauses allow for the financing of large-scale natural gas projects. However, they also create important obligations for purchasers, who may refuse to grant third-party access to their transmission network, if such access would cause them serious economic and financial difficulties.[41]

It should be observed that such a refusal is legitimate only if a third-party access exemption has been granted under article 48 of Directive 2009/73/EC. Despite the existence of take-or-pay obligations in contracts, an undertaking cannot automatically refuse access rights to its transmission network, but it must seek approval from the responsible

---

38 *Distrigaz*, Case COMP/B-1/37.966.

39 Talus and Hunt, 2011, p. 274.

40 Verde, 2008.

41 Article 35 of Directive 2009/73/EC.

regulator. Upon granting the third-party access exemption, the responsible Member State or NRA has to take into consideration several conditions, including the objective of achieving a competitive gas market, the need to ensure security of supply, the market position of the undertaking in cause and the level of competition in the market. The procedure is of strict interpretation owing to the fact that, under certain circumstances, take-or-pay clauses could qualify as market foreclosure facilitators. In this sense, when associated with territorial sale restrictions, such clauses are brought under scrutiny for their capacity to preclude potential third-party access to the market, stop the development of market liquidity and favour market division.[42]

Sale restrictions prevent the buyer from reselling the contracted gas or condition such resale on the original seller's acceptance. When employed together with a take-or-pay clause they distort the equitability of risk division between suppliers and purchasers.[43] Territorial restriction clauses are also known as destination clauses, which generally take the form of a prohibition imposed on the buyer from reselling gas in countries other than those for which it was originally intended. More subtle forms of destination clauses include profit-splitting mechanisms,[44] use restriction[45] and market sharing.[46] In the case of profit splitting, the buyer is obliged to give the seller part of the profits. Use restrictions limit the capacity of the purchaser to make decisions over the use of its profit. Market sharing refers to the historical segmentation of the EU gas market both vertically and horizontally between large producers and national incumbent suppliers.[47] Limitations imposed on the buyer allow the producer to maintain different price areas for the same product. In consequence, different customers may be charged different prices for the same product distributed at different delivery points.

While acknowledging the important role played by long-term contracts in the European gas markets,[48] the Commission severely condemns restriction clauses,[49] considering them a breach of European competition

---

[42] Boussena and Locatelli, 2011, pp. 27–46.

[43] Talus and Hunt, 2011.

[44] Case COMP/37 811 *Territorial restrictions 1) Algeria gas transport contracts, 2) Expansion of TAG pipeline*.

[45] *Gas Natural + Endesa*, COMP/37.542.

[46] *E.ON/GDF collusion*, COMP/39.401.

[47] *ENEL/ENI/Gaz de France*, Case COMP/38.662.

[48] *British Gas/British Gas Trading/Wingas/*, Case COMP/36.559.

[49] Cameron, 2007, p. 325. However, quite a different approach is followed in the case of LNG contracts where destination clauses ensure that LNG exports

law, a barrier to the accomplishment of the internal gas market and a threat to the security of supply.[50] The Commission's position is reflected in a settlement following an investigation into an agreement between NLNG and one of its European customers.[51] According to the investigation, the long-term agreement prevented the latter from re-selling gas outside its national borders. The investigation was settled after NLNG agreed to delete the territorial restrictive clause.

The exclusion of territorial restrictions from long-term supply contracts has been made the subject of a vast jurisprudence.[52] The recent GDF and E.ON gas foreclosure[53] cases concern upstream contracts used to import significant gas volumes into the Union. Both cases reveal persistent obstacles to gas competition owing to long-term reservation of transport capacity. In the case of GDF, the Commission also observed strategic underinvestment in import infrastructure. In both cases the incumbents were suspected of concerted practices constituting an abuse of dominant market position in breach of article 102 TFEU. Their behaviour was deemed detrimental for potential competitors and domestic and industrial gas consumers. The commitments accepted by the Commission in both cases are in line with its policy and entail a release of an important share of long-term reservations of pipeline capacities by both GDF and E.ON.

4.1.2.2 The impact of sector regulation upon the advancement of long-term gas contracts

The recent jurisprudence regarding long-term contracts indicates stricter competition law enforcement to correct present market failures caused by past behaviour of market participants. Complementary to this, the provisions of the new gas legislation are certain to affect the future of long-term contracts. Directive 2009/73/EC acknowledges the fact that long-term contracts are an intrinsic characteristic of the natural gas

---

will not be suddenly redirected to more competitive markets. See Roggenkamp *et al.*, 2007, p. 213.

[50] *Territorial restrictions – Austria (GAZPROM/OMV)*, Case COMP/38.085 and *Territorial restrictions in Germany (Ruhrgas/Gazprom)*, Case COMP/38.307 regard the removal of clauses infringing the Treaty rules on restrictive business practices (article 101 TFEU).

[51] Case COMP/37 811 *Territorial restrictions 1) Algeria gas transport contracts, 2) Expansion of TAG pipeline.*

[52] *Gas Natural + Endesa*, COMP/37.542; *GFU*, Case COMP/36.072.

[53] *GDF foreclosure*, Case COMP/39.316; *E.ON gas foreclosure*, COMP/39.317.

market.[54] However article 41(1) conditions their use to their compatibility with EU law and policies. Already in its recital, the Directive hints at significant market changes in aspects closely linked to the manner in which gas is contracted and traded throughout the EU. Such aspects concern an enhanced consumer protection, an increased transparency through information sharing and collaboration among regulators, a strengthened role for the NRAs and the development of network codes.

Under the new Directive, the relationship between consumer protection and long-term contracts should "deliver real choice for all consumers among natural gas suppliers".[55] This provision generates certain limitations to long-term contracts, which in general presume the locking of a significant segment of consumers into supply arrangements with main incumbents. In order for all consumers to be able to enjoy the benefits of competition and fair prices in the sense of recital 47 of Directive 2009/73/EC, it is necessary that they are able to freely switch among gas providers. Such a development would not immediately and directly affect the flux and volume of gas traded between external producers and internal suppliers. However, in time it could exert an indirect impact on this by increasing the volatility of demand, which constitutes an essential part of suppliers' import strategies.

Furthermore, the new energy package provides for increased transparency and information sharing between different regulators to which European energy undertakings are accountable. To this end, ACER and the Committee of European Securities Regulators investigate together whether transactions in gas supply contracts should be subjected to pre- or post-trade transparency requirements. In complement to this, the Commission and Member States are expected to cooperate in exchanging meaningful information regarding energy relations with third countries, including long-term supply arrangements.[56] The importance of increasing the transparency of relevant information in the case of long-term contracts has two important consequences. On the one hand, it helps detect market misconduct in the shape of competition distortion. In this sense, article 41(k) of Directive 2009/73/EC requires NRAs to monitor the introduction of exclusivity clauses in supply contracts, which may prevent nonhousehold customers from contracting simultaneously with more suppliers. On the other hand, enhanced transparency contributes to preventing such market misconduct through decreasing the incentives of

---

[54] Recital 37.

[55] Recitals 1 and 3 of Directive 2009/73/EC.

[56] Recital 47 of Regulation (EU) 994/2010.

market players to behave in an inappropriate manner by introducing restrictive conditions or by influencing wholesale gas prices.

Wholesale prices are defined daily on the traded market through the input created by a variety of producers, utilities and speculators. As they change on a daily basis, such prices are generally more volatile than long-term set prices. However, they also reflect better the dynamics of the market, especially with regard to supply and demand fundamentals. The level of gas wholesale prices may be influenced by different political, environmental or social–cultural factors. On a noneventful day, however, a low level of wholesale prices is secured by energy supply companies buying and reselling gas constantly.[57] Such an observation has an important consequence for the design of long-term contracts. Contractual restrictions preventing large industrial users from reselling gas to the market artificially trigger the increase of gas wholesale prices above the contract price.[58] As a result, the value of long-term contracts (interpreted as the difference between the value of gas in a long-term contract and the sale price in a more competitive market) is also artificially preserved. In the absence of use restrictions, the number of wholesale gas buyers and sellers in the market allegedly increases. This should determine a decrease in the wholesale price of gas. In the sense of article 41(i) of Directive 2009/73/EC, NRAs are responsible for monitoring the level of transparency of wholesale prices and thus ensuring that no restrictive contractual practices are affecting their definition.

NRAs' attributions have a notable impact on the design of long-term contracts. Along with the transparency oversight presented above, national regulators have to ensure the respect of contractual freedom with regard to both interruptible supply contracts and long-term contracts.[59] For consumers this means that they can buy their gas from the supplier of their choice. For sellers it means that they can dispose of their gas in the manner they see fit. Consequently, this constitutes an additional regulatory measure adding *a priori* control in support of competition requirements regarding restrictions in long-term contracts.

Last but not least, the TEP introduces network codes. The purpose of such codes is to ensure network security and reliability, network connection rules, third-party access and rules regarding harmonized transmission tariff structures.[60] Network codes have the potential to stimulate an

---

57 Essex, 2010.

58 http://www.oxera.com/cmsDocuments/Agenda_Oct%2007/The%20long%20and%20short%20of%20it_Agenda.pdf

59 Article 41(1) of Directive 2009/73/EC.

60 Article 8(6) of Regulation (EC) 715/2009.

increase in short-term gas trading, which could bring additional limitations to long-term gas contracts. The affirmation is verified in the case of network codes on gas balancing, as provided by article 6(2) of Regulation (EC) 715/2009. In 2011 ACER suggested a new balancing regime, encouraging TSOs to procure flexible gas on a short-term basis.[61] Under the current balancing rules, TSOs have been tempted to lock capacity in long-term contracts owing to the latter's proven ability to compensate for imbalance charges. It is expected that, in conformity with the new balancing regime which should be adopted by the end of 2013, flexible gas will be released to the wholesale market, thus enhancing competition and trading in markets that currently lack liquidity.[62]

With regard to the above, it appears that long-term gas supply contracts are in a process of reinvention at EU level. To this end, Talus maintains that such contracts will continue to exist in the market, albeit reshaped by market forces in a different direction than that suggested by the traditional model.[63] Challenges to this type of contracting come from both regulators and some of the industry stakeholders. In the EU, sector-specific regulation and competition policy converge towards reframing the conditions under which gas is contracted in the long term. Additionally, the reinforced development of LNG production and unconventional gas discoveries, as well as the preference granted to spot markets' trade arrangements, exert pressure upon the current form of long-term contracts. A redefinition of the contractual terms might have a positive effect upon the level of market interconnection, which would ultimately benefit the achievement of an integrated European gas market.

4.2 THE DEGREE OF MARKET INTERCONNECTION

In contrast to its predecessors, the Third Energy Package establishes a complex system of cooperation between regulators and participation of stakeholders in developing interconnection capacity between Member States, in view of achieving a well-functioning, efficient and open internal market.[64] The construction and maintenance of network infrastructure is focused on cross-border interconnection priorities supporting the diversification of gas routes and sources of supply, which should ensure a stable gas flow towards and within the Union.

---

61 ACER, 2011: Framework Guidelines on Gas Balancing … .

62 *Ibid.*

63 Talus and Hunt, 2011.

64 Recital 13 of Regulation (EC) 715/2009.

There is a direct link between the persistence of market foreclosure strategies and poor market interconnection. In this respect, the Energy Sector Inquiry observed that major incumbents, who preserved their dominant position on their respective national markets, rarely enter other national markets as competitors.[65] Such a situation is favoured by the insufficiency of cross-border capacity and the coexistence of different gas market designs across Member States.

As noticed by the Bundeskartellamt, one of the major negative effects of long-term contracts on the gas market has been the full booking of entry and exit capacity at major market interconnection points.[66] The main incumbents controlling the majority of the pipelines have shown little incentive to increase the transmission capacity or to manage it in a manner allowing third-party access. The Second and Third Energy Packages have established strict rules regarding the elimination of barriers to entry to the energy market. Specifically, national and European regulators have been assigned with defining technical rules applicable to the cross-border flow of gas in the Union and with regulating the behaviour of market players. Despite such developments, recent infringement cases referred to the CJEU concern the failure of several Member States to transpose the new gas legislation so as to increase gas market interconnection.[67] The conclusion derived from the case law has been that, in the absence of reverse flow capacity and short-term service provisions, the maximum interconnection level between Member States and between gas transmission systems cannot be achieved.

Failures in market interconnection have negative effects both on market competition and on final consumers. A weak interconnected market does not allow transport capacity to reach its full potential. As a result, barriers to entry stemming from capacity congestion issues are maintained. At consumer level, this translates into a poor choice among suppliers and services.

---

65 DG Competition Report on Energy Sector Inquiry, /*SEC/2006/1724/, p. 8.

66 Bundeskartellamt, 2009: Summary.

67 Case C-406/13 – *European Commission v Romania*; Case C-253/13 – *European Commission v Republic of Bulgaria*; see also IP/12/52 Commission refers Ireland and the United Kingdom to Court to ensure European law is properly implemented.

4.2.1 Legal Aspects Contributing to Gas Market Segmentation

4.2.1.1 Cross-border capacity

Article 2(3) of Regulation (EC) 715/2009 defines capacity as "the maximum flow … to which the network user is entitled in accordance with the provisions of the transport contract". In order for market participants to be able to fulfil their contractual obligations and establish their position in the market, they require access to the volume of network transport capacity, to which they are contractually entitled. Access to network transport capacity is regulated through various allocation mechanisms, which either favour the supplier or encourage a more consumer oriented approach.[68] These mechanisms are expected to show regard to market conditions and to be regularly updated by the responsible TSO.[69]

Prior to the liberalization waves, the preferred capacity allocation method has been the first-come-first-served model, which defines the allocation of transport capacity to the first shipper that demands transport services. The method continues to be employed as the basis of a significant amount of capacity allocation decisions, despite the potential negative impact it might exert on new market entrants and the provisions of the Capacity Allocation pilot guideline of the European Regulators' Group for Electricity and Gas (ERGEG).[70] Long-term contracts represent a relevant example of this. The employment of the first-come-first-served capacity allocation mechanism in such contracts justifies the long-term granting of capacity rights to a specific market player, to the detriment of potential competitors.

Other capacity allocation mechanisms such as the auction method and the "capacity goes where consumer goes" approach allow a more transparent competition between market players and a better protection of consumers' interests. The former method provides market signals for investment in transport capacity, through facilitating access to published available capacity before submitting it to a transparent and nondiscriminatory tender procedure. The latter method is based on the Swedish market model and seems to specifically match markets with oversupply of capacity.

---

[68] Haase, 2009, p. 160.

[69] ACER, 2011: Framework Guidelines on Capacity Allocation Mechanisms … .

[70] It should be observed that the formulation is missing from ACER 2011 Framework Guidelines on Capacity Allocation Mechanisms, which took upon the lead in ERGEG's document.

The granting of gas transmission capacity rights based on distributive (e.g. first-come first-served) allocation methods, rather than on market based ones (e.g. auctioning)[71] may have negative effects on EU gas market integration. This happens while, at the level of cross-border interconnections, unavailable or limited cross-border capacities contribute to market segmentation and limit the scope for cross-border competition. The lack of transport capacity sets barriers to the normal access to the transmission network. This should be correlated to the fact that anti-competitive practices and excessive tariff charges through vertically integrated network operators persist on interconnection pipelines.[72] In order for such market integration obstacles to be overcome, transparent and efficient methods are required to direct unused pipeline capacity towards secondary market and unused flexibility in existing long-term contracts towards wholesale markets.[73] The offering of unused capacity on a secondary market is essential in correcting contractual congestion,[74] which occurs when demand for long-term allocated capacity exceeds what can be made available.[75] Owing to pre-existing capacity contracts, some of which have emerged through transit agreements concluded by vertically integrated operators with themselves, such practice is not without controversy. Under the TEP, NRAs are required to draft cross-border and congestion management capacity allocation procedures.[76] Network codes are employed when defining the standardized content of transmission capacity contracts and the general terms and conditions for capacity allocation and capacity services.[77] Article 13 of Regulation (EC) 715/2009 mentions several factors[78] that should be considered by TSOs and NRAs when deciding how to allocate capacity. Two types of capacity can be granted: long-term capacity for one year or longer and short-term capacity for less than one year. When allocating capacity, the right

---

[71] http://newsletter.epfl.ch/mir/newspaper-article?np_id=75&np_eid=21&cat id=0

[72] Cameron, 2005a, p. 47.

[73] van der Linde *et al.*, 2006, pp. 53–54.

[74] Article 16(3) of Regulation (EC) 715/2009.

[75] Commission Staff working document on capacity allocation and congestion management for access to the natural gas transmission networks regulated under article 5 of Regulation (EC) No. 1775/2005 … /*SEC/2007/822*/.

[76] Article 41(6) of Directive 2009/73/EC.

[77] ACER, 2011: Framework Guidelines on Capacity Allocation Mechanisms … .

[78] For example, cost-reflectivity; the need of a market based system which reflects the market value instead of costs; the need to provide investment incentives; mandatory use of entry-exit point.

balance must be struck between short-term capacity services required for new competition to emerge and long-term capacity serving investments and security of supply.[79]

The choice of an appropriate capacity-allocation mechanism must be accompanied by accurate balancing rules. Such rules determine who exerts control upon primary capacity in transit lines. Poor balancing rules allow for the limited capacity in transit pipelines to be concentrated in the hands of incumbents in accordance with contracts signed prior to liberalization. When applying such rules, information transparency is essential owing to its direct impact upon market integration. A lack of transparent and reliable information regarding network capacity may severely hamper the integration of the EU gas market. Regular information during balancing periods enhances new entrants' ability to balance their portfolios and decreases their exposure to charges issued from the financial settlement of balancing their inputs and off-takes from the gas transmission system.[80] TSOs play an instrumental role in the sharing of such information, as well as in the enforcement of balancing rules.[81] In this sense, they are responsible for publishing on time relevant information regarding the available capacity and for undertaking the balancing of cross-border capacity-allocation endeavours.

The difference in network access models between neighbouring markets and within the same market represents an additional cause for significant misallocation of cross-border transit capacity and, hence, for the maintenance of market segmentation. In general, network access is regulated through either an entry–exit system or a point-to-point system. The parallel development of the two systems divides the market into two distinctly regulated areas. According to article 13(1) of Regulation (EC) 715/2009 the point-to-point system should have been abandoned by September 2011. The practice presumes a coupling of entry and exit capacities, limiting the flexibility of network users and reducing the liquidity of the local market.[82] Despite the provisions of the new gas

---

[79] Talus and Hunt, 2011, pp. 267–268.

[80] ACER, 2011: Framework Guidelines on Gas Balancing … .

[81] According to article 12(2) of Regulation (EC) 715/2009 "Transmission system operators shall promote operational arrangements in order to ensure the optimum management of the network and shall promote the development of energy exchanges, the coordinated allocation of cross-border capacity through nondiscriminatory market-based solutions, paying due attention to the specific merits of implicit auctions for short-term allocations and the integration of balancing mechanisms."

[82] KEMA International B.V., 2009, p. 49.

regulation, in February 2012 national legislation in eight Member States was still not in line with EU rules.[83]

In relation to capacity allocation, the Madrid Forum deemed contractual congestion on gas transmission pipelines and interconnection points as one of the key barriers to cross-border trade and market interconnection.[84] Contractual congestion may be caused by capacity hoarding or by capacity reservation in view of meeting fluctuating demand. This type of congestion results in an inefficient and suboptimal use of the technical capacity of the network where it occurs and prevents market participants from enjoying the benefits derived from using the maximum pipeline capacity available.[85] Congestion management procedures are expected to tackle efficiently the allocation of capacity through the application of the use-it-or-lose-it principle[86] or through the building of new capacity. The need for a unified approach to national and cross-border congestion management has been emphasized in a case law regarding Swedish interconnectors.[87] The cross-sectoral application of the findings in this case law is also relevant for the role performed by TSOs in congestion management. In this sense, the case underlines the fact that the trade-off between managing domestic congestion and limiting cross-border capacity trade is in opposition to the advancement of the European energy transmission network, which aims at interconnecting the EU energy market.

4.2.1.2 The deficiencies of the integrated transmission network

The first requirement for having a truly integrated market is to have an interconnected network throughout the Member States. The causal relationship between the two is reflected in recital 57 of Directive 2009/73/EC, which deems it as one of the main goals of the new legislative package. Recital 27 of Regulation (EU) 994/2010 acknowledges the

[83] European Commission, 2012: Internal energy market: national legislation in 8 Member States still not in line with EU rules.

[84] Mercados, 2008, p. 67.

[85] Commission Staff working document on capacity allocation and congestion management for access to the natural gas transmission networks regulated under article 5 of Regulation (EC) No. 1775/2005 … /*SEC/2007/822*/.

[86] The use-it-or-lose-it principle refers to the loss of network capacity from the possession of a network user, if such capacity was unused over a certain period of time when contractual congestion existed. The said network user may be deprived temporarily or definitely of the respective network capacity. The principle is used to prevent anticompetitive hoarding of capacity on the gas market, which would otherwise hamper the participation of competitors.

[87] *Swedish Interconnectors*, Case COMP/39.351.

priority of the infrastructure projects favouring gas market integration. Most network interconnections among Member States are insufficient and serve only transit purposes.

According to the International Energy Agency, the absence of network interconnection slows down the transformation of national markets into more regional ones.[88] As previously observed, the European gas network is divided into three segments. Gas flows from the production site to the consumption site under the supervision of transmission system operators for high-pressure transportation and of distribution system operators for low pressure transportation, involving LNG and storage facilities.[89] Only few pipelines support bidirectional flow.[90] Reverse flow is an important element of gas market interconnection, owing to the fact that it increases the flexibility of the rigid pipeline network and the number of supply routes to the market.[91] According to article 6 of Regulation (EU) 994/2010, network operators are obliged to ensure bidirectional flow on all cross-border interconnections between Member States by December 2013 at the latest. The new legislative package gives way to interpretation regarding the conditions justifying exemption demands. Article 7 of the 2010 Regulation provides that the proposal for an exemption should be based on a multitude of factors including:

- an assessment of market demand;
- projections of demand and supply;
- technical feasibility;
- the costs of reverse flow capacity; and
- the required infrastructure standards.

Such criteria allow a large margin of discretion for TSOs responsible for the bidirectional flow proposal or the corresponding exemption request.

As a complement to structural interconnection obstacles, incumbents do not enter each other's markets. A relevant example of this is provided by case COMP/B-1/39.401, regarding ownership issues and gas transportation rights across Southern Germany between France on the one hand and the Czech Republic and Austria on the other. The case revealed

---

88 Cronshaw *et al.*, 2008, p. 75.

89 Keyaerts *et al.*, 2008.

90 The Belgian–UK interconnector, the majority of the Belgian grid, and parts of the Czech network.

91 European Commission, 2012: Commission refers Ireland and the United Kingdom to Court to ensure European law is properly implemented, http://europa.eu/rapid/press-release_IP-12-52_en.htm

that E.ON and GDF committed not to sell gas from the MEGAL pipeline in each other's territories. Their understanding contributed to preventing the integration of the French and German gas markets and it distorted competition and consumers' free choice of suppliers.

Bearing in mind the above presented context, one should ask whether and how the new energy package can determine deeper network integration. An answer appears to be given by the provisions regarding regulatory oversight and cooperation between various regulators. Such provisions are not entirely novel. To a certain extent, cooperation and coordination between market stakeholders regarding transnational infrastructure development have existed since the nineteenth century.[92] The new formulation allows for more transparency and less discretion. Both primary and secondary European legislation require Member States to demonstrate regional solidarity in identifying and developing or upgrading natural gas interconnections.[93] Article 7 of Regulation (EU) 994/2010 and article 42(2)(a) of Directive 2009/73/EC underline the importance of cross-border cooperation between competent authorities, when deciding upon interconnection capacity allocation, gas flows, network management and balancing regimes. In the absence of such a cooperation, market integration would be hampered by the coexistence of conflicting national specificities on the same interconnection knot. The procedure for fostering transmission network integration requires ENTSO for gas to adopt every two years a plan regarding cross-border interconnection in the EU.[94] ACER is responsible for monitoring that such a plan provides a sufficient level of cross-border interconnection open to third-party access.[95] The two platforms for regulatory cooperation play an important role in the cross-border interconnection of gas transmission systems and are expected to cooperate with Member States and national competent authorities in the development of new cross-border infrastructure.[96] Competent authorities are entrusted with the development of joint preventive action plans, which are meant to increase the inter-regional infrastructure connection and provide essential information on existing or future interconnections, cross-border flows, cross-border access to storage facilities and bidirectional capacity.[97] As suggested by their name, joint action plans require competent authorities to consult with each other

[92] Schipper and van der Vleuten, 2008.
[93] Article 194 TFEU and article 6(2)(b) of Directive 2009/73/EC.
[94] Article 8(10)(b) of Regulation (EC) 715/2009.
[95] Article 9(2) of Regulation (EC) 715/2009.
[96] Recitals 15 and 18 of Regulation (EU) 994/2010.
[97] Article 5(1)(d)(f) of Regulation (EU) 994/2010.

prior to their adoption. These plans are based on ERGEG's 10 year network development plan and should be renewed every two years.[98]

4.2.1.3 Isolated markets

The concept of "isolated gas markets" is employed to define those Member States whose pipeline infrastructure is not directly connected to any other Member State and which have only one main external supplier. Article 49 of Directive 2009/73/EC details the derogations applicable to isolated Member States[99] and the conditions determining the expiration of such derogations. A reading of the article reveals that in fact Member States who most require the reinforcement of competition conditions on their gas markets in order to increase investment in their pipeline systems and connect to the rest of the EU are the most exempted under the new legislative package.[100] Such exemptions, albeit designed to protect gas consumers in the isolated markets, are liable to promote measures that further isolate such markets. The affirmation is evident in the case of Finland and the Baltic States, whose main external gas supplier is Gazprom. An exemption from the application of article 9 of Directive 2009/73/EC allows Gazprom's subsidiaries active in these Member States to escape the certification procedure provided in article 11 of the same Directive. In line with practices observed above, the application of exemptions which benefit the Russian incumbent may help the latter preserve its market position and may favour the lack of interconnection between isolated Member States and the rest of the EU.

The persistence of isolated markets forming gas islands prevents the achievement of an interconnected gas market and additionally poses challenges to the security of energy supply and to the sale of gas on equal terms.[101] It could be asked at this time whether regional integration, as a preceding step to full market integration, is indeed a correct approach. A 2010 ERGEG Press release maintained that regional initiatives deliver a more integrated Europe by improving the use of the existing gas interconnection and by enhancing the transparency of information.[102] Furthermore, regional cooperation between Member States and system operators, with the inclusion of isolated systems, is provided in annex IV

---

[98] Article 5(4) of Regulation (EU) 994/2010.

[99] Authorization procedures, unbundling of transmission systems and TSOs, market opening and reciprocity, direct lines.

[100] Dreyer *et al.*, 2010.

[101] Regulation (EC) 715/2009.

[102] ERGEG, 2010: Success in regional and inter-regional energy market integration. Press Release, Brussels, 10 November 2010.

of Regulation (EC) 715/2009, article 194 TFEU and recital 56 and article 7 of Directive 2009/73/EC. According to the latter, Member States are expected to promote and facilitate the cooperation of transmission system operators at regional level. This should include cross-border issues in view of creating a competitive internal market in natural gas and facilitate the interconnection of isolated systems forming gas islands.

In contrast to what the latest energy package suggests in terms of market interconnection, it should be observed that the regional cooperation currently foreseen follows the existent grid fragmentation lines. Indeed the three potential regional gas markets in Europe are addressed by the North-West initiative, the South-South East initiative and the South initiative. There are six Member States who are not present as participating countries in any of these regional initiatives. Among them, Malta and Cyprus are obvious bystanders. The other four are the Baltic States and Finland. The absence of these latter Member States from gas regional initiatives raises important questions, especially when looking at regulatory provisions demanding their integration and interconnection with the rest of the Union.

Taking into consideration that, so far, there has been little progress in respect to inter-regional gas projects,[103] it appears that the efficiency of market interconnection measures envisaging gradual regional integration is still to be determined.

4.2.2 Mechanisms Correcting Market Interconnection Failures

4.2.2.1 Balancing rules

Balancing plays an important role in correcting market interconnection failures, specifically with regard to cross-border capacity arrangements. In example, virtual balancing allows gas to be exchanged at a single price irrespective of where it is injected into or withdrawn from the network. This is essential considering that gas is traded at multiple international border-crossing points and major junctions between pipeline systems,[104] which may be submitted to various regulatory regimes. Currently, gas markets are fragmented by the coexistence of several balancing zones in the EU, to which correspond different balancing arrangements. Often, such diversity occurs across neighbouring markets. This increases the regulatory complexity of cross-border trade and obliges network users to

---

[103] CEER: Gas Regional Initiative (GRI), http://www.energy-regulators.eu/portal/page/portal/EER_HOME/EER_ACTIVITIES/EER_INITIATIVES/GRI

[104] Harris, 2006.

reserve capacity at each border point. As a result, the market power of incumbents is increased and there are more difficulties to entry. The small size of balancing zones contributes significantly to the complexity and excessive costs of transporting gas across EU territory.[105]

As a consequence of network unbundling, transmission and distribution system operators, as well as shippers, had to rearrange the sharing of their responsibilities in a suitable manner for the fair distribution of opportunities and obligations between market incumbents and new entrants.[106] According to Regulation (EC) 715/2009, regulatory overview over the progress of harmonization of rules regarding adjacent balancing zones has been entrusted to the ENTSO for gas. ACER's proposal to modify the role of TSOs and to increase the voice of stakeholders in regulating imbalances[107] has been in line with the elaboration of network codes by sector regulators, aiming at the use of gas balancing to encourage cross-border trade and to develop competition in the EU.[108]

Prior to ACER's proposal, TSOs have been the main responsible market actors for balancing activities. In this sense, they have been able to decide upon the trade-off between the capacity offered in the market and the transaction costs and risks imposed on shippers by a complex hourly or zonal system. TSOs could decide whether there was a need to limit transaction costs and risks imposed on shippers in order to promote new entry and competition in gas supply.[109] It has been observed that in the past TSOs' balancing method of choice has been the blocking of capacity through long-term contracts, instead of releasing such capacity on the wholesale market. The tendency has been explained as a consequence of the vertical integration of gas undertakings. Incumbents controlling primary capacity in transit pipelines have generally manifested little interest in serving the needs of new entrants. This fact is reflected in the behaviour of TSOs under their ownership.

TSOs have also been traditionally responsible for dealing with imbalances in the market. As a rule, imbalances occur owing to rapid demand variations, pipeline disruptions, upstream problems and downstream

---

[105] Dreyer *et al.*, 2010.

[106] Haase, 2008, p. 52.

[107] ACER, 2011: Framework Guidelines on Gas Balancing … .

[108] Article 8(6) of Regulation (EC) 715/2009: specific network-codes objectives in relation to balancing with regard to network-related rules on nomination procedures, rules for imbalance charging and rules for operational balancing between TSO systems.

[109] Hough *et al.*, 2005.

interruptions. Major market imbalances are due to the lack of synchronization between variations in demand and supply and the timeframe for commercial action.[110] Owing to the fact that imbalance charges do not reflect the cost of TSOs balancing the gas network, the latter could find incentives to perform their tasks inefficiently and to allow cross-subsidies between network users.

In order to prevent TSOs from abusing their balancing role, ACER's proposal focuses on information transparency and information sharing.[111] Nontransparent processes do not guarantee nondiscriminatory rules and are aimed to place barriers to new entrants for the benefit of incumbent suppliers.[112] The lack of information available for network users during balancing times results in new entrants being unaware of whether their portfolios are in balance. Depriving them of such information leads to an increase in their exposure to imbalance charges, which deflate only when the portfolios are maintained balanced. Under ACER's proposal network users are consulted by regulators and TSOs prior to the adoption of balancing rules. In this sense, they receive up-to-date information on their own balancing position and the system's balancing status during the balancing period, which is supposed to enable them to balance their portfolio efficiently.

Under the TEP, TSOs exert an important role in cross-border interconnection through their cooperation in the framework of ENTSO. Recital 15 of Regulation (EC) 715/2009 underlines the increased cooperation among TSOs in order to promote the creation of interconnection capacities. Cross-border TSOs' cooperation in order to integrate EU gas markets takes place through merging entry and exit balancing zones. In consultation with stakeholders, TSOs elaborate the impact assessment of expected costs and benefits of the balancing regime. The results of the consultation are shared with ACER and NRAs, which take the final decision regarding the necessary steps for integrating the market.

4.2.2.2 Trans-European energy networks

Gas pipelines and compressors represent the physical backbone of any natural gas market.[113] Even prior to the liberalization waves, it has been claimed that the transnational infrastructure, as a means of transporting goods, providing services and disseminating information, determines

[110] Cronshaw *et al.*, 2008, pp. 54–55.
[111] ACER, 2011: Framework Guidelines on Gas Balancing … .
[112] Hough *et al.*, 2005.
[113] Keyaerts *et al.*, 2008, pp. 865–876.

socio-economic integration.[114] Indeed, transport and communication revolutions have preceded formal European integration and have naturally sparked incipient infrastructure integration among European countries.[115] Recital 22 of Directive 2009/73/EC maintains the high importance of the transmission system for the EU gas sector and the necessity of additional safeguards to preserve the security of European energy supply. The performance and reach of gas pipelines influence the market's security of supply and interconnection. In this sense, article 170(1) TFEU underlines the importance of energy transportation networks in achieving the Union's strategic security of supply objectives. Additionally, article 170(2) TFEU envisages the integration of landlocked and peripheral regions and market pipeline interconnection to the benefit of gas islands and countries at the edge of the European supply networks.

Member States' gas transmission infrastructure has a cross-border nature and importance. Owing to this fact, its control goes beyond the scope of national regulations, which have too limited a range to address the international regulatory aspects that it raises. Consequently, energy interconnection development can be best achieved when planned with a European perspective.[116] European primary and secondary legislation confirms this assumption. Article 194(1)(d) TFEU requests the interconnection of energy networks. At the same time, Regulation (EU) 994/2010 emphasizes the link between a sufficient and diversified gas structure within and between Member States and the connection of isolated systems to the gas network.

In the framework of Europe 2020 Strategy for smart, sustainable and inclusive growth, the European Commission suggested the development of a European "smart super grid".[117] In this sense, the Commission's 2011 report to the Energy Council anticipated large investment in high-pressure gas transmission pipelines coming to the EU and between Member States' storage, LNG terminals and reverse flow infrastructure. Gas investment volumes are expected to increase by 30 per cent between 2011 and 2020, with a significant impact upon trans-European energy networks (TENs-E). The European supergrid presumes the interconnection of three gas transmission corridors. The Southern Corridor should

---

[114] http://ec.europa.eu/ten/index_en.html

[115] Schipper and van der Vleuten, 2008.

[116] Proposal for a Regulation of the European Parliament and of the Council on guidelines for trans-European energy infrastructure … /*COM/2011/0658 final–2011/0300/COD*/.

[117] Communication from the Commission Europe 2020 A strategy for smart, sustainable and inclusive growth /*COM/2010/2020 final*/.

transport gas from the Caspian basin, Central Asia and the Middle East. The North–South Corridor is foreseen to link the Baltic, Black, Adriatic and Aegean Seas and to ensure the North–South connection in Western Europe. These corridors follow the traditional setting of European gas pipelines. The turning point of the interconnection plan is given by the Baltic Energy Market Interconnection corridor, which has the potential to end the isolation and single supplier dependency of the three Baltic States and Finland.[118]

Apart from Europe 2020 Strategy, a proposal for a regulation on guidelines for TENs-E infrastructure was adopted in 2011.[119] The proposal is based on articles 171(1) and 172 TFEU regarding the establishment and adoption of guidelines covering objectives, priorities and broad lines of measures related to trans-European networks. The initiative has a two-fold justification. From a structural point of view, the existent EU energy infrastructure requires improvement and innovation, as well as more interconnection across borders in order to be able to meet EU's core energy objectives. Additionally, from a regulatory point of view, new energy infrastructure policy is required to coordinate and optimize network development on a continental scale. The aim is to ensure the operability of Member States' solidarity in energy matters, the completion of the internal market in energy and the linking of isolated regions including the gas islands. Three directions are suggested for the gas market interconnection: bidirectional pipelines, enhanced storage capacities and flexible supply including LNG and compressed natural gas. Additionally, the proposal aims at the materialization of alternative supply and transmission routes and diversified imports for the European gas market.

In completion to the strategic trans-European energy infrastructure elaborated by Europe 2020 Strategy, the 2011 proposal details the procedure for their implementation through projects of common interest (PCIs). Regional expert groups and ACER are responsible for monitoring and evaluating the implementation of PCIs. Each Member State has to have a national competent authority, which oversees and coordinates the permit-granting process for PCIs. This authority (NRA or other) is also responsible to set minimum standards for transparency and public participation and to fix the maximum allowed duration of the permit granting process. Projects of common interest can be implemented under

[118] Proposal for a Regulation of the European Parliament and of the Council on guidelines for trans-European energy infrastructure … /*COM/2011/0658 final – 2011/0300/COD*/.

[119] *Ibid.*

certain conditions for reasons of overriding public interest. The measures are expected to be proportionate, considering that they aim at a minimum alignment of national administrative procedures necessary to facilitate the implementation of – mostly cross-border – PCIs. NRAs and ACER are responsible for allocating costs across borders for PCIs in gas sectors, according to the benefits received by Member States, which are directly or indirectly concerned with the PCIs in question. NRAs are requested to grant appropriate tariff incentives for the implementation of PCIs facing justified higher risks.

The impact of the two initiatives presented above upon the interconnection of the EU gas transmission network seems promising, but is yet to be assessed. As resulting from the overview of the gas market structure, several obstacles have prevented the achievement of an integrated market. In general, the nature of these obstacles can be perceived as structural (linked to the fixed, rigid pipeline infrastructure, which segments the EU gas market into different areas), functional (current market arrangements drawing on former monopolies) or regulatory (the absence of a real European regulator, unequal national regulatory authorities). In this sense transmission networks are unequally distributed on EU territory. They cross countries with different unbundling rules and different capacity balancing principles and are further segmented by public and private ownerships. In their large majority, owing to the specificity of the pipeline grid,[120] energy networks continue to function as monopoly businesses under the control of traditional TSOs. Each of the observed obstacles could be referred to as unsatisfied market needs, in an extended sense of the general economic definition. Such an association facilitates the interpretation of the first category of obstacles as corresponding to "basic gas market needs" and of the latter two categories as issuing from "elevated gas market needs". The slow advancement of gas market integration may be explained while drawing upon this economic metaphor. The three liberalization waves addressing the EU gas market have aimed at satisfying the elevated gas market needs before accomplishing the basic ones. As a result, the current gas market continues to be divided by poor West–East and North–South connections, owing to the regulators focalization on national markets' liberalization, rather than on the balancing of supply and cross-border trade in a single gas market.

---

[120] For example, structure of costs characterized by high capital intensity and low variable costs, life span ranging up to 60 years and slow technical progress. See http://www.energypolicyblog.com/2007/08/22/no-regulation-immunity-for-german-gas-pipelines/

4.3 SUMMARY OF THE EU GAS MARKET STRUCTURE

The Third Energy Package established a certification procedure for all TSOs active on the EU gas market, new unbundling provisions aimed at increasing the market competitiveness and a clear mechanism to protect investments,[121] benefiting both new entrants and network development. Furthermore, it increased the role of TSOs, by cutting the vassalage links between producers or suppliers and network operators. Independent TSOs are intended to reduce the level of market concentration, by allowing unbiased third-party access and an investment regime protecting and encouraging new investment, and by prohibiting cross-subsidies between network users. At the same time, they are also meant to participate in strengthening the EU gas market's interconnection. TSOs are responsible for distributing capacity on pipelines and thus their decisions constitute an important component of the capacity congestion management process. As seen above, it is important that the trade-off between managing domestic congestion and limiting cross-border capacity trade does not hamper the interconnection of the EU gas market.

Notwithstanding the positive developments, under the TEP, different market unbundling models continue to coexist. This allows the maintenance of divergent regulatory regimes across Member States, which could ultimately result in further barriers to market integration and interconnection. Gaps in the current legislative framework, such as those favouring bad balancing rules, contribute to the maintenance of poorly interconnected national gas markets. It is precisely the weak interconnection among Member States' gas markets and the lack of an external energy representation of the EU that render the bidimensional European gas market problematic.

Certainly, measures regarding cooperative regulatory oversight among various accountability instances could exert a positive impact upon the deeper integration of the EU gas market and transmission network. Their implementation is aimed at blurring conflicting national regulatory specificities regarding cross-border issues. However, a further weakness of the TEP lies in the effectiveness with which it is implemented in the Member

[121] Based on exemptions from certain access rules and a setting of the access tariff that would take into consideration investment incentives – e.g., returns.

States. Recent case law of CJEU[122] reveals that several Member States have yet to transpose the provisions of the TEP, contributing thus through their action to the continuation of failures in market interconnection.

---

[122] *European Commission v Estonia*, Case C-241/2013; *European Commission v Poland*, Case C-55/2013; *European Commission v Slovenia*, Case C-9/2013; *supra* note 67.

5. Defining and assessing the current EU gas market design

Each organized market has a specific configuration, which derives from the setting of its institutions, the arrangements regulating the interactions between market actors, the number of market participants, the mechanics of trade, the sequence of market models on the wholesale market and the methodologies of establishing market access tariffs.[1] Such a configuration is generically perceived as market design. There are different ways in which market design may be defined, based on the market features listed above and the importance assigned to each of them. The common thread of all the possible definitions is the construction and matching of a set of institutions and legal tools to meet pre-defined market objectives. By default, the concept of *design* entails a purposefully creative or constructing action. In the case of natural gas, the EU market design has been specifically conceived in order to satisfy energy policy goals associated with the said market, while forging market integration. In its development, EU gas market design has been influenced by traditional principles of market reform (unbundling, competition, access to resources and transit lines) and has been modulated by specific energy objectives and tangential policies.[2] Based on this premise, market design is defined, in the understanding of the present chapter, as the pattern of collaboration and distribution of responsibilities between legislative and regulatory tools operating in the EU gas market, in view of achieving predetermined energy policy objectives.

The assessment of EU gas market design relies on two complementary sets of arguments. The first set of arguments ponders over the accommodation between energy policy goals in the gas market. The second one looks at the confrontation between envisaged market normative goals and legal tools at hand.

---

[1] The enumeration is not exhaustive. Its purpose is to hint at the variety of elements that could be taken into consideration when discussing the gas market design.

[2] van der Linde *et al.*, 2006, p. 22.

5.1 OPERATING ENERGY POLICY GOALS APPLICABLE TO THE EU GAS MARKET

The main goals of EU energy policy are the achievement of a competitive and efficient energy sector, the guarantee of the security of energy supply and energy sustainability.[3] The formulation of these goals draws – in a not completely coherent manner – upon internal market principles, environmental and climate change policy, regional integration requirements and trans-European networks initiatives. The variety and multitude of factors influencing EU energy policy explain the complex institutional architecture of this sector, as well as the sometimes overlapping application of different legal tools, pursuing different market objectives. In general, the three energy policy goals are regarded as interconnected and complementary. It appears that the Commission assumes that the pursuit of any of the three goals hints at the realization of the other two.[4]

However, as shown in the following, in the case of the EU gas market this observation is verified only to a certain extent. The particular nature of the gas market puts under scrutiny the compatibility between the three energy policy goals. This means that the pursuit of one goal might contradict the fulfilment of another.

5.1.1 Identifying and Defining the Policy Goals

5.1.1.1 Requirement of competitiveness

The first goal of EU energy policy is the completion of a competitive and efficient market, characterized by lower and increasingly similar energy prices across national borders, more efficient energy-using industries, increased employment and economic growth. The objective of competitiveness has been defined in a three-fold manner by referring to the benefits brought to consumers by energy market opening, the importance of keeping energy prices stable and the need to develop new and clean energy technologies at EU level.[5]

In the Third Energy Package, the achievement of a competitive gas market is realized through access to storage and primary capacity rights, regional cooperation between Member States in view of market opening and effective unbundling between the monopolistic and the competitive segments of the gas chain. The degree of realization is, however,

---

3 Green Paper /*COM/2006/105 final*/.

4 *Ibid.*

5 Green Paper /*COM/2006/105 final*/.

heterogeneous across Member States. The particularity of this market goal is that it allows both the Commission and national competition authorities to ensure that market liberalization is properly implemented at the level of undertakings. Consequently, competent authorities can make use of antitrust tools to ensure the opening of the market and the prohibition of price fixing, territorial restriction clauses and abuses of dominant market positions by undertakings. At the same time, the Commission is able to enforce gas market liberalization principles in those situations where a Member State has failed to do so. Already before the end of the Energy Inquiry, the Commission had started infringement proceedings under article 258 TFEU, as a result of the failure of Member States' authorities to withdraw certain national measures that created unjustified restrictions to mergers with a Community dimension.[6]

The competitiveness goal can be perceived as directly deriving from the application of internal market principles to the energy sector. Concerns about energy security have played a major role in the original establishment of the EU and have continually subsisted since that moment and until the present day, in a more or less pressing manner. Adversely, the employment of competition for the shaping up of the EU energy market was introduced only at the end of the 1980s. As shown in the preceding chapters, natural monopolies over grid industries started to be challenged only with the demise of the regulatory state. The introduction of competition in the gas sector was based on its previous success in forging internal markets in other sectors. In the absence of a Treaty basis for energy, EU competencies stemming from single market provisions and competition policy enabled the Commission to regulate in the gas sector.[7] The reliance on competition law and state aid for dismantling dominant market structures was maintained despite, and complementary to, the adoption of EU secondary law, aimed at harmonizing the regulatory frameworks addressing anticompetitive industrial behaviour in liberalized gas markets.

However, the results of applying competitive pressure upon market efficiency in the Member States have not always been very conclusive.[8]

---

[6] *E.ON/Endesa*, Case M.4110; *ENEL/ACCIONA/ENDESA*, Case M.4685.

[7] Egenhofer, 1997.

[8] According to Eurostat's observations, the level of gas prices was the same for households and slightly different for industry in UK and Estonia, in the first half of 2011. The UK was the first Member State to advocate and put in practice the liberalization of its network industries and has extensively made use of competition tools to decrease its gas prices. Estonia is one of the four Member States catalogued as gas islands. According to article 49 of Directive 2009/73/EC

The requirement that European operators use the most cost-efficient methods of production has not guaranteed an immediate reduction of gas prices. In Lithuania and Estonia an economic assessment of unbundling incentives for vertically integrated entities appears to indicate instead a significant temporary increase in prices for end consumers.[9] This might have occurred as a result of the traditional *competition dilemma*. Accordingly, a correct balance should be forged between ensuring maximum short-term competition and determining increased competitiveness of the industry in the long run.[10] Although consumers might be able to choose from different providers and contract schemes, such a choice could generate substantial switching costs that add up to the original gas supply costs.

Furthermore, the absence of gas-to-gas competition, the indexation of gas prices to oil and the malfunctions and complexity of gas liberalization reforms have complicated the appreciation of the effects exerted by competition upon EU gas market arrangements.

5.1.1.2 Security of energy supply

Energy represents an essential component of the stable and prosperous economic and social development of both the EU and the Member States. Despite the role played in the European integration, the concept of security of energy supply lacks a real definition at EU level.[11] Certainly, different energy resources require a different focus with respect to their varying exposure to security of supply threats. These might derive from unequal conditions of storage, transportation infrastructure, substitutability and environmental sustainability. In this respect, a unified security of energy supply definition might not be possible. However, in its absence,

---

this status allows it to derogate from a significant part of the liberalization directives applied to the national gas markets. The most recent Eurostat figures show similar gas prices for one of the most competitive and one of the least competitive EU gas markets. What is even more surprising is that in the two preceding years household gas prices in Estonia were smaller than those in the UK. European Commission, Eurostat: File: Half-year electricity and gas prices, first half of year, 2009–2011 (EUR per kW h), http://epp.eurostat.ec. europa.eu/statistics_explained/index.php?title=File:Half-yearly_electricity_and_gas_prices,_first_half_of_year,_2009-2011_(EUR_per_kWh).png&filetimestamp=20111124164017

9 Seventeen per cent in the case of Estonia. See Kanapinskas and Urmonas, 2011, p. 245. A similar case appeared in the Czech Republic. See Slabá, 2008.

10 Nyssens and Schnichels, 2007, p. 1363.

11 Cameron, 2007, p. 518.

discussions are raised regarding the scope of the concept, its compatibility with other energy policy goals and the mechanisms and subpolicies required for its achievement.[12]

The Commission's 2006 Green Paper presents a series of steps meant to ensure the satisfaction of security of energy supply, but omits to give an explanation regarding what is included in the concept.[13] An association is made with tackling the EU's rising dependence on imported energy. This confirms the view that the Union perceives energy import dependency as the major cause of energy security unease. Article 194 TFEU does not automatically exclude other potential sources of supply disruptions, but makes clear reference to gas supply crisis determined by third-country suppliers. Such an approach is maintained in the Third Energy Package. The recitals of Directive 2009/73/EC link security of supply and solidarity between Member States with the Commission's oversight upon the certification of TSOs, controlled by a third-country national.

The EU's approach to energy security has not been without critique. Such an approach brings important limitations to the identification of causes that can have an impact upon EU's security of supply and further dictates the direction taken by the corresponding legislation. Talus observes that the EU's perception of risk in relation to energy supply does not correspond either to the reality of the risk or its geographical origin.[14] In 2006 Stern advanced the argument that increasing import dependency should not be automatically equalled to decreasing supply security.[15] His observation was based on the fact that the latter did not constitute at the time the principal security threat to European gas supplies, in either the short or the longer term.[16] The value of Stern's reasoning for energy security discussions in Europe consists of indicating other sources of threats for EU energy supply besides import dependency.[17]

---

[12] See Cameron, 2011; Haase and Bressers, 2008, pp. 15–17.

[13] Green Paper /*COM/2006/105 final*/, p. 18.

[14] Talus, 2008, p. 125.

[15] Stern, 2006.

[16] The most important EU energy supply disruptions of the time were due to accidents that occurred at gas storage facilities in the UK.

[17] This has become obvious in 2011 in the aftermath of the Fukushima Daiichi nuclear accidents, which turned the EU's attention towards the potential effect of natural disasters, man-made accidents, and terrorist and malevolent attacks upon the security of energy supply.

The absence of a unified definition for the European security of energy supply is accompanied by the coexistence of various definitions of the concept at Member States' level. In Member States which are relatively highly dependent on gas imports the security of gas supply presumes that under normal circumstances gas is delivered at reasonable prices.[18] In the few EU countries which have been able to rely on their national energy resources, energy supply security refers also to the resilience of the supply infrastructure in case of attacks or natural disaster and the capacity to maintain self-reliability in energy provision.[19] Natural gas represents different percentages in the energy mix of Member States. Each Member State preserves its autonomy in designing national energy policies.[20] The latter dispose of different availabilities of potential substitutes for gas, should its supply be disrupted. Furthermore, each Member State is differently equipped with storage facilities and benefits from specific levels of access to energy supplies, in the form of either long-term capacity or short-term flexible volumes. These elements contribute to different understandings of the security of gas supply in various Member States. Recital 22 of Regulation (EU) 994/2010 indicates that there is a need for a comprehensive and effective common approach to security of supply among Member States. Stability and clarity of security of supply standards are required in order to provide legal certainty. Member States are expected to abstain from imposing unreasonable and disproportionate burdens on natural gas undertakings, while applying their specific criteria for the maintenance of the security of supply.

Having regard to the specificities of gas as an energy resource, its increasing exposure to geopolitical pressures and the unequal vulnerability of Member States to supply disruptions, it is believed that any definition of security of gas supply should also aim at clarifying the idea behind the solidarity concept.[21] To this end, the following definition is suggested. Security of gas supply represents the capacity of the Union to prevent or react in a cooperating and coordinated manner to supply

---

[18] Mileczarski, 2007.

[19] House of Commons Energy and Climate Change Committee, 2011: UK Energy Supply Security or Independence: Eighth Report of Session 2010–2012, Volume 1, http://www.publications.parliament.uk/pa/cm201012/cmselect/cmenergy/1065/1065.pdf

[20] Despite solidarity demands in primary (article 194 TFEU) and secondary legislation (Communication from the Commission – Second Strategic Energy Review … /*COM /2008/0781 final*/).

[21] To date CJEU has not delivered any judgment regarding the interpretation of solidarity action of Member States on energy markets.

threats, such that it would ensure the maintenance of the balance between gas consumption and supply or, respectively, the immediate restoration of such balance, should a disruption of gas supply occur.

5.1.1.3 Sustainability

The third affirmed energy policy objective regards sustainability. As in the case of the security of supply, the definition of the term is vague and not unitary. The most accepted understanding of sustainability is the "development that meets the needs of the present without compromising the ability of future generations to meet their own needs".[22] Natural gas has been considered to be a sustainable energy resource owing to the fact that it is the cleanest, most efficient and versatile of all fossil resources and more available than renewable energy.[23] As such, it is expected to contribute more efficiently to the reduction of greenhouse emissions than its potential substitutes.

The sustainability impetus is a hybrid goal, common to both energy and environmental policies. A look at the three-fold definition of the concept in the Commission's 2006 Green Paper supports the affirmation that energy policy has in fact imported and transposed in totality the notion of sustainability formulated by environmentalists. Accordingly, energy sustainability aims at the development of alternative energy resources, the decrease of energy demand in Europe and the halt of climate change.[24] As demonstrated in the next subsection, there are certain problems that derive from such a transfer of policy goals. These refer in general to the risk of inconsistencies between original and imported goals and to the potential conflict between applicable legal tools, envisaging the accomplishment of different goals.

The interdependency between the environmental and the energy sector has not always been dealt with comprehensively in the existent legislation. With the adoption of SEA, the area of shared competencies between the EU and the Member States widened to include environmental concerns. After the introduction of majority voting on environmental issues, ex-article 130-R EEC (article 191 TFEU) constituted a basis for the increase of the European Commission's powers to propose and get environmental regulations adopted. In its paragraph 5 (article 191(4)

22 Report of the World Commission on Environment and Development: Our Common Future, http://www.un-documents.net/ocf-02.htm

23 Gas Infrastructure Europe: Contributing to a sustainable society: Gas as an integral part of the Roadmap 2050, http://www.gasnaturally.eu/uploads/Modules/Publications/gie_roadmap2050.pdf

24 Green Paper /*COM/2006/105 final*/, p. 17.

TFEU), the article provided expressly that EU could participate in international environmental agreements. At the same time, it pushed for the integration of environmental protection requirements as a component of other common European policies. Of course, at the time of the adoption of the SEA, a European energy policy was still far from becoming a reality. This would explain the first interpretation of the limitations of ex-article 130-R EEC formulated by the 1985 Conference of the Governments of Member States. According to this, the EU's activities in the sphere of environmental protection should not interfere with national policies regarding energy resources.[25] The Member States' position could be further observed in their initial refusal to accept a hybrid tax regrouping energy and carbon emissions that the Commission proposed in September 1991.[26] It is difficult to say whether such a refusal was entirely the outcome of Member States' initial decision to keep the two sectors separated, or if it represented a sign that the latter continued to be reluctant to allow the EU to use tax matters in order to regulate issues under sole state sovereignty. It could, however, be observed that the success of hybrid carbon taxes has been in general connected to the ability of their initiators to make them politically acceptable for the targeted actors.[27] Notwithstanding the original tendency of limiting EU's influence upon national energy policies through the enforcement of environmental measures, the beginning of the 1990s witnessed a change in the Member States' approach to energy–environment relationships. A growing awareness of the impact exerted by energy consumption upon the environment determined several Member States to treat the two departments together, through a combined approach. As a consequence, despite the failure of the Union-wide energy/carbon tax project, by 2000 seven of the current Member States had already adopted a similar taxation policy.[28]

Since the Maastricht Treaty, the EU's commitment to environmental protection has increased and environmental policy has continued to develop its links with other departments.[29] Several initiatives have been taken by the Commission in the initial years of the European energy policy to integrate environment-based provisions into the latter. In 1998 a

[25] Final Act and Declarations of the Single European Act, 1985.
[26] Eikeland, 2004.
[27] Vollebergh *et al.*, 1997.
[28] Andersen, 2010.
[29] Wilkinson, 1992, pp. 221–239.

Communication from the Commission[30] identified the three main objectives of the European energy policy with an environmental dimension:

- the promotion of energy efficiency/savings;
- the increase of the share of production and use of green energy;
- the reduction of the environmental impact exerted by the production and use of energy resources.

Article 11 TFEU provides a treaty basis for the energy–environment interdependency. Accordingly, environmental protection requirements must be integrated in other policies of the Union, especially in view of promoting sustainable development. In contrast to article 130-R EEC, article 11 TFEU underlines the sustainability factor, which is one of the advocated objectives of EU energy policy.

At both Member State and EU levels, environmental concerns have determined a re-arrangement of the institutional setting, which has also had consequences for energy matters. According to article 4(e) TFEU, environment continues to be a shared responsibility of EU and its Member States. However, article 21(2)(f) TEU provides that the Union is entrusted with defining and pursuing common policies and actions in order to "develop international measures to preserve and improve the quality of the environment and the sustainable management of global resources in order to ensure sustainable development". This article represents an example of the effect exerted by the creation of more environmental competencies for the Union upon the energy dynamics between the EU and its Member States. Indeed, the EU's gas relationships take place within a global context, in which sustainability concerns require a decrease in the use of fossil fuels susceptible to augment the emissions of CO_2 into the atmosphere. An integrated EU gas market is more likely than any Member State acting alone to succeed in regulating such emissions through the progressive establishment of an efficient use of energy resources both at the level of industry and that of households.

5.1.2 Inconsistencies and Inherent Problems

There is a considerable trade-off between the Member States' right to be solely responsible for their energy mixes and the harmonious coexistence

[30] Communication from the Commission: Strengthening environmental integration /*COM/1998/0571*/.

of the three main energy policy goals.[31] Without being mutually exclusive, each of these goals determines a specific direction for EU gas and energy policy. The increase in the share of renewables in a decarbonization scenario, which would benefit the goal of sustainability, might require targeted subsidies generating competitiveness distortion in other energy resources. Similarly, the application of competition principles to the gas market might endanger the realization of the security of gas supply. From this perspective an integrated EU gas market would open the path for the emergence of an external EU energy policy and representation, which would be better suited than any Member State acting alone to ensure the accomplishment of energy policy priorities.[32] For the present time though, the inconsistencies between the latter cannot be ignored, owing to the effect that they exert upon the market design. The context is further complicated by the existence of additional energy policy goals that result from market liberalization packages. Among these, public service obligations and consumer protection occupy distinctive positions, while they play a specific role in defining gas policy and gas market arrangements. The relationship between these additional goals and the main energy policy objectives is complicated by the fact that, although all are constitutionalized in the TFEU,[33] they are differently pursued by the existent legal tools, which stems from both primary law and secondary legislation. As explained in the following section, this gives way to several concerns regarding the hierarchical satisfaction of energy policy goals, deriving from the constituency of the EU legal order.

5.1.2.1 Competitive normative goals

The complementarity between the three main energy policy objectives advocated by the Commission has not always found support in the jurisprudence of the Court of Justice or even in the formulations of the specific sector regulations. The *Almelo* judgment[34] has been generally quoted to justify competition law limitations in certain cases of provision of public service obligations. This book considers, however, that there is an additional interesting finding of CJEU in the judgment, which has not

---

[31] Fischer and Geden, 2012.

[32] McGowan, 2008, pp. 90–106.

[33] See article 194 TFEU for energy policy goals, article 12 TFEU for consumer protection, article 14 TFEU for services of general economic interest and protocol 26 to the TFEU for the specific pursuit of services of general interest in the utilities sector.

[34] *Municipality of Almelo and others v NV Energiebedrijf Ijsselmij*, Case C-393/92.

been the focus of discussions around the case law. In delivering its decision, the Court signalled the tension between *environmental provisions and the competitiveness requirement.* According to the Court, when deciding upon the necessity of applying restrictions to competition, the environmental legislation regulating the economic conditions in which the undertaking operates had to be taken into consideration. It could be legitimately assumed that, in the absence of such legislation, the burden imposed on the undertaking entrusted with public service obligations would decrease to the point where the necessity of derogating from competition principles would be reassessed. The position of the Court in this particular case law is clearly in favour of maintaining strict environmental provisions to the detriment of competitiveness. Such an approach confirms the Court's prior jurisprudence, according to which limitations on the application of competition have been considered as lawful measures in cases where the protection of the environment is at stake.[35] The incompatibility between internal market rules emphasizing free and fair competition and the sustainability requirement became obvious during the first drafting of the electricity and gas Directives. The Commission had to accept derogations from competition provisions if Member States decided to adopt measures leading to an increased environmental protection that demanded such derogation.[36] Keeping in mind that the sustainability impetus cannot justify the maintenance of inefficient energy policies, the argument should, however, be directed towards a substantive change in the articulation of the normative goals. In their current form, the latter cannot be pursued simultaneously in an efficient and noncontradictory manner. Furthermore, their prioritization and mutual balancing should be clearly regulated. The order should be first increasing the energy efficiency, then regulating the environment.[37]

The conflict between *competitiveness and security of supply* is more subtle. *A priori* more competition in the gas market is expected to reinforce the security of gas supply. In a competitive gas market, in

---

35 *Commission v Belgium*, Case C-2/90.

36 Eikeland, 2004.

37 At the moment, EU Impact Assessments are used to elaborate *ex ante* upon the possible effects of the adoption of a certain policy in economic, social and environmental terms. However, the ability of Impact Assessments to act as guidelines, as well as their real impact upon the legislative proposal are questionable. See, e.g., House of Commons European Union Committee, 2010: Impact Assessments in the EU: room for improvement? Chapter 2: Production of Impact Assessments, http://www.publications.parliament.uk/pa/ld200910/ldselect/ldeucom/61/6105.htm; see also Renda, 2006.

which barriers to entry have been set aside, the overall efficiency of the gas system is improved. An increased number of market players is equivalent to a decrease in the dependency on single undertakings and monopolies. At the same time, it arguably offers alternative supply routes for end consumers. The argument is only partly valid as long as at the other side of the supply flow remains the same limited number of exporting producers. As previously shown, the principal side effect of introducing competition in the gas sector has been the diminution of the attractiveness of investments in gas projects, which demand large capital inputs and entail long payback periods. Traditional incumbents might find little stimulation to invest in transmission networks that will have to be shared with third-party entrants, while the latter might prove reluctant to engage the required capital amounts in the first stages of accessing the market. Investments in the gas transmission network and storage facilities are mandatory for the security of supply. This is confirmed by article 36 of Directive 2009/73/EC, which provides for the exemption from competition rules of investments generated by major new infrastructure, in case the security of supply might suffer prejudice should the former be applied.[38] Furthermore, the conflict between competition and security of supply is acknowledged by the Commission,[39] which observes that the pressure exerted by the former upon gas prices could result in an increase in the gas demand to the disadvantage of price supply. A similar reasoning opposes the increase of gas demand as a result of applying competition rules to sustainability requirements, which might favour the emergence of different energy mixes.

The inconsistencies between *sustainability and security of energy supply* derive from the inherent conflict between environment protection and energy consumption. The energy sector is regarded as a major source of distress for the local, regional and global environment. Energy consumption determines the release of gaseous emissions in the atmosphere, which can affect human health or produce acid rain or the greenhouse effect. A stable and continuous supply of energy is equivalent to a constant and sustained threat to the environment. Such an incompatibility is reduced to a certain extent in the case of natural gas, which is considered to be the most environment-friendly fossil resource. Regulation (EU) 994/2010 attempts to establish a balance between the two policy goals. Recital 15 conditions investments in gas infrastructure, required to enhance the security of gas supply, to a prior appropriate

---

[38] Szydło, 2009, p. 254.

[39] Green Paper /*COM/2000/0769 final */.

environmental impact assessment. Article 5(3) and annex II of the Regulation provide that national and joint Preventive Action Plans have to consider the impact that they might produce on the environment and opt for the least harmful solution. None of the three quoted legal sources explains what happens in those cases where the investment or action plan is liable to produce environmental side effects. An answer could come, however, from recital 14 of the Regulation. This provides that preference should be granted as far as possible to demand- and supply-side measures adopted in view of ensuring energy security which produce the least environmental impact. The interpretation of the text suggests that priority should be given to a balancing exercise between environmental needs and energy security to the extent that such a possibility exists. In its absence, concerns related to security of energy supply appear to prevail.

The clashes between the three main policy goals – both in pairs and, even more complicated, all three together – are determined by the fact that EU energy policy does not clearly prioritize among them.[40] As a result of this, it is often the role of the sector regulation to determine which goal will prevail. In cases in which the formulations of the specific secondary legislation give way to interpretation, the CJEU is expected to pronounce upon their meaning. Accordingly, of most relevance for the EU gas market is the jurisprudence accommodating public service obligations with the achievement of regulation-for-competition.

5.1.2.2 Tensions created by the accommodation of public service obligations

Respect for public service obligations has been a fundamental requirement in all three energy legislative packages, in accordance with the Commission's policy of gradually opening up the market, while protecting the general interest.[41] Public service obligations (PSOs) can be defined as obligations imposed by Member States upon public or private undertakings in view of providing a service, bringing public benefit or being of general interest. Regarding the gas market, PSOs are related to energy security (supply security, regularity and quality), as well as environmental protection (energy efficiency). A particularity of the definition of PSOs applicable to gas is the absence of reference to the "universal service" concept. In contrast to electricity, gas is a substitutable energy resource, which could be replaced by oil or coal, should its

---

[40] Haase and Bressers, 2008, pp. 15–17.

[41] Paragraph 32(1) of Commission of the European Union, 2004: Green Paper on Services of General Interest. OJ C 76. 25.03.2004.

supply incur disturbances.[42] As a consequence, the dependence of end consumers on the provision of gas is considered to be less intense than in the case of electricity. Despite the substitutability of gas, public gas service obligations are not less valuable than other PSOs. There are two natures attached to the object of public service obligations: one which entails public nonprofit interest and the other which presumes an economic component. Concerning the former, there is no straightforward definition of public interest either at national or at EU level. In general, public nonprofit interest is incorporated in national legislation that refers to shared responsibility, autonomy and the ideological or nonprofit-making nature of the operations. Its provision is not submitted to the review of EU law.[43] Regarding the latter nature of the object of PSOs, this provides the context in which tensions created by the accommodation of PSOs appear. This happens owing to the fact that Member States appointing PSOs for services of general economic interest are required to apply EU law regarding competition and public procurement.

Article 14 TFEU underlines the role played by PSOs in the fulfilment of the shared values of the Union and the promotion of social and territorial cohesion. Both concepts of "shared values" and "social and territorial cohesion" are vague and give way to a large margin of interpretation. Hence, they should be read in correlation with the *lex specialis* provisions of Directive 2009/73/EC. The major implication of article 14 TFEU lies in the emphasis placed upon the Union's and Member States' shared competence over PSOs, which constitutes one of the sources of conflict in accommodating gas PSOs and competitiveness requirements. It is clear from the article that both parties have to take care that such services operate under conditions that enable them to fulfil their missions. At the same time, the establishment and functioning of the internal market in gas remain exclusive competences of the Union. Both purposes are well in line with the competitiveness requirement, which generally presumes the institution of a level playing field among Member States, in view of achieving a common gas market.

The same cannot be said in the case of public service obligations. In conformity with a well-established jurisprudence,[44] Directive 2009/73/EC is very specific when providing that PSOs should be determined at the level of Member States, in conformity with their specific national

---

[42] Prosser, 2005, p. 191.

[43] Ministry of Employment and the Economy: Services of General Economic Interest, http://www.tem.fi/index.phtml?l=en&s=2017

[44] *Criminal proceedings against Paul Corbeau*, Case C-320/91; *Commission v France*, Case C-159/94; *Almelo*, *supra* note 34.

environment. Hence, it seems sensible to assume that such obligations may differ considerably from one Member State to another and potentially constitute barriers to trade with direct repercussions upon the common market. In an attempt to prevent the normative conflict between PSOs provisions and general market harmonization rules, the three successive gas legislative packages have introduced and maintained reference to the necessity that Member States respect common minimum standards, when constituting PSOs.[45] However, this provision has not precluded the appearance of obstacles to common market values in the fulfilment of particular PSOs. For example, article 35 of Directive 2009/73/EC provides that third-party access to the network may be refused by natural gas undertakings, where such an access would prevent the undertakings from carrying out public service obligations assigned to them. It is obvious that, in this case, following certain conditions, such as the obligation of duly substantiated reasons for the undertaking's refusal to grant access, the regulatory provisions give priority to PSOs.

The balance between the pursuit of market-based values and public welfare has been the object of an evolving jurisprudence. Over the years, the CJEU has demonstrated a fluctuating tendency towards limiting the application of competition rules in cases regarding PSOs. After an initial series of cases from the early 1990s proclaiming the triumph of competitiveness above public service obligations in regulated sectors,[46] the Court of Justice changed its approach in favour of the former with particular relevance for the EU gas market. In the so-called *Gas and Electricity monopoly* cases,[47] the Court found that Treaty rules, like those regarding competitiveness, are not to be applied to an undertaking entrusted with public service obligations, if such rules obstruct the performance in law or fact of the specific obligations that the said undertaking is deemed to fulfil. The interpretation of the Court brings a further limitation to the competitiveness requirement facing public service obligations. Two additional case laws are important for the present discussion. In the *Corbeau*[48] and *Almelo*[49] rulings, the Court found that restrictions on

[45] For example, objectives of common protection, security of supply, environmental protection and equivalent levels of competition.

[46] *Elliniki Radiophonia*, Case C-260/89; *Klaus Höfner and Fritz Elser v. Macrotron GmbH*, Case C-41/90; *Merci convenzionali porto di Genova SpA v Siderurgica Gabrielli SpA*, Case C-170/90.

[47] *Commission v Netherlands*, Case C-157/94; *Commission v France*, *supra* note 44; *Commission v Italy*, Case C-158/94.

[48] Case C-320/91 *Corbeau*, *supra* note 44.

[49] *Almelo*, *supra* note 34.

competition are allowed where the guarantee of the economic stability of an undertaking entrusted with PSOs would be prejudiced. Specifically, in *Almelo* the Advocate General maintained that "competition rules may be disapplied not only where they make it impossible for the undertaking in question to perform its public service task, but also where they jeopardize its financial stability".[50] Such an approach appears to have provided the basis for article 48(3)(b) of Directive 2009/73/EC. According to this article, derogations from third-party access obligations are allowed in those cases in which these would provoke serious economic and financial difficulties to undertakings which entered take-or-pay contracts in view of fulfilling public service obligations and ensuring the security of supply. In this case, as in the case provided by article 35 of the Directive, priority seems to have been granted to the accomplishment of PSOs over the achievement of the competitiveness goal. Such an approach is, however, not uniform in the text of the Directive, which produces further confusion.

It could be argued that articles 35 and 48 of Directive 73/2009/EC represent the exception from the general rule of competitiveness. Such an argument would be nevertheless contradicted by primary law provisions, such as those issued from article 106(2) TFEU. This article establishes a negative obligation for competition rules of not obstructing in law or fact the accomplishment of public service obligations. The relevance of this article for the opening of EU energy markets derives from the jurisprudence associated with it, which has introduced the principle of proportionality in competition issues regarding the assessment of the legality of public monopolies and privileged undertakings.[51]

National regulatory authorities play an important role in balancing competitiveness and public service obligations. In case *Federutility*[52] CJEU found that setting rates for energy companies is not inconsistent with achieving competition in the energy market, but that not all price interventions by regulators are legally permissible. According to the judgment, creating and setting rates for energy companies involves a balance between the justified economical interest of energy companies regarding their profit maximization and the general interest of protecting

---

50 *Ibid.*, para. 146.

51 *Federutility*, Case C-265/08; *Altmark Trans GmbH, Regierungsprasidium Magdeburg and Nahverkehrsgesellschaft Altmark GmbH*, Case C-280/00; *Ferring SA v Agence Centrale des Organismes de Sécurité Sociale*, Case C-53/00; *Enirisorse SpA v Sotacarbo SpA*, Case C-237/04; *Ministre de l'Economie, des Finances et de l'Industrie v Gemo SA*, Case C-126/01.

52 *Federutility*, *supra* note 51.

end users or businesses from unjustified price levels. In the view of the CJEU, article 23(1) of Directive 2003/55/EC did not prevent national regulators from influencing price levels for supply of natural gas by setting "reference prices". The Court, however, established three cumulative conditions that should be fulfilled in order for such intervention to be justified. First, the intervention must be based on "a general economic interest consisting in maintaining the price of the supply of natural gas to final consumers at a reasonable level", having regard to Member States' obligation to reconcile liberalization and the protection of final consumers, deriving from public service obligations. Secondly, such intervention coming from the state should be delimitated in time and scope, in order to circumscribe its discretionary application. The intervention is permissible only in as much as it is required in order to achieve the protection of final consumers and within the margins provided by the general economic interest. The duration of such measures is limited to the period necessary for the accomplishment of the objective. The third condition set by the Court regards the rules for the administrative regulation of prices. Such rules must be "clearly defined, transparent, nondiscriminatory and verifiable". At the same time they must guarantee energy companies in the EU equal access to consumers in the given Member State. Only upon the cumulative meeting of the conditions set by the Court is a price intervention of a public authority legal.

Additional frictions between public service obligations and competitiveness as gas market goals stem from their different legal sources, aiming at different market directions. PSOs are generally governed by public service law. This aims at protecting social rights and social solidarity. In performing PSOs, undertakings become accountable to citizens, through the scrutiny performed by regulators and governments. Regarding competitiveness, this stems from the concerns of competition law, which aim at opening the market. In this case, gas undertakings become accountable to consumers and shareholders.[53]

Both primary EU legislation and specific gas sector regulation attempt to blur the demarcation lines between the origins of the two market goals by linking the realization of one to the nondisturbance of the other. Such an approach has not always brought the expected results. Article 3(2) of Directive 2009/73/EC conditions the discretion of Member States in defining PSOs to the respect of EU law provisions and especially those regarding state aid. It has been argued that the new legislative package has strengthened the prescriptiveness of PSOs by introducing new

---

[53] Prosser, 2005, p. 2.

mandatory obligations for Member States, in view of increasing cross-border regulatory harmonization.[54] This is confirmed by article 3(4) of the Directive, which establishes an obligation for Member States to notify the Commission upon each adopted measure, which could have an impact on the quality of public service in the EU. The Commission is supposed to publish regularly a report on the measures taken by Member States and compare their efficiency. Member States are additionally required to notify the Commission upon any measure constituting state aid under article 107(1) TFEU taken in order to fulfil the respective PSOs. The notification obligation is aimed at moderating the differences between the levels of social intervention in the Member States, which may act as a barrier to competition in the EU gas market.

Further clashes between market-based values and public interest requirements resulting from the poor articulation of EU energy goals in the European energy policy are reflected in the enforcement of proper customer protection. The 2009 Directive emphasizes the close connection between public service obligations and consumer protection, especially in relation to the category of vulnerable consumers and their access to gas resources. Such a connection is confirmed by the recent codification of consumer protection in the utilities sector in Protocol 26[55] annexed to the TFEU. In light of the new Energy Package, the clarification of the vulnerable customers' concept and the introduction of the idea of energy poverty in Directive 2009/73/EC might lead to a rusty articulation with market competitiveness objectives. At the same time, a better equipment of consumers with effective tools to enforce his rights might finally allow them to take advantage of the liberalized market benefits. This could result in an increase in energy consumption, which in turn could conflict with energy sustainability principles.

5.1.2.3 The implications of reinforcing consumer protection requirements

One of the main objectives behind the gas market liberalization and integration process has been the increase of the benefits that European consumers can draw from the market. The entire liberalization discourse has been founded upon the advantages that gas competition may create for the ultimate consumer, in terms of energy supply affordability, quality and continuity. The critics have, however, noticed two major failures of the seemingly perfect partnership between more gas market competition

---

[54] McHarg *et al.*, 2010, pp. 384–385.

[55] Protocol 26 on services of general interest.

and better gas consumer protection. The first is related to gas supply prices, which contrary to expectations have not decreased along with the liberalization of the gas market. In fact, the transition from regulated gas prices to market-based ones has resulted in the final quote reflecting the marginal cost of production instead of the average one.[56] Consequently, as indicated by several economic studies, in the aftermath of gas policy reforms, household prices have either stagnated or increased.[57] A potential explanation for this development stems from the second major failure of the competition–consumer protection partnership. This regards the inability of European energy consumers to take advantage of market liberalization benefits and to ensure an accurate reflection of their interests in the market.[58]

The Third Energy Package attempts to correct the effects of the former inadequate and insufficient European consumer legislation by favouring a higher consumer protection than its predecessors, more transparency regarding the formation of prices and switching procedures, as well as an easier access to dispute settlement mechanisms. Consumer protection measures are reinforced by a detailed explanation of penalties that can be applied to a transmission system operator or a vertically integrated undertaking, in cases of noncompliance with the existent legislation. These can amount to 10 per cent of the annual turnover of the offender, which can result in a considerable financial loss for the undertaking or the system operator.[59] The amount of penalties[60] appears to have inspired a turn in the Commission's approach to energy case law. Indeed in October 2009, only a few months after the adoption of the TEP, the Commission imposed its first financial sanctions upon gas undertakings engaged in concerted practices that did not comply with existent legislation.[61]

Such a development inspires the assumption that the strengthening of consumer protection measures in the EU gas market is likely to generate a positive impact on the competitiveness of the market. At the same time, a reinforced gas consumer protection policy presumes an increased coordination between purpose and means. If the purpose is to augment the advantages of gas consumers in the gas market, this would imply that

---

56 Spector, 2008, pp. 50–56.

57 For example, Brau *et al.*, 2010, pp. 167–182.

58 Cseres, 2008, pp. 77–121.

59 Article 41(4) of Directive 2009/73/EC.

60 Guidelines on the method of setting fines imposed pursuant to Article 23(2)(a) of Regulation No. 1/2003, para. 28.

61 *E.ON/GDF collusion*, COMP/39.401.

the means have to target a stable and secure flow of energy, of a certain quality and at competitive prices. The latter condition entails the presence in the market of a multitude of participants, competing for the largest share of gas consumers. Accordingly, article 3(2) of Directive 2009/73/EC underlines the guarantee of equal access for natural gas undertakings to national consumers. On the one hand, the provision grants consumers with a greater choice between energy suppliers. On the other hand, it aims at dismantling the remaining barriers in the market preventing third-party access. Articles 41(4) and 3(2) of the Directive lead to the conclusion that the potentially anticompetitive, consumer-harming behaviour of gas undertakings is expected to be deterred by the amplitude of potential fines, correlated with actual examples of their enforcement.

Despite an apparent perfect compatibility between regulation-for-competition and consumers' interest, certain consumer protection provisions in TEP seem to have brought more complexity to the relationship. These provisions refer to energy poverty and vulnerable customers. In general, the direction adopted in the Package seems to incline towards a balance between ensuring a continuous and safe supply of energy to consumers and maintaining a level playing field for all suppliers, domestic and foreign. This derives especially from article 40(d) and (g) of the Directive, regarding the tasks of NRAs. Along with other attributions, the latter are required to simultaneously ensure consumer protection and promote effective competition.

While it is apparent from the Directive that the legislator continues to abstain from setting up a hierarchy between market goals,[62] the question that emerges is to see to what extent, if any, the balance between the two market objectives is challenged by energy poverty concept. Energy poverty has been described as "the absence of sufficient choice in accessing adequate, affordable, reliable, high quality, safe and environmentally benign energy services to support economic and human development".[63] Member States that are affected by energy poverty are obliged to develop national action plans or other appropriate frameworks to tackle this matter.[64] The concept is used to emulate a situation in which the security of gas supply to a certain group or groups of consumers is endangered. Subsequently, the employment of the concept serves to identify a special category of consumers, deemed vulnerable, to

---

62 In this specific case between competitiveness and consumer protection.

63 Clancy *et al.*, 2003.

64 Article 3(4) of Directive 2009/73/EC.

which the uninterrupted supply of gas becomes a national public service obligation. Hence, it appears that, in cases of energy poverty, the balance between consumer protection and competitiveness requirements discussed above ceases to remain a matter of their sole interest. Instead it becomes a dynamic relationship between competitiveness on the one side and consumer protection, public service obligations and security of supply requirements on the other. As seen in the preceding subsections, such a relationship determines in general a hierarchical framework for the market objectives, which varies on a case-by-case basis.

As seen in Chapter 2, under Directive 2009/73/EC all consumers are equally protected and are expected to equally benefit from the advantages of a competitive market.[65] Additional guarantees are, however, foreseen with regard to vulnerable customers. These represent a particular category of consumers, which is generally exposed to energy poverty and whose disconnection from the supply of gas in critical times is expressly prohibited by the text of the 2009 Directive.[66] According to article 3(3), each Member State enjoys discretion over defining the parameters of the said category. The article follows the logic that, in the same way as PSOs may vary according to specific national contexts, the measures taken by Member States to protect final consumers may differ according to the segment of consumers they are aimed at: households or small and medium-sized enterprises. In France vulnerable customers are considered to be people in precarious situations[67] or consumers in payment difficulties.[68] In the UK, customers of pensionable age, chronically sick, disabled consumers and those who are in payment difficulties are deemed vulnerable customers.[69] In Greece the supplier is obliged to provide uninterrupted natural gas in case of emergency to consumers who are not

[65] That is, recitals 1 and 3 of Directive 2009/73/EC do not differentiate between industrial consumers and households.

[66] Although the concept of vulnerable consumers does not represent a novelty of Directive 2009/73/EC, the latter has certainly increased its importance. For example, the text of Directive 2003/55/EC indicated a mere obligation for Member States to develop measures to help vulnerable consumers avoid disconnections. No express prohibition of disconnection was stipulated in the 2003 gas Directive.

[67] Décret no. 2001-531 du 20 juin 2001 relatif à l'aide aux personnes en situation de précarité pour préserver ou garantir leur accès à l'électricité.

[68] Décret no. 2005-971 du 10 aout 205 relatif à la procédure applicable en cas d'impayés des factures d'électricité.

[69] OFGEM, 2010: Notice of Proposed Modification in Relation to the Disconnection of Vulnerable Consumers … .

major customers.[70] Furthermore Greece has adopted administrative measures facilitating gas supply to disabled and elderly customers and prohibiting their disconnection from the service during periods of peak demand in case of nonpayment.

Despite the discretion allowed for concept definition purposes, article 3(3) of Directive 2009/73/EC establishes a real obligation for each Member State to take appropriate measures to protect final consumers in remote areas.[71] This refers to the possibility of appointing a supplier of last resort for vulnerable customers connected to the gas system. It is specifically such a provision that can create potential tensions between consumer protection and the competitiveness requirements. The appointment of a supplier of last resort might temporarily interfere with the principle of free competition between gas market players. There are two important arguments here. First, suppliers of last resort are generally appointed by public authorities. There is no tendering procedure that could facilitate competition between potential suppliers. Consequently, in at least half of the cases the major gas incumbent is the one assuming such a position.[72] Secondly, the service of a supplier of last resort is not time-limited. Article 3(3) of the Directive provides only that Member States have to ensure that eligible customers are able to easily switch to another supplier. However, there are no clarifications regarding the steps that should be followed if the consumers do not wish to change their providers.

The 2009 Directive attempts to avoid such a situation by educating the consumer with regard to their actual gas consumption and the costs that this generates. In the past few years, the relationship between energy sustainability and consumer protection has been reinvented by new secondary legislation adopted in the two directions. As a result, the awareness of the inter-conditioning between energy consumption and environmental sustainability has been increased. Changes in the patterns of consumption towards more energy efficiency could have a positive impact upon energy sustainability. As stated in EU's Energy Efficiency Plan, energy consumption should be diminished by 20 per cent by

---

[70] Article 24(3) of Greek national Gas Law 3428/2005.

[71] In this respect, the Directive suggests an integrated approach that could use the framework of the social policy, as well as the formulation of national policies in favour of the vulnerable customers.

[72] ERGEG, 2009: Status review of the definitions of vulnerable customer, default supplier and supplier of last resort, p. 31.

2020.[73] There are various ways through which the Commission intends to achieve such an objective, one of which is a better reflection of environmental concerns in energy prices.[74] The reasoning followed is that a rise in energy prices is expected to determine a shift towards lower energy consumption, which would ultimately result in a reduction of environment-harming emissions.

The balance between sustainability and consumer protection is further influenced by the direction taken by competitiveness in the gas market. If competitiveness is pursued to increase the total welfare in the market, then at least in the short term this would lead to higher prices for consumers. This happens owing to the fact that an increase in total market welfare is generally achieved by pushing market efficiency upwards, which generates new costs (technology innovation, network infrastructure modernization, investments in new transmission infrastructure) that are reflected in the end prices. From an energy efficiency point of view such an outcome is desirable, because it could lead to avoiding overconsumption.[75] As a consequence, the strain put on environment could theoretically decrease. If on the other hand, competitiveness is pursued to increase consumer welfare, the targeted outcome would be the maximization of consumer surplus at the expense of energy end-prices. This could determine an upward evolution of gas demand, which would result in increased energy consumption and an added stress upon the environment.[76]

In the case of the EU gas market, the competition-based approach seems to oscillate between the two strands of welfare. To a considerable extent, this is due to ambiguous signals provided by Commission officials, which at times give the impression of a perfect correlation between the efficient allocation of resources and consumer welfare, while at other times state that the latter should be given priority.[77] While the European legislator appears to have finally succeeded in bringing the

---

[73] Communication from the Commission: Energy Efficiency Plan /*COM/ 2011/109 final*/.

[74] *Ibid.*, recital 5.

[75] Spector, 2008, pp. 50–56.

[76] There is an important limitation to this discussion in that it is highly based on economic theoretical assumptions. The Commission's Eurostat report from 2011 shows a fluctuating evolution of gas demand, which cannot be solely attributed to competitive efforts on the market. Moreover, as previously explained, the introduction of competitiveness in the gas markets of Member States has not necessarily resulted in a decrease in gas prices.

[77] Kroes, 2005; Monti, Mario, 2004.

interests of EU energy consumers to the core of the EU energy policy,[78] the substantive rules of EU competition law focus on different aspects of market liberalization when applied. This will be further developed in the next section.

5.2 LEGAL ROUTES TO ACHIEVE THE POLICY GOALS

Taking into consideration the definition of market design put forward at the beginning of this chapter, the present section aims to look at the legislative and regulatory tools operating in the EU gas market and at the patterns of sharing responsibilities and assuming collaboration between them, in order to achieve the gas market goals. As seen in Chapter 2, the legal basis for a common European action in the EU gas market has evolved in recent years from single market provisions and competition policy, to sector regulation and treaty codification. Despite the position of various agents manifesting different market beliefs,[79] the EU gas market stands as an example that competition law and regulation are not and should no longer be regarded as "rhetorical friends and deadly enemies".[80] Instead, it appears that regulatory authorities in the EU gas market are committed to apply the regulatory tools at hand in order to ensure market competitiveness.[81]

The provisions of the Third Energy Package indicate a divergent approach to the original drive of the gas market reforms. Until the period of time immediately prior to the adoption of the Second Energy Package, the advocated view was that competition law was better equipped than state regulation to order and organize any economic sector, including the gas one. Hence, it appeared that competition law and policy should be extensively used to control and remove adverse behaviour of market players in the EU gas market, while the regulation of the sector should be reduced.[82] Nevertheless, already in the Second Energy Package, the powers and attributions of NRAs have been increased and a series of accompanying regulatory measures regarding unbundling have been

---

78 Commission Staff Working Paper: An Energy Policy for Consumers, /*SEC/2010/1407 final*/.

79 Eikeland, 2004.

80 Stigler, 1975, p. 183.

81 Jordana and Levi-Faur, 2004, pp. 1–2.

82 OECD Reviews of Regulatory Reform: Regulatory Reform in the United Kingdom Regulatory Reform in Gas and Electricity and the Professions.

introduced, in support of the market liberalization process. In the TEP, the regulatory grip appears to have been even further reinforced through the introduction of effective unbundling and complementary requirements of transmission system operators' certification, as well as a further increase in the powers of NRAs and the establishment of European bodies responsible for a supranational supervision of the market and the cross-border coordination of national regulators and system operators.

The increased role of regulation in the EU gas market has coincided with a more effective enforcement of competition law in the gas sector.[83] As a result, the relationship between the two has evolved and is possibly in need of re-definition. As Cameron observes, "the principal role of competition law in energy market reform has been to provide *support* to the development of a sector-specific regime".[84] The affirmation appears to hold even truer in the present day, especially owing to the strengthening of competition law enforcement in the field. However, along with the reinforcement of both legal routes used to achieve the gas policy goals, the instances of overlapping rules between the two are also likely to multiply. This could potentially influence the oscillation of the relationship between competition policy and regulation from complementarity to conflict and back.

In the following, the complexity of this relationship will be analysed bearing in mind three main questions:

- what are the instances of complementarity between the action of competition law and that of regulation in the EU gas market;
- what are the factors that might fuel a conflicting application of the two; and
- to what extent – if any – are they still mutually exclusive?

5.2.1 The Ability of Competition Law to Satisfy Gas Policy Goals

Competition law tools, the expected success of their application evidenced by the remedies that they entail, as well as their potential limitations, are elements that allow an assessment of the ability of competition law to satisfy EU gas market goals. Two questions are of interest. These are to what extent competition law is equipped to tackle the said goals and whether its application is likely to contradict them in certain instances.

---

[83] *E.ON/GDF collusion*, COMP/39.401; *ENI SpA*, Case COMP/39.315, Case COMP/39.402 *RWE AG* 2009.

[84] Cameron, 2007, p. 564.

Competition policy has traditionally served as one of the main pillars for Commission action in economic fields.[85] Article 119 TFEU provides that the activities of Member States and of the Union shall include the adoption of an economic policy based on the close cooperation between national economic policies in the internal market and on the definition of common objectives. The conduction of such an economic policy is expected to occur in accordance with the principle of an open market economy with free competition. The scope of application of competition policy stretches over anticompetitive agreements between undertakings, abuses of dominant position in the market, mergers and anticompetitive state behaviour, favouring one or some market players above others.

In the EU gas market the use of competition policy tools is justified by the market structure, where main EU and non-EU gas incumbents have historically engaged in anticompetitive, market-foreclosing practices. The anticompetitive behaviour of a gas undertaking might replicate monopolistic conditions, for example, when it involves collusion with another company to set prices above the competitive level and reduce output.[86] Such behaviour leads not only to a redistribution of income, but also to a misallocation of resources.[87]

Of high relevance for applying competition law and policy tools to satisfy gas market goals are the findings of a 1999 White Paper of the European Commission. In the said document, the Commission identified a double purpose for EU competition law and policy: forging integration of national markets into a single economic area and ensuring that competition is developed as the main driving force of the economy.[88] The first purpose is unique to EU competition law, which acts to prevent the frustration of its realization by actions of private undertakings.[89] The two identified objectives correspond to one of the main EU gas policy goals – which is the requirement for market competitiveness – and the ultimate goal of the gas market – which is market integration. In this context, it seems legitimate that the Commission has used competition tools to open up and bring together the formerly monopolistic European gas markets, as well as the rest of the network utilities' sectors.

---

85 Report from the Commission – XXXth Report on competition policy 2000, /*SEC/2001/0694 final*/.

86 *E.ON/GDF collusion*, COMP/39.401.

87 Jones and Sufrin, 2004, p. 18; Keske, 2009.

88 Commission White Paper on Modernisation of the Rules implementing articles 85 and 86 of the EC Treaty.

89 Craig and de Búrca, 2011.

Despite the apparent lack of contradiction between competition law and EU gas market goals, it should be observed that competition law has not been limited to the achievement of the two-fold purpose resulting from the 1999 Commission's White Paper. Instead, there are a multitude of goals pursued through competition law. Their exact interpretation and internal compatibilities have raised fervent discussions, notwithstanding the undisputable role played by competition in the internal EU legal order.[90] Several studies seem to agree that the primary objective of competition policy is economic efficiency.[91] The goal is codified in article 102 TFEU. In subsidiary to economic efficiency, competition law has been applied to protect competitiveness and competitors, public welfare, security of supply and market integration. In the specific case of EU competition law, efficiency refers both to the maximization of consumer welfare and to the achievement of an optimal allocation of resources. This interpretation of efficiency represents the logic behind the liberalization of network industries.

When applying competition law to the EU gas market, one of the first concerns should be the accommodation of consumer welfare with the efficient allocation of resources. It has been shown above that, despite the liberalization of the EU gas market, energy prices have demonstrated a tendency to increase in the short run. Although such a development has most likely benefited the dynamic efficiency of production, it has not necessarily brought about a simultaneous increase in consumer welfare.[92]

The second issue when applying competition law to achieve efficiency in the gas market is that both the concept of energy efficiency in EU competition law and the term efficiency itself in general economics lack a true definition.[93] The appreciation of efficiency requires a balancing test, applied on a case-by-case basis. Within the framework of the test, the negative effects of an agreement, transaction or merger upon market competitiveness are balanced with the "efficiencies" that they bring in the market. Such efficiencies may refer to innovation, the realization of economies of scale or scope, security of supply, ensuring public service obligations or guaranteeing a high level of protection for vulnerable

---

90 Roggenkamp *et al.*, 2007, p. 250; Craig and de Búrca, 2011, p. 950.

91 Jones and Sufrin, 2004; Geradin, 2004; Goyder *et al.*, 2009.

92 At least in the sense of the concept which entails the maximization of consumer surplus and prohibits excessive prices. If consumer welfare is understood in a more extensive approach to encompass long-term gains, rather than short-term ones, then the achievement of economic efficiency could be regarded as truly serving the interests of consumers at all stages. See more in Keske, 2009.

93 Geradin, 2004.

consumers. Following this reasoning, it appears very likely that, when the balance is tilted in favour of one of the ends of competition, the accomplishment of another will be frustrated. In consequence, it could be concluded that, much like in the case of EU gas policy, the goals of competition policy are at times at odds with one another. This has a direct effect on the pursuit of gas market goals through competition law tools.

5.2.1.1 Instruments of competition law

5.2.1.1.1 Antitrust measures Articles 101 and 102 TFEU seem to perfectly capture the essence of the gas market prior to liberalization. The purpose of antitrust measures is to prevent restrictive agreements between undertakings (article 101 TFEU) and abuses of market power (article 102 TFEU). As seen above, article 101 TFEU refers to restrictive agreements between undertakings, which produce anticompetitive effects within the internal market and which may affect trade between Member States. This latter condition entails the scope of application of EU competition law. In conformity with article 3(1) of Regulation (EC) No. 1/2003, in matters affecting trade between Member States article 101 TFEU prevails, where the application of national competition provisions in the Member States is concerned. In the sense of article 101(1) TFEU, a restriction or distortion of competition is specifically expected to occur in cases of price-fixing, limitation or control of production, markets, technical development or investment, sharing of market or sources of supply, application of dissimilar conditions to equivalent transactions with other trading parties and tying.[94] Article 101(2) TFEU provides that such agreements are automatically void. A horizontal agreement between major gas producers regarding price fixing in gas supply contracts with the EU might distort market competition and lower the protection of consumers.[95]

Article 101(3) TFEU provides for several exemptions regarding agreements between undertakings. These are in general individual exemptions, although the Commission may also grant block exemptions. In order for individual exemptions to apply, four cumulative conditions must be met:

---

[94] Tying refers to subjecting the conclusion of contracts to acceptance of supplementary obligations by other parties which have no connection with the subject of such contracts.

[95] *GFU*, Case COMP/36.072.

- the agreement must contribute to improving the production or distribution of goods or to promoting technical or economic progress;
- it must allow consumers a fair share of the resulting benefit;
- it must not impose on undertakings restrictions that are not indispensable to the attainment of this objective; and
- it must not give such undertakings the possibility to eliminate competition in respect of a substantial part of the products in question.

Block exemptions have been covered by different European regulations in accordance with the nature of the agreement or the market sector envisaged.[96] In the gas market, such exemptions have been raised under the "specialisation block exemption"[97] by undertakings attempting a joint production or a joint selling of gas.[98] In general, joint production ventures have been regarded as less threatening to competition than joint selling, owing to their limited effect on prices.[99] However, in a series of investigations into cooperative agreements in the EU gas market regarding specific cooperation between producers, the Commission has been of the opinion that such practices could harm the creation of a competitive market in the Member States concerned and in the neighbouring countries.[100]

Article 102 TFEU prohibits abuses of dominant position by one or more undertakings, within the internal market or a substantial part of it. As in the case of article 101 TFEU, the defining condition for the application of article 102 TFEU is that the abuse affects trade between Member States. An abuse of dominant position occurs in the same cases as described above for article 101(1). It should be observed at this stage that article 102 TFEU sanctions the abuse of a dominant position, not the dominant position itself. This is a very important distinction for the organization of the EU gas market and the potential limitations of competition policy.

The concept of dominance has been coined by the Court of Justice as "a position of economic strength enjoyed by an undertaking, which

---

96 For example, Commission Regulation (EC) No. 2790/1999/EC; Council Regulation 19/65/EEC.

97 Commission Regulation (EC) No. 2658/2000.

98 *Irish Corrib*, Case COMP/37.708.

99 Roggenkamp *et al.*, 2007, p. 256.

100 *GFU*, Case COMP/36.072; *Corrib*, *supra* note 98; *DONG/DUC*, Case COMP/38.187.

enables it to prevent effective market competition being maintained on a relevant market, by affording it the power to behave to an appreciable extent independently of its competitors, customers and ultimately of the consumers".[101] According to the Court "such a position does not preclude some competition, which it does whether there is a monopoly or a quasi-monopoly, but enables the undertaking which profits by it, if not to determine, at least to have an appreciable influence on the conditions under which that competition will develop, and in any case to act largely in disregard of it so long as such conduct does not operate to its detriment".[102]

Relevant for the EU gas market is the fact that third-party access refusal may qualify as an abuse of dominant position. Access to the transmission and distribution network, as well as storage and LNG facilities, represents an indispensable component of functioning competition in the EU gas market. As shown in the previous chapters, since the initiation of the gas market liberalization process, major gas incumbents have directed all their strategies towards maintaining their dominant position in the market. This included refusing third-party access to their gas systems. In the *Marathon* cases[103] the systematic refusal of five gas companies in continental Europe to grant access to their gas pipelines to a Norwegian subsidiary of an American oil and gas producer was considered by the Commission as a potential abuse of a dominant market position and restrictive concerted practice.

The legitimacy of using competition rules in cases regarding access to gas infrastructure derives from the *essential facilities* doctrine.[104] This doctrine applied to the case of the EU gas market refers to establishing whether or not dominant firms should be compelled to allow third-party access to their essential assets that are not available to their competitors. In *Bronner*[105] the CJEU enunciated three cumulative conditions under which access to facilities deemed essential had to be granted:

- the refusal of access should be likely to eliminate all competition in the market of the entity requesting the service;
- there is no objective justification for the refusal of access;

---

101 *Hoffmann–La Roche and C. AG v. Commission*, Case 85/76, para. 38.

102 *Ibid.*, para. 39.

103 *Marathon/Ruhrgas/GDF et alia*, Case COMP/36.246.

104 More on the essential facilities doctrine in Geradin, 2005.

105 *Oscar Bronner GmbH & Co. KG v Mediaprint Zeitungs, Case* C-7/97, para. 41.

- access to the facilities is indispensable for carrying on that entity's business.

Should these conditions not be met, the refusal of access cannot be qualified as abuse of a dominant position under article 102 TFEU.

5.2.1.1.2 Merger Regulation (EC) No. 139/2004 Regulation (EC) No. 139/2004 provides the rules for the control of concentrations between undertakings at EU level. The purpose of the Regulation concerns the delimitation of merger cases with an EU dimension from those with a predominantly national focus. In this respect, the Merger Regulation is based on the principle of subsidiarity, whereby any merger under scrutiny should be examined by the judicial authority best positioned to do so. Under the Merger Regulation, Member States preserve the right to have sovereign control over concentrations between undertakings, when the said concentrations are below a certain level of economic importance. The Commission has control over mergers with a Union dimension.

The concept of market concentration refers to a transaction between two or more undertakings. The transaction must bring about a lasting change in control, joint or sole, over the whole or part of one or several undertakings or businesses.[106] At the same time a concentration can entail the acquisition by one or more persons already controlling at least one undertaking, or by one or more undertakings, whether by purchase of securities or assets, by contract or by any other means, of direct or indirect control of the whole or parts of one or more undertakings.[107] Control can be constituted by rights, contracts or other means which separately or in combination confer the possibility of exercising decisive influence on an undertaking.[108] Such a control can refer to ownership or decisive influence upon the internal organization of an undertaking. Additionally, it is considered as the creation of a joint venture performing on a lasting basis all the functions of an autonomous economic entity.[109]

For the application of the merger regulation in the EU gas market it is important to define concentrations with a Union dimension. According to recital 10 of Regulation (EC) No. 139/2004, a concentration with a Union dimension should be deemed to exist where the aggregate turnover of the undertakings concerned exceeds a given threshold, irrespective of whether or not the said undertakings have their seats or principal fields of

---

[106] Article 3(1)(a) Council Regulation (EC) No. 139/2004.
[107] Article 3(1)(b) of Council Regulation (EC) No. 139/2004.
[108] Article 3(2) of Council Regulation (EC) No. 139/2004.
[109] Article 3(4) of Council Regulation (EC) No. 139/2004.

activity in the Union, provided they have substantial operations there. This latter aspect is all the more important for the gas sector, where very often undertakings initiating concentrations which affect the EU are neither established nor have their principal seats in the Union. The conditions for a concentration with a Union-wide scope are inscribed in article 1(2) of Regulation (EC) No. 139/2004 and include:

- a higher than €5000 million combined aggregate turnover of all the undertakings concerned;
- a higher than €250 million aggregate Union-wide turnover of each of at least two of the undertakings concerned.

A concentration is considered to have a Union dimension as well, where

- combined aggregate worldwide turnover of all undertakings concerned is more than € 2500 million;
- in each of at least three Member States, the combined aggregate turnover of all undertakings concerned is higher than € 100 million;
- in each of at least three Member States, the aggregate turnover of each of at least two of the undertakings concerned is more than € 25 million;
- the aggregate Union-wide turnover of at least two of the undertakings concerned is more than € 100 million.

There is an exception to both situations described above. This occurs when each of the undertakings concerned achieves more than two-thirds of its aggregate Union-wide turnover within one and the same Member State.

The powers of the Commission in relation to a concentration with Union dimension are described in article 8 of the Merger Regulation. According to the article, the Commission may declare a concentration compatible with the common market, it may require certain commitments from or set conditions on the undertakings concerned so as to clear the concentration, it can declare the concentration incompatible with the common market and it may dissolve an already installed concentration or impose commitments. Article 8(5) provides for the possibility of the Commission adopting interim measures. Article 11 stipulates the right of the Commission to request information and article 13 confirms the Commission's powers of inspection. Article 14 establishes that the Commission may impose fines on the concerned undertakings in specific cases.

The application of the European merger control to concentrations on the gas sector is relatively recent. Before the gas market liberalization, it was almost impossible to demonstrate the EU impact of a merger accomplished within the national borders of Member States. Following the opening up of Member States' gas markets and the recognition of the cross-border nature of the EU gas market, merger rules have been increasingly used to ensure better access to essential infrastructure for all market players and to limit the ability of major incumbents to strengthen their already dominant position in the market.[110] Recital 25 of the Merger Regulation has been relevant for the regulation's application in the original stages of opening up the EU gas market. According to the recital, in oligopolistic market structures, concentrations that set aside important competitive constraints between the merging parties and decrease the competitive pressure on the remaining competitors may result in significant impediments to effective competition. The justification of the continuation of the use of merger control in the EU gas market derives from recital 3 of Regulation (EC) No. 139/2004. This confirms that the completion of the internal market coupled with the lowering of international barriers to trade and investment and the multiplication of membership in the EU will result in major corporate reorganization. The Commission's right to approve significant mergers has been aimed at determining necessary structural changes in the market, such as increased transparency and better access to the grid.

The importance of the Merger Regulation in opening up the EU gas market became visible in a series of cases where Member States continued to shelter national champions from foreign takeover. Case COMP/M.4197 E.ON/Endesa[111] concerned the decision of the German company E.ON to acquire the entire share capital of the Spanish energy company Endesa. The bid launched by E.ON was competing with a hostile bid launched some months earlier by the Spanish traditional gas incumbent Gas Natural. Following E.ON's bid, Spain adopted by Royal Decree a legislative measure increasing the supervisory powers of the local NRA.[112] According to the new measures, E.ON's bid became subject to the prior approval of the Spanish NRA. The latter imposed a series of conditions creating unwarranted obstacles to a merger judged by both the Commission and the CJEU as falling under the Commission's sole competence. In its 6 March 2008 decision the Court of Justice found

110 Monti, Giorgio, 2002.

111 See also *Commission v. Spain*, Case C-207/07.

112 Comisión Nacional de Energía (CNE).

that Spain had failed to comply with two of Commission's decisions pursuant to article 21 of the Merger Regulation, which required Spain to withdraw the conditions it imposed on the E.ON acquisition. The judgment of the Court confirms the Commission's position that Member States cannot create unjustified obstacles to mergers which fall under the Commission's exclusive jurisdiction. As a consequence, any disproportionate state measures restricting or having a negative impact on mergers with a Union dimension are considered to violate EU law. The fact that the concentration under discussion had a Union dimension allowed the Commission to intervene and prevent local authorities from distorting competition on the Spanish market.

When analysing a concentration with a Union dimension, the Commission takes into account the need to maintain and develop effective competition within the common market, the structure of all the market concerned and the actual or potential competition from undertakings located either within or without the Union.[113] At the same time it evaluates alternatives available to suppliers and users, their access to supply or markets, any legal or other barriers to entry, supply and demand trends for the relevant goods and services and the interest of the intermediate and ultimate consumers.

Market definition plays an essential role in the application of merger control rules to the gas sector. Specifically, the definition of the relevant market is necessary in order to assess the market position of the undertakings concerned and their economic and financial power.[114] Two features are important when defining the relevant market: the product concerned and the geographical scope. The product concerned regards activities in a particular segment of the market. In the case of gas these can mean either upstream or downstream activities. Upstream activities refer to the exploration,[115] development,[116] production and sale,[117] transmission (through gas pipelines or LNG ships) and processing of gas.[118] Downstream activities refer to sales,[119] transport or transmission via the

---

[113] Article 2(1)(a) and (b) of Regulation No. 139/2004.

[114] In *BPAmoco/Arco*, the Commission observed that, while the exploration market was worldwide in scope, from a European-demand perspective, the geographic market for gas included only the European Economic Area (EEA), Algeria and Russia.

[115] *Exxon/Mobil*, Case M.1383.

[116] *Ibid.*

[117] *Shell/Enterprise Oil*, Case M.2745.

[118] *Ibid.*

[119] *Gaz de France/Suez*, Case COMP/M.4180.

high-pressure pipeline grid,[120] storage[121] and blending.[122] Regarding the geographic market, the European Commission considers a triple conditioning.[123] First, this is an area where the undertakings concerned are implicated in the supply and demand of products and services. Second, the area has to present sufficiently homogenous competition conditions. Third, it has to distinguish itself from neighbouring territories through specific competition conditions.

5.2.1.1.3 State aid In the sense of article 107(1) TFEU, state aid in national gas markets can be defined as a measure designed to help a gas undertaking that fulfils four cumulative conditions:

- the aid must be granted by a State or through State resources;
- it must distort or threaten to distort competition in the gas market;
- by selectively favouring certain gas undertakings or the production of certain goods (e.g. gas of certain quality);
- it must affect trade between Member States.

The application of state aid measures in the EU gas market is of very high relevance owing to the continued high level of involvement of national governments in the regulation of the gas sector and their traditional allegiance to major gas companies.[124] Following EU gas market liberalization, this fact has been visible in cases involving state compensation for "stranded costs".[125] Under the TEP, Member States continue to play an important role in the market opening to true competition, as well as in forging market integration and interconnection. According to article 13 of Regulation (EC) 715/2009, Member States –

---

[120] For example, "swap arrangements" that allow gas of a certain quantity and quality to be inserted at point A of a gas network and withdrawn at point B of the gas network.

[121] *Gaz de France/Suez*, *supra* note 119.

[122] This refers to the transformation of H-gas into L-gas. H-gas is a "high" calorific natural gas with a high fuel value, while L-gas is a "low" calorific gas. See Case *RWE Gas/Lattice*, Case M.2744.

[123] Commission Notice on the definition of relevant market for the purpose of Community competition law, para. 7.

[124] Either as a result of ownership or partial ownership, or as a consequence of the specificities of national policies for security of gas supply.

[125] Stranded costs result from commitments that gas undertakings have entered into prior to the liberalization waves and which, under the current conditions, they are unable to fulfil without help from the state. *Italy*, Case S.A. 15876, NN 27/2005.

with the approval of NRAs – may decide if tariffs for network access are to be determined through market-based arrangements, such as auctions. Furthermore, it is the Member States which can grant authorization if a licence, permission or consent is required for the construction or operation of natural gas facilities.[126] Member States are also responsible for the implementation of a nondiscriminatory system of third-party access to transmission, distribution and storage[127] and for ensuring that the undertaking refusing third-party access makes the necessary economic enhancements to correct the lack of network capacity of connection.[128] At the same time, there are still examples of Member States backing up the dominant position of their gas market incumbents, to the detriment of potential supply competitors, consumer protection and market competition in general.[129] Hence, the role of state aid control in the EU gas market appears to be that of guaranteeing that Member States do not contribute to the foreclosure of their national markets through granting discriminatory aid in favour of a limited number of market players.

State aid rules are relevant in regulating network industries specifically with regard to cross-subsidization, preferential tariffs and long-term contracts. In the case of cross-subsidization, the element of state aid occurs if the granting of a favourable tariff by the monopolistic part of an undertaking to its competitive segment is based on the decision of the state to grant the particular tariff. Concerning preferential tariffs in natural gas, these were challenged in a series of cases regarding the Dutch gas supplier, *Gasunie*.[130] In the oldest case of the series,[131] the CJEU held that the level of gas prices practised by *Gasunie* was too low, despite the Dutch government's argument that such tariffs were necessary to prevent consumers switching from gas to coal heating, which would have increased air pollution. However, in a latter case,[132] the Court found that the tariffs practised by *Gasunie* were justified on commercial grounds.

---

[126] Article 4 of Directive 2009/73/EC. The granting of such an authorization is however submitted to pre-established criteria, which are objective and nondiscriminatory.

[127] Articles 32 and 33 of Directive 2009/73/EC.

[128] Article 35 of Directive 2009/73/EC.

[129] For more details, see Chapter 4.

[130] *Kwekerij Gebroeders van der Kooy BV and others v Commission*, joined cases 67, 68 and 70/85; *Société CdF Chimie azote et Fertilisants et al v Commission*, Case C-169/84; *Belgium v Commission*, Case C-56/93.

[131] *Kwekerij Gebroeders van der Kooy*, *supra* note 130.

[132] Case C-56/93 *Belgium v Commission*, *supra* note 130.

The application of state aid rules to long-term energy contracts is of particular interest owing to the fact that such measures may be declared illegal even if they were granted before the country which infringed competition law joined the EU.[133] In complement to this, Member States are required to notify measures that originally did not constitute state aid, but which owing to the evolution of the market have transformed into an aid.[134]

In order for state aid rules to apply, the first condition is that the measure under scrutiny has been granted by a state or through state resources. CJEU jurisprudence on energy matters generated extensive discussions on the interpretation of the three elements of the condition.[135] It has been established that the concept of aid includes not only financial aid, but also any type of intervention through state resources that creates an economic benefit for a particular undertaking.[136] In this sense, aid could refer to direct subsidies, tax exemptions,[137] preferential interest rates, public supply contracts, state guarantees, state participation in capital tax concessions and infrastructure aid.[138] Specifically, the latter form of aid is of considerable relevance for the EU gas market, where Member States have often participated in the operation of the infrastructure, through public–private partnerships.[139] According to a note of the European Commission, the financing of any type of infrastructure –

---

[133] In *Budapesti Erőmű 'Zártkörűen Működő Részvénytársaság' v Commission of the European Communities*, joined cases T-80/06 and T-182/09, the General Court decided that an electricity procurement agreement between Budapesti Erömü and the Hungarian public undertaking MVM was illegal under EU rules on state aid, even though it was concluded before Hungary joined the Union.

[134] Council Regulation (EU) No. 734/2013.

[135] See for example, Hancher *et al.*, 2006; Roggenkamp *et al.*, 2007, p. 256.

[136] *Enirisorse SpA v Sotacarbo SpA*, Case C-237/04, para. 42; *Banco de Credito Industrial SA (Banco Exterior de España SA) v. Ayuntamiento de Valencia*, Case C-387/92; *Sloman Neptune*, joined cases C-72/91 and C-73/91.

[137] There is a very interesting exception to considering tax exemption as state aid. If the said tax exemption derives from an EU measure that does not leave scope for discretionary application by the Member States, then the measure under scrutiny cannot be imputable to the state and does not amount to a state aid.

[138] This is not an exhaustive list. For more examples see Craig and de Búrca, 2011, p. 1087.

[139] For example, "Gasgebouw", a public–private partnership between the Netherlands and a private joint-venture called NAM, which was formed of Shell and ExxonMobil, was put in place in 1963 to manage the exploration of the Groningen gas field.

except that related to security – that will be commercially exploited in the future must be considered under the scope of state aid rules.[140] The notion of state refers not only to governmental organs, but also to public or private bodies established by the state. This is a very important observation when considering the scope of the concept of "state resources". In a case regarding requirements set by a national law for electricity supply undertakings to purchase electricity produced from renewable resources at minimum fixed prices, CJEU decided that the measure did not constitute state aid, as it did not involve any transfer of public resources.[141] The decision of CJEU in the case received a considerable amount of critique, owing to an apparent creation of a loophole.[142] This allowed Member States to circumvent state aid rules by devising schemes of mandatory payments from one individual to another, instead of distributing revenue through state organs or state controlled bodies.

The second condition for the application of article 107(1) TFEU requires that the aid distorts or threatens to distort competition in the market. Unlike in the case of antitrust and merger rules, the Commission is not expected to perform a complete assessment of the product and geographical market.[143] What has to be established is whether through the disputable measure certain undertakings have received economic advantages that place them in a better market position than their competitors.[144] This condition is closely linked to the third requirement of article 107(1), which asks for the aid to create a selective benefit. If the aid does not result in the benefit of only one or several undertakings in the market, but is of general application, then the measure does not constitute state aid. In *Adria-Wien*,[145] an Austrian undertaking applied for rebates granted through a state tax measure. While the measure concerned the awarding of rebates for companies producing goods and *Adria-Wien* qualified rather as a supplier of services, its application was discarded. Called to express its opinion in the case, the CJEU maintained that the tax measure – which granted a rebate on energy taxes charged on supply of natural gas and electricity to undertakings active in the

140 European Commission Note to DG REGIO Application of State aid rules to infrastructure investment projects.

141 *PreussenElektra and Schleswag*, Case C-379/98.

142 Hancher *et al.*, 2006.

143 *Philip Morris Holland v Commission*, Case 730/79.

144 Roggenkamp *et al.*, 2007, p. 256.

145 *Adria Wien Pipeline GmbH v Finanzlandesdirektion für Kärtnen*, Case C-143/99.

manufacture of goods and in so far as energy taxes exceeded 0.35 per cent of the production value – was equivalent to state aid.

The fourth and last condition for a measure to qualify as a state aid is that it affects trade between Member States. In case N 183-2007 *Spain*, the Commission's evaluation established that the prolongation of a state aid would create an economic advantage to beneficiaries, enabling them to reduce investment costs for electricity and gas distribution. The potential effect of the prolongation of the aid upon market competition was considered to be a probable impact on trade between Member States. The Commission took into consideration the liberalization of the construction of infrastructure for gas and electricity in Spain. This allowed it to conclude that, should non-Spanish undertakings decide to invest in the Spanish infrastructure, they could incur unfair competition from subsidized national undertakings. It was also observed that the aid was selective, as it was only applicable to undertakings investing in a specific Spanish region. Finally, the Commission noted that public resources were involved, as the aid grants were given through state funds. Nevertheless, the Commission decided to accept the prolongation of the state aid based on the fact that this did not alter substantially the grounds of the Commission's previous decision approving the existing scheme of aids. An important facet of the Commission's decision was the fact that the new granting scheme attempted to give a smaller intensity of aid to Spanish major energy players than to smaller energy undertakings on the Spanish market.

Article 107(2) TFEU provides for automatic exemptions from the rules of state aid in cases of measures having a social character and granted to individual consumers, if the aid is provided without discrimination with regard to the origin of the products or services concerned. Furthermore, the aid granted to repair damage caused by natural disasters or extraordinary occurrences is automatically exempted. This exemption appears to cover the case of vulnerable gas consumers exposed to energy poverty and certain cases of security of gas supply concerns. Nevertheless, in the case of the EU gas market, the article has not been invoked, while the same gas market objectives seem to be better protected by the third paragraph of article 107 TFEU.[146]

Article 107(3) provides for discretionary exemptions from the application of state aid rules. For the EU gas market the first three paragraphs

---

[146] Energy case law based on article 107(2) concerns renewable energy resources – *HAMSA v Commission*, T-152/99 and *Austria – Aid to renewable energy installations*, Case N 241/2004 [2005], JOCE C/2206/2005.

of the article are the most relevant. Article 107(3)(a) provides that aid to promote economic development in areas where the standard of living is abnormally low or where there is serious underemployment may be considered compatible with the common market. In case N 435/2008 *Poland*,[147] the state envisioned the construction of gas distribution networks in areas not supplied with gas, as well as the modernization of existing distribution networks. The aid offered incentives for investments in the construction of gas distribution networks on territories which at that time were not supplied with gas. Having consideration to the regional aid element of the case, the Commission considered that the state measure was covered by the exemption contained in article 107(3)(a) TFEU.

Article 107(3)(b) TFEU considers that an aid may be compatible with the common market if it promotes the execution of an important project of common European interest. This is an interesting point for the sustainability objective pursued by energy policy in the gas market. According to the guidelines on environmental aid, under certain conditions[148] state measures which "promote the execution of important projects of European interest – which are an environmental priority – may be considered compatible with the common market according to article 87(3)(b) of the EC Treaty (107(3)(b) TFEU)".[149]

In general, environmental concerns seem to have tilted the application of state aids in energy matters rather towards the objective of market sustainability than that of market competitiveness.[150] In case *Preussen-Elektra* the CJEU underlined the importance of state measures for protecting the environment through their contribution to the reduction of

---

147 *Poland – Regional aid scheme for construction and reconstruction of the gas distribution network*, Case N 435/2008.

148 The aid proposal is clearly defined in regarding its implementation: it must be in the common European interest, be necessary and present an incentive for the execution of the project, which is of great importance with regard to its volume.

149 Notices from European Union Institutions and Bodies Commission Community Guidelines on State Aid for Environmental Protection.

150 In the Union's guidelines on state aid for environmental protection, one of the main underlined issues has been that there must be a balance between environmental protection and sustainable development, on the one hand, and competition and economic growth, on the other. In case *PreussenElektra*, the Court underlined the importance of the state measures for protecting the environment through their contribution to the reduction in emissions of greenhouse gases, the fact that such protection was amongst the priority objectives and obligations of the Union and the Member States and that Treaty provisions

emissions of greenhouse gases. The Court emphasized the fact that such protection was amongst the priority objectives and obligations of the EU and its Member States and that Treaty provisions demanded that environmental protection requirements must be integrated into the definition and implementation of other Union policies.[151] In case N/660a/2000 *Northern Ireland*,[152] the local authorities proposed a five-year exemption from a climate change tax for natural gas consumption by all businesses in Northern Ireland. The exemption was aimed at encouraging the substitution of coal and oil consumption with natural gas, which from an energy sustainability point of view had the positive effect of reducing CO_2 emissions. It was established that the tax exemption was selective, owing to the fact that it favoured undertakings located in a specific region. More specifically, it was noticed that the position of the undertakings in Northern Ireland would be reinforced by comparison to that of their competitors. Furthermore, it was found that the tax exemption would have positioned natural gas in a more competitive position price-wise than the traditional energy sources used in Northern Ireland. However, the environmental benefits of the measure were considered more important and in line with the Commission's approach to sustainability.[153] Hence, it was considered that the distortion of competition through the application of the tax exemption was necessary in order to realize the environmental benefits. The state measure was analysed both on the basis of the Union's guidelines on state aid for environmental protection[154] and in the light of article 107(c) TFEU.

Article 107(3)(c) TFEU provides that aid to facilitate the development of certain economic activities or areas is compatible with the common market, if it does not adversely affect trading conditions to an extent contrary to the common interest. In a 2011 case,[155] the Commission observed that the national measures that amounted to state aid promoted regional development in areas designated in accordance with articles 107(3)(a) and (c) TFEU. In order to assess the compatibility of the aid measure with the internal market, the Commission referred to the

---

demand that environmental protection requirements must be integrated into the definition and implementation of other Union policies.

151 *PreussenElektra*, *supra* note 141, paras 73–77.

152 *Exemption from climate change levy for natural gas in Northern Ireland*, Case N/660a/2000.

153 Communication from the Commission: Environment 2010: Our future, our choice, /*COM/2001/31 final*/.

154 *PreussenElektra*, *supra* note 141, paras 73–77.

155 *Greece*, Case SA.31225 (2011/N).

Guidelines on national regional aid for 2007–2013.[156] The fact that the projects proposed contributed to a coherent regional development strategy and that the aid was calculated in reference to expenditures was taken into consideration. The decision of the Commission was not to raise objections to the notified measure. The exemption under article 107(3)(c) TFEU has been also invoked in cases regarding the improvement of the security of gas supply. In N 660/2009 *Poland*,[157] the Commission decided to approve €390 million in aid for an increase in gas storage capacity in Poland, owing to the positive effects of the measure upon the increase of the security of gas supply to the benefit of all customers.[158] Following the same logic of Directive 2004/67/EC, it was found that the maximization of the gas storage capacity would allow the stabilization of the supply of gas, which would result in an increase in the security of supply of consumers. The case acknowledged security of supply as one of the pillars of the European energy policy[159] and referred to the texts of specific sector legislation as a supporting argument for allowing the selective aid in order to enhance the security of supply.[160]

5.2.1.2 Possibilities and effects of using competition measures in the EU gas market

The success of using competition tools in gas cases resides in their effective enforcement. The application of articles 101 and 102 TFEU is governed by Regulation (EC) No. 1/2003, which provides for behavioural and structural remedies,[161] interim measures,[162] acceptance of commitments[163] and fines for the infringement of competition rules by undertakings in the market.[164] Similar remedies derive from Regulation (EC) 139/2004, which advocates the possibility to accept commitments,[165]

---

[156] Guidelines on National Regional Aid for 2007–2013.

[157] *Poland Aid to PGNiG for underground gas storage*, Case N 660/2009.

[158] State aid: Commission approves € 90 million aid for increase of gas storage capacity in Poland, IP/10/786.

[159] *Poland Aid to PGNiG for underground gas storage*, Case N 660/2009, para. 52.

[160] For example, Council Directive 2004/67/EC and Decision No. 1364/2006/EC of the European Parliament and of the Council.

[161] Article 7 of Regulation (EC) 1/2003.

[162] Article 8 of Regulation (EC) 1/2003.

[163] Article 9 of Regulation (EC) 1/2003.

[164] Article 23 of Regulation (EC) 1/2003.

[165] For example, recitals 30, 31, 35 and articles 6(2) and 8(2) of Regulation (EC) 139/2004.

adopt interim measures[166] and impose fines[167] or periodic penalties[168] in cases in which a concentration causes or is likely to cause significant impediments to effective competition.

5.2.1.2.1 Investigations into alleged breaches of articles 101 and 102 TFEU In the beginning of the liberalization process of network industries, the application of competition policy to the newly liberalized sectors did not occur as a result of carefully planning the optimal regulation or regulatory bodies that should be in place in order to achieve the promised advantages of liberalization.[169] Nevertheless from the early years on, the logic followed by the Commission has been that competition was necessary to ensure the creation of a level playing field in previously monopolistic sector industries for business to thrive and operate effectively.[170] A stronger enforcement of competition law in the gas sector was envisaged however only in the aftermath of the adoption of the competition "modernization" Regulation (EC) 1/2003.[171]

EARLY SETTLEMENTS The Commission's early settlements in the gas sector have largely been dealt with in the above sections. In general, they involved vertical relationships between energy producers and European supply incumbents or horizontal agreements between external producers or internal suppliers. The elements hindering the liberalization of the EU gas market concerned the introduction of restrictive clauses in long-term contracts,[172] price fixing in supply agreements,[173] joint marketing and market-sharing between producers or suppliers[174] and third-party access.[175]

---

166 Article 8(5) of Regulation (EC) 139/2004.

167 Article 14 of Regulation (EC) 139/2004.

168 Article 15 of Regulation (EC) 139/2004.

169 Pelkman, 2001, pp. 432–456.

170 Schaub, 2000.

171 Roggenkamp *et al.*, 2007, p. 251.

172 *Gas Natural + Endesa*, COMP/37.542; *Territorial restrictions 1) Algeria gas transport contracts, 2) Expansion of TAG pipeline*, Case COMP/37.811.

173 *GFU*, Case COMP/36.072.

174 *Irish Corrib*, Case COMP/37.708.

175 *UK Gas Interconnector*, Case COMP/38.075; *Marathon/Ruhrgas/GDF et alia*, Case COMP/36.246.

In a number of cases involving Algerian and Russian gas suppliers, the Commission intervened against the inclusion and maintenance of restrictive clauses and mechanisms in supply contracts with EU counterparts.[176] The reasoning behind this was that such clauses were a threat to the realization of a more integrated EU gas market by setting up barriers to the supply of gas among Member States and by favouring gas market segmentation. In case COMP/37.542 *Gas Natural + Endesa*, a gas supply contract concluded between two Spanish undertakings provided for Spanish gas market segmentation, through the introduction of a destination clause and several use restrictions in favour of the dominant undertaking. The original agreement resulted in *Gas Natural* securing an important gas contract in terms of quantities sold and duration regarding the supply of gas for electricity generation. *Endesa*, the other undertaking in the agreement, was not allowed to resell the gas purchased from *Gas Natural*. Among the restrictive clauses, the agreement also raised questions regarding third-party access. According to some of the clauses, *Endesa* was supposed to receive better treatment than other future clients of *Gas Natural*. This led to the conclusion that the long-term contract signed between *Gas Natural* and *Endesa* would have raised barriers to entry into the liberalized Spanish gas market. The commitments proposed by the two undertakings following the Commission's investigation included: the freeing of a part of *Endesa*'s purchasing capacity, the decrease of the supply contract by one-third, the removal of use restriction clauses and the removal of clauses that could favour *Endesa* above other potential gas customers.

What is interesting to observe for this initial level of the application of competition rules to the EU gas market is the fact that the Commission seems to have been reluctant to initiate formal infringement proceedings basing on article 102 TFEU. Instead it opted for what Commissioner Monti in 2003 called "a commercial solution for a competition problem".[177] The reasoning behind the Commission's reluctance could be explained by drawing on a multitude of elements. These include the still premature stage of liberalization process, the reluctance of Member States to cede their sovereign right over national energy policies to the Commission, the need to "educate" traditional incumbents to behave competitively in the market and the pressure of different interest groups at EU level. However, it seems legitimate to ask whether such a cautious

---

[176] *Territorial restrictions (1) Algeria gas transport contracts, (2) Expansion of TAG pipeline*, Case COMP/37.811.

[177] Monti, Mario, 2003.

approach best served the liberalization of the Spanish and European gas markets, which at the moment of the contested agreements found themselves "in a particularly crucial moment".[178] Not at all surprising, as will result from the following, it should be noted that the use of commitment decisions in energy cases has led to vigorous discussions.[179]

COMMITMENTS FOLLOWING ARTICLE 9 OF REGULATION (EC) NO. 1/2003

The manner in which the Commission has dealt with "abuse" cases under article 102 TFEU in the energy sector during the Energy Sector Inquiry and in the aftermath of the adoption of Regulation (EC) No. 1/2003 has stirred the interest of analysts and scholars[180] and attracted critique.[181] Article 17 of the Regulation has empowered the Commission to launch inquiries in sectors and types of agreements where there were concerns that competition might have been distorted. The Commission did not hesitate to use its new powers for the revision of energy sectors that seemed to resist market liberalization at the level of both undertakings and governments.[182] During and in the years immediately following the Energy Sector Inquiry, the Commission completed six complex investigations in the gas sector, out of which five dealt with alleged "abuse" under article 102 TFEU and were concluded with commitments from the parties.

Of interest for the present discussion, regarding the application of competition tools to satisfy gas market goals, is the potential antagonizing role that commitment decisions exert upon the realization of such goals. Specifically, this is observed in relation to the pursuit of competitiveness and consumer protection in the gas market. Article 9 of Regulation (EC) No. 1/2003 gives the Commission the ability to close antitrust investigations without finding any infringement or imposing a

178 European Commission, 2000: Commission closes investigation on Spanish company Gas Natural, http://europa.eu/rapid/pressReleasesAction.do?reference=IP/00/297&format=HTML&aged=1&language=EN&guiLanguage=en

179 Koch and Gauer, 2011, p. 242.

180 See Tapia and Mantzari, 2012; Scholtz and Purps, 2010, pp. 37–51; Koch and Gauer, 2011, p. 242.

181 Sadowska, 2011.

182 The Commission initiated in 2006 its first round of infringement procedures against 25 Member States for noncompliance with EU energy legislation. Letters of formal notice were sent to 17 Member States, out of which 11 regarded the functioning of the EU gas market. European Commission, 2006: Infringement procedures opened in gas and electricity market sector, by Member State, http://europa.eu/rapid/pressReleasesAction.do?reference=MEMO/06/152&language=en

fine. Following the economics of the article, in case of an alleged abuse of a dominant position in the market, neither the abuse nor the dominant position must be proved. Instead, the Commission may accept commitments from the undertakings under scrutiny and make them legally binding. For consumers, however, this means a further barrier to private enforcement. In general, commitment decisions expressly state that they do not represent a conclusion whether or not there has been or still is an infringement under article 101(1) or 102 TFEU. The lack of a legally binding finding on a competition infringement makes it very difficult for consumers to claim back damages caused by the anticompetitive market behaviour of the undertaking.

In case COMP/B-1/37.966 *Distrigaz*,[183] the Commission raised important competition concerns regarding the long-term gas supply contracts between *Distrigaz* and its customers. According to the agreements, *Distrigaz* customers were required to acquire a certain minimum amount of gas from *Distrigaz*. In return, they had the right to source all their gas from the Belgian incumbent. This led to the foreclosure of customers' supply options, while it appeared that the latter did in fact outsource all their gas from Distrigaz, instead of looking for alternative suppliers. The situation was not surprising. Generally EU consumers have shown a low tendency to switch energy suppliers either owing to the limited benefits of switching or to favourable contract terms offered by a supplier, coupled with the quality of service.[184] The decision in *Distrigaz* was considered a breakthrough in establishing guidance on the appropriate form of compliance of downstream bilateral long-term supply agreements with European competition law. The commitments offered by the Belgian undertaking regarded the noninclusion of any use, resale or destination restriction in any new supply contract and the removal of such clauses where appropriate, no tacit renewal clause, no long-term contract with resellers longer than two years and no long-term contract with customers longer than five years. The settlement resulted in the encouragement of competition among suppliers, which certainly led to an increase in consumer choice. However, no private enforcement case was brought in front of the European Courts to claim back the consumers' damages resulting from the market behaviour of the Belgian incumbent.

Private damage actions are directed to obtain compensation for the harm caused by antitrust infringements. There is no actual private damage practice regarding commitment decisions in energy cases. Partly,

---

183 *Distrigaz*, Case COMP/B-1/37.966.

184 Cseres, 2008, pp. 77–121.

this is due to the existent gaps in the European legislation protecting consumers' interests. Nevertheless, important limitations derive from the lack of transparency of information regarding the Commission's investigations into alleged competition distortions. Commitment decisions related to energy sector cases and listed on the DG Competition website make reference only to the text of the commitments decision and European press releases, without providing a detailed presentation of data and information that has been taken into consideration by the Commission when starting the investigation. Very often the text of the commitments decision is also absent. According to article 12 of Regulation (EC) 1/2003, information exchanged between the Commission and national competition authorities (NCAs) regarding a specific case and including confidential information can be used as evidence only for the purpose of applying articles 101 and 102 TFEU. In commitment decisions, neither of these articles is applied. Hence, information obtained by the Commission in such situations cannot serve as evidence in the case. Furthermore, article 27 of the Regulation recognizes the right of access to the file only for the defending parties. Such a right does not extend to confidential information and internal documents of the Commission. The article remains mute with regard to collateral parties to the case, for example, consumers. In a recent speech, the Vice-president of the European Commission[185] underlined the need to regulate access to the evidence held by NCAs so as to allow private damage actions. It remains to be seen to what extent Commission's Staff Working Document regarding collective redress[186] will be able to ensure the effective enforcement of the rights of every consumer.

A common critique to the procedure envisaged by article 9 of Regulation (EC) No. 1/2003 is that the latter has considerably lowered the standard of proof by allowing the Commission to adopt decisions based solely on concerns. In a case concerning ENI SpA's[187] conduct on the gas supply market in Italy, the Commission initiated an inspection on the grounds that the Italian incumbent might have abused its dominant position in the Italian gas transportation market to the detriment of competitors, competition and ultimately downstream consumers by foreclosing access to the network. At the moment of the investigation, ENI was a fully integrated gas company controlled by the Italian state, with activities in the production, import, transmission and storage of gas,

---

185 Almunía, 2011.

186 Commission Staff Working Document public consultation: Towards a Coherent European Approach to Collective Redress, /*SEC/2011/173 final*/.

187 *ENI SpA*, Case COMP/39.315.

while being active in the downstream gas distribution business. It was the Commission's view that ENI might have abused its dominant position by capacity hoarding[188] or capacity degradation[189] towards its competitors and by maintaining strategic underinvestment in ENI's international transmission pipelines. ENI's commitments concerned the divesting of its share in three international transmission pipelines. However, owing to the fact that the undertaking chose to settle the case with the Commission, no additional proof was required in relation to the alleged infringement.

From the point of view of market competitiveness, the lowering of the standard of proof has a double consequence. On the one hand, it has been argued that commitment decisions have allowed a greater improvement in the functioning of energy markets than what could have been achieved by a *stricto sensu* application of antitrust rules.[190] On the other hand, it has been pointed out that some of the downsides of negotiations regarding commitments are that they are often lengthy and not always successful.[191] In regards to the first strand of argumentation, it has been observed that settlement procedures have allowed for a more expeditious finalization of Commission's investigations than the usual "case output" of DG Competition when applying article 102 TFEU.[192] At the same time, it was shown that parties were eager to propose remedies to remove Commission's concerns and avoid the establishment of an infringement under article 7 of Regulation (EC) No. 1/2003, which could have levied significant fines.[193] It is interesting to see that such commitments have managed in certain cases to impose remedies that go beyond what could have been achieved under article 7 of Regulation (EC) No. 1/2003.[194] This was specifically the case of an electricity settlement decision regarding the German incumbent E.ON.[195] Commitments offered by E.ON to close the Commission's investigation included the divestiture of 5000 MW of generation capacity and its high-voltage transmission grid for electricity supply in Germany. The size of the commitments makes them the most significant structural remedies ever imposed on the EU energy market. Both the appropriateness and the proportionality of the

---

[188] Capacity hoarding refers to the refusal to grant access to unused capacity on the transmission network.
[189] Granting access in a less useful manner.
[190] Monti, Giorgio, 2008, pp. 123–145.
[191] Koch and Gauer, 2011, p. 242.
[192] Rab, 2012.
[193] Tapia and Mantzari, 2012.
[194] Scholtz and Purps, 2010, pp. 37–51.
[195] *E.ON electricity*, Case COMP/39.389.

settlement have been questioned,[196] as well as Commission's legitimacy of using article 9 of Regulation (EC) No. 1/2003 in order to obtain such commitments.[197] However it could not be denied that network divestiture ensured the ownership unbundling of the transmission network of the German incumbent, which at least in theory should allow for increased market competitiveness.

A similar, but less contested decision was adopted in a case regarding the suspected infringement of EU competition law by RWE AG.[198] The commitments were offered by the German company RWE in response to the Commission's concerns regarding an abuse of dominant position on the German gas transmission markets. In fact RWE, which operates the second largest high-pressure gas transmission network in Germany, pursued a strategy of systematically keeping transport capacities for itself, preventing its competitors from access to its gas network. In the aftermath of Commission's investigation, RWE offered a structural remedy, regarding the selling of its existing German high-pressure gas transmission network – including the necessary personnel and ancillary assets and services necessary for a viable gas transport business – to an independent purchaser, who did not give rise to competition concerns. The remedies offered by RWE were considered by the Commission as necessary and proportionate to address the concerns identified during its Preliminary Assessment.

Despite the advantages of commitment decisions for competitiveness in the EU gas market, there are several concerns that could be formulated. First, as observed above, commitments do not always lead to fast decisions, but could stretch over several years from the initiation of proceedings until the actual adoption of a decision.[199] Second, they could cause significant procedural delays, if an undertaking decides to withdraw its commitment proposal.[200] Third, the fact that neither the dominant position nor its abuse are actually investigated or proved in settlement decisions leads to "weak cases, based on far-fetched allegations".[201] All of these arguments point to the question whether the Commission's recourse to article 9 of Regulation (EC) 1/2003 instead of article 7 of the

---

196 See Sadowska, 2011; Tapia and Mantzari, 2012.

197 Monopolkommission, 2009.

198 *RWE gas foreclosure*, Case COMP/39.402.

199 That is, in *ENI SpA*, Case COMP/39.315 it took four years from the initial concerns of the Commission until the adoption of the commitments decision.

200 Koch and Gauer, 2011, pp. 241–243.

201 Sadowska, 2011.

same Regulation – which provides for an actual finding and termination of infringements – is not in some cases detrimental to the competitiveness requirement in the EU gas market.

The question seems particularly pertinent in cases involving the alleged abuse of a dominant position in the market by non-EU gas incumbents. This regards specifically recent investigations into alleged hindering of competition in Central and Eastern European gas markets by the Russian incumbent, Gazprom.[202] Investigations in the case have been carried through in a similar manner to investigations into alleged market abuse of EU incumbents, following surprise dawn-raids of Gazprom-linked offices and its trade counterparts in September 2011.[203] Most of energy-related investigations initiated during the Sector Inquiry share as a common feature the Commission's practice of basing its investigations on information and data collected during surprise inspections at the premises of the investigated undertakings.[204] However, none of the divestiture commitments adopted until the present has concerned a non-EU gas undertaking. Hence, it seems appropriate to ask to what extent far-reaching settlements of the amplitude registered in cases involving EU gas incumbents could occur in cases regarding the alleged abusive behaviour of non-EU incumbents. First of all, it must be observed that commitments under article 9 of the Regulation (EC) No. 1/2003 are offered voluntarily by the undertakings concerned.[205] In September 2012, Russia adopted a presidential decree that provided that, in cases where foreign authorities require specific actions from Strategic Russian Enterprises, the latter must obtain the consent of the Federal Executive Authority prior to performing such actions.[206] In addition to this, the presidential decree provided that the Federal Executive Authority is obliged to refuse permission for an undertaking to conduct cooperative activities with

202 *Gazprom*, Case COMP/39.816.

203 For example, Polish company PGNiG.

204 The most vivid example in this direction stems from the case *E.ON Seals*, COMP/39.326, where the Commission fined the German company for tampering with seals applied by European officials during a surprise raid on the premises of the undertaking. The fine was contested by E.ON in case T-141/08, where the General Court decided in favour of maintaining it. This technique, which has been mostly associated with Commission's cartel investigations, appears to have been transformed into common practice of abuse investigation in energy sectors. A recent example is provided by *Čez and others Commitments to the European Commission*, Case COMP/39.727.

205 Koch and Gauer, 2011, pp. 241–242.

206 Goltsblat BLP, 2012.

foreign authorities where such activities have the potential to damage the economic interests of Russia.

With regard to investigations in the case COMP/39.816 *Gazprom*, the presidential decree could effectively block Gazprom – which is one of the 137 Russian strategic enterprises – from cooperating with EU antitrust officials if it was found that such a cooperation was contrary to the interests of the Russian Federation. Bearing in mind the fact that *Gazprom* has already manifested its discontent with the TSO certification procedure under the Third Energy Package,[207] it seems very unlikely that it could or would commit to an unbundling of its transmission network to stop the Commission's investigations. At the moment it is still premature to consider the validity of such a reasoning. However, should it be supported by a future decision regarding the *Gazprom* case, it could be concluded that commitment decisions in alleged abuse cases have serious limitations regarding the establishment of far-reaching commitments with non-EU incumbents. This would be an important constraint considering the fact that the EU gas market competitiveness is still directly influenced by the behaviour of foreign gas suppliers. Hence it is advisable to see what has been the practice of the Commission in using other remedies in the gas market.

FIRST FINES In 2009, for the first time in an antitrust case regarding the EU gas market, the Commission imposed a fine totalling €1,106,000,000 on EU undertakings.[208] The case regarded market-sharing between E.ON AG and its subsidiary E.ON Ruhrgas AG (of Germany) and GDF Suez SA (of France). Market sharing constitutes a breach of the Treaty rules on cartels and restrictive business practices in the sense of article 101 TFEU. In 2009, the Commission found that the two companies had secretly agreed not to sell gas transported from Russia over the jointly owned MEGAL pipeline into each other's home markets. The agreement was concluded in 1975, when European energy markets were still far from being liberalized. It was observed that the market-sharing agreement helped the German and French incumbents to maintain strong positions in their respective national markets after the gas market liberalization process was initiated. It was further stated that the companies have deliberately denied French and German gas consumers the benefits of more price competition and a larger choice of suppliers, as promised by the liberalization directives. The agreement was completely

---

207 Ćwiek-Karpowicz and Gawlikowska-Fyk, 2012.

208 *E.ON/GDF collusion*, COMP/39.401.

abandoned only as late as 2005. Having consideration for the gravity and the duration of the competition law infringement, the Commission imposed an extensive fine in accordance with article 23 of Regulation (EC) No. 1/2003.

The adoption of the fine seems to have anticipated a new direction in the application of competition law tools to the gas market.[209] As EU officials observed, the Commission intended to use "the full gamut of competition enforcement tools at [its] disposal to pursue individual cases that could significantly improve the level of competition in the market".[210] The decision of the Commission to apply fines in an energy case involving an alleged violation of article 101 TFEU and the proof of the violation represent a significant advancement from the prior practice in similar cases. In case COMP/38662 – *GDF*[211] the Commission found in accordance with article 7 of Regulation (EC) No. 1/2003 that the concerned undertakings violated article 101(1) TFEU, but decided not to apply any fines. The Commission's motivations were related to the specificities of the gas sector, where at the moment of the decision the commercial practices of market players were undergoing an important change owing to liberalization.[212]

In the joined cases T-360/09 *E.ON Ruhrgas/Commission* and T-370/09 *GdF Suez/Commission*, the General Court partly annulled the decision of the Commission under article 101 TFEU, inasmuch as the infringement concerned a period of time when competition was *de facto* absent from both Germany and France. The fines were adjusted accordingly. However, the Court considered that, if it applied the same method as that used by the Commission in setting up the initial fines, the final amount of the fine imposed on each undertaking would be too low in light of the gravity and the duration of the infringement. Instead it imposed a final fine of €320 million on each undertaking. Despite an only partial recognition of the Commission's decision, the judgment of the Court regarding the final amount of fines for the infringement of competition

---

209 Dreyer *et al.*, 2010.

210 Lowe, 2008.

211 Décision de la Commission du 26 octobre 2004 relative à une procédure d'application de l'article 81 du traité CE (Affaire COMP/38662 – *GDF* – décision *GDF/ENI*) and Décision de la Commission du 26 octobre 2004 relative à une procédure d'application de l'article 81 du traité CE (Affaire COMP/38662 – *GDF* – décision *GDF/ENEL*).

212 Case COMP/38662 – *GDF* – *GDF/ENEL* para. 163; Case COMP/38662 – *GDF* – *GDF/ENI*, para. 133.

law rules by gas undertakings indicates support for the pro-active action of the Commission in the sector.

It has been argued that fines are not capable of restoring the competitive process that was affected through the action of one or several undertakings in the market.[213] Quite the contrary, while the accepted commitments must be suitable to remove competition concerns expressed by the Commission, they have been perceived both as a means to restore effective competition and as an obstacle to the infringer continuing to benefit from the advantages of a past violation.[214] However, it should be observed that fines in the energy sector, albeit limited, have amounted to high pecuniary sanctions.[215] This has served the aim of article 23 of Regulation (EC) No. 1/2003 of exerting a preventative effect not only on future actions of the undertaking concerned, but also on those of other undertakings in the market that have engaged or are planning to engage in similar practices.[216]

As seen in *Gas Insulated Switchgear* (*GIS*), the application of antitrust fines may occur extraterritorially under the effects doctrine. In *GIS* the Commission imposed fines on five Japanese companies and on several European ones owing to the impairment of competition rules in the EU. As argued above, it seems hardly likely that in the recent *Gazprom* investigation[217] the Commission will be able to receive extensive structural commitments from the Russian incumbent. Instead, drawing on the *GIS* precedent it appears that fines are a reliable competition enforcement tool in a market where the cross-border activities of private-sector companies constitute the majority of market transactions and the relationships between trade partners occur at the crossroads of sometimes conflicting legal systems.

Furthermore, the application of fines increases the legal certainty and coherence of the Commission's findings during preliminary investigations in alleged anticompetitive conduct of an undertaking in the gas market in a manner that escapes the reach of commitments. This would potentially allow an enhancement of competition in the gas market by

---

213 Hellström *et al.*, 2009, pp. 43–63.

214 In fact in general, the commitments offered by the parties have amounted to structural remedies. See *RWE gas foreclosure*, Case COMP/39.402, paras 48 and 50.

215 *E.ON/GDF collusion*, Case COMP/39.401; *Gas Insulated Switchgear*, Case COMP/38.899.

216 Guidelines on the method of setting fines imposed pursuant to Article 23(2)(a) of Regulation No. 1/2003, para. (4).

217 *Gazprom*, Case COMP/39.816.

providing the rest of the gas market stakeholders with the transparency, due process and predictability necessary for ensuring the promotion of the private enforcement of EU competition law. For gas consumers, fines – unlike commitments – may constitute a ground for individual or collective redress,[218] provided that the right system is in place for private actions in EU energy matters. The initiation of a 2008 White Paper regarding damages actions for breach of EU competition law[219] shows the Commission's efforts to enhance private enforcement of European competition rules. However, the White Paper is still under discussion and no clear implementation steps have been taken at either European or Member State levels.[220]

5.2.1.2.2 Structural remedies under the Merger Regulation Merger control in the light of Regulation (EC) 139/2004 has also been used extensively by the European Commission in its efforts to create a competitive EU gas market. Notably, under article 2(3) of the Merger Regulation, the Commission is entitled to declare a concentration that causes significant impediments to effective competition incompatible with the internal market, particularly if it concerns the strengthening of a dominant position in the market. As seen in the preceding subsections, mergers between energy undertakings have occurred either within the confines of the same sector[221] or cross-sector between gas and electricity undertakings,[222] and have had at times the potential effect of eliminating or distorting market competition.

In this context it appears surprising that the Commission seems to have been reluctant to use its merger prohibition powers under article 8(3) of Regulation (EC) 139/2004.[223] In the contested EDP/ENI/GDP notified merger,[224] the Commission considered that the remedies offered by the parties were insufficient to remove the identified competition concerns regarding the reinforcement of dominant positions of EDP and GDP on their respective markets. Following the prohibition of the merger, the parties brought an action of annulment of the Commission's decision in

---

218 OECD, 2006: Remedies and Sanctions in Abuse of Dominance Cases.

219 White paper /*COM/2008/165 final/.

220 Moodaliyar *et al.*, 2010, pp. 141–162.

221 *Enel/Endesa*, Case M.5494; *EdF/British Energy*, Case M.5224.

222 *Vattenfall/Nuon*, Case COMP/M.5496; *E.ON/Endesa*, Case M.4197.

223 It has been observed that in the period of time 1990–2003 only 18 mergers in the European gas and electricity markets have been blocked. See Eikeland, 2004.

224 *EDP/ENI/GDP*, Case M.3440 [2004].

front of the General Court.[225] The Court found several errors in the Commission's decision, regarding the appreciation of a dominant position in the market. Thus it was stated that the markets under scrutiny at the time of the concentration were not yet open to competition. Consequently, the undertakings concerned could not "be criticised for significantly impending effective competition where that competition [did] not exist as a result of national and Community legislation".[226] Notwithstanding such findings, the General Court upheld the prohibition decision considering that the concentration would have caused an important potential competitor to disappear from all the electricity markets, while another's dominant position would have been reinforced to the detriment of effective competition.[227] It appears from its decision that the General Court has given the Commission its support in liberalizing the EU gas market through formal merger control.

Despite the support of the General Court as regards formal merger control, the Commission seems to be more supportive of an informal control of concentrations on energy markets, through making the commitments offered by the concerned undertakings legally binding.[228] The purpose of the commitments is to remove impediments to, restore or maintain effective competition, while being "comprehensive and effective from all points of view".[229] One of the most frequent commitments offered by undertakings in energy sectors has been divestiture. The Commission considers the divestiture of viable businesses as the best way to eliminate competition concerns which result from horizontal overlaps or vertical or conglomerate concerns.[230]

In case COMP/M.4180 *Gaz de France/Suez*, the Commission identified various threats to competition on the Belgian and French markets prior to the commitments offered by the concerned undertakings. The threats referred primarily to the strengthening of the dominant position of Suez on the Belgian gas market and the strengthening of GDF's dominant position on the French gas market. Furthermore, it was found that extensive barriers to access to gas and infrastructure that existed in the market risked increasing further the horizontal effects caused by the combination of market shares through the merger. The Commission

---

[225] *Energias de Portugal (EDP) SA v Commission*, Case T-87/05.

[226] *Ibid.*, *supra* note 230, para. 130.

[227] *Ibid.*, paras 186, 189, 195, 197, 236.

[228] Article 6(2) of Regulation (EC) 139/2004.

[229] Commission notice on remedies acceptable under Council Regulation (EC) No. 139/2004 and under Commission Regulation (EC) No. 802/2004.

[230] *Ibid.*

approved the merger in exchange for the divesting of Suez's gas supply business Distrigas, giving up control over the Belgian gas transmission network operator Fluxys and the divesting of GDF's shareholding in the Belgian electricity supplier SPE. Similar commitments have been previously offered and accepted in *E.ON/MOL*[231] and *DONG/Elsam/E2*.[232] In the former case, the remedies included full unbundling of the ownership of the gas production and gas activities bestowed upon MOL and the wholesale and storage activities acquired by E.ON. In the latter case, DONG offered to divest the larger of its two Danish gas storage facilities and to carry out a gas release programme through auctions, thus making large quantities of gas available to competitors. The commitments offered by DONG constituted the first structural unbundling remedy in a merger case with Union dimension, which envisaged the separation of assets of the incumbent wholesale operator and infrastructure facilities.[233] In an earlier case[234] the Commission was satisfied by behavioural remedies offered by the undertakings concerned, which referred to the sharing of facilities between incumbents, new entrants and others. At the time, such commitments were considered sufficient to guarantee secure access to the network.

In the case of merger control, the CJEU has ruled that the Commission's acceptance of commitments under the merger regulation must respect the principle of proportionality.[235] It has also been noted in relation to the use of antitrust measures in the energy sector that, during and after the Sector Inquiry, the Commission has shown a tendency to push for far-reaching commitments resulting in the ownership unbundling of the undertakings concerned. This tendency has generated the concern that competition tools have been employed not only to remedy a specific infringement of EU competition rules, but also to restructure markets according to the Commission's competition goals and regulatory strategies.[236] The difference between antitrust enforcement practice and merger control concerns the moment of intervention of the Commission. Antitrust measures occur *ex-post*, to correct a certain competition infringement in the market. In contrast, merger control following the notification of the parties concerned occurs usually *ex-ante*, to prevent a

231 *E.ON/MOL*, Case M.3696.

232 *DONG/Elsam/E2*, Case M.3868.

233 di Jole, 2008.

234 Case COMP/M3410, *Total/GdF*, 8 October 2004.

235 *Cementbouw Handel & Industrie BV v Commission of the European Communities*, Case C-202/06.

236 Scholtz and Purps, 2010, p. 50.

potentially anticompetitive concentration. Hence, far-reaching commitments in merger cases have been received more favourably than their antitrust counterparts, where the Commission has been suspected of following political aims beyond the reasonable interest of a sustainable remedy against anticompetitive practices.[237] Nevertheless, it has been observed that, despite the forward-looking perspective of merger control, this remains a competition law tool and its application as a regulatory instrument could increase disparities between the competition levels achieved so far in energy markets.[238]

5.2.2 The Regulatory Approach

Regulatory tools addressing EU gas market goals comprise three Directives and afferent Regulations, which refer to rules of access, measures to guarantee the security of gas supply and the creation of new Europe-wide bodies to ensure the facilitation of cross-border cooperation between national regulatory authorities. The following will look at the necessity of market regulation in order to attain gas market goals and the respective potential success and limitations of its enforcement in the EU gas market.

5.2.2.1 The role of regulation in the EU gas market

There are many ways in which market regulation can be defined.[239] In the case of the EU gas market, a fitting definition refers to regulation as a set of specific and intentional rules aimed at steering Member States' markets towards harmonization, full opening and cross-border integration.[240] The regulatory process is accompanied by independent regulatory authorities entrusted with monitoring and enforcing compliance and encompasses additional rule-making deriving from EU energy policy objectives.[241]

In the EU gas market, regulation has been initiated and developed to achieve economic competitiveness. Although such a purpose does not automatically contradict the main role of regulation – which is the

---

237 Monopolkommission, 2009, para. 496.

238 di Jole, 2008, p. 368.

239 There has been an extensive discussion of what could be understood under the general concept of regulation. This discussion escapes the scope of the present thesis. For further detail see Jordana and Levi-Faur, 2004.

240 The definition refers to the specific features of the EU gas market and is based on the definition of regulation in Baldwin *et al.*, 1998, p. 3.

241 For example, regarding taxation, public ownership, subsidies, etc.

promotion of public good[242] – the accommodation of gas policy goals has revealed a potential tension between the two. The general requirement has been that gas regulators apply competition law, while observing their specific social duties.[243] It has been maintained that regulation provides an "interface between the public interest, the interests of consumers, the interests of those providing regulated services under monopolistic conditions and the interest of those using the monopolistic infrastructure".[244] This is an important feature, which distinguishes the priorities of gas market regulation from those of competition law on the same market. Market competitiveness is pursued through regulation only to the extent that it serves public welfare. As an example of this, in the matter of deciding between market competitiveness requirements and third-party access refusal,[245] regulation is expected to perform a balancing test, which should ensure the least onerous decision for the public good.

Regulation entails a delegation of power from national governments to regulatory authorities that are experts in a certain field. Majone underlines the fact that regulation does not equal the simple passing of a law, but requires regulatory agencies to have a detailed knowledge and to be involved in the regulated activity.[246] Such an activity demands not only legal knowledge, but also economic, technical and social information. The multidisciplinary approach of regulation in the EU gas market allows it to address gas market failures in a manner that escapes the reach of competition and competition authorities. In order to fulfil its purposes, regulation relies on deeper monitoring, assessment of whether the market is competitive or not and decision-making based on specialized expertise. The desirability of *ex-ante* regulatory intervention, as opposed to the *ex-post* competition application, is related to the increased level of certainty of market conditions that regulation is able to provide.[247]

As deriving from its definition, gas market regulation has a three-fold objective, which *a priori* does not contradict the gas market goals. Already in the Second Energy Package sector regulation has been entrusted with the provision of "structural changes in the regulatory

---

242 Glachant, 2009.

243 Prosser, 2005, p. 13.

244 Vasconcelos, 2005.

245 For example, owing to risks of constructing or maintain an existing or new capacity that could satisfy security of supply concerns or customer protection.

246 Majone, 1994, pp. 77–101.

247 Geradin and O'Donoghue, 2005.

framework, to tackle remaining barriers to the completion of the internal market in particular regarding the trade of gas".[248] In contrast to this, Majone underlined an important constraint to forging harmonization of national laws, policy objectives and national institutional design through international market regulation.[249] Such a constraint stemmed from internal limitations of EU law with regard to the possibility of harmonizing the internal legal systems of the Member States.[250] As shown in Chapter 2, article 114 TFEU has been used extensively for the harmonization of the EU gas sector. However, its text conditioned the adoption of a specific harmonization measure to prove that the intention of such a measure was to improve the conditions and functioning of the market.

The limitations posed by article 114 TFEU explain to a certain extent the Commission's failure to impose its preferred model of unbundling on the EU gas market. In the immediate aftermath of the Energy Sector Inquiry, the Commission maintained that ownership unbundling was the most efficient way to eliminate incentives for preferential treatment, improve efficiency leading to lower prices for consumers, ensure choice for energy users and encourage investment, and that it was crucial in order to create a level playing field.[251] However, ownership divestiture could not be homogenously enforced across Member States owing to remaining divergent opinions on the model's suitability to remove obstacles to gas trade,[252] in comparison to its competitors – the ISO and ITO models. Hence, it appears that the simple favourable position of the Commission for a specific regulatory approach is not enough. Unless its benefit for the conditions and functioning of the market is demonstrated, the afferent market regulation is prevented from fulfilling its harmonizing role.

This does not mean, however, that regulatory convergence of Member States' gas regulatory regimes is unlikely to occur. Haase maintains that the more specific European law is in setting targets or prescribing regulatory instruments, the more convergent national gas market regimes will become.[253] In contrast to the manner in which EU gas market design

---

[248] Common position (EC) No. 5/2005 adopted by the Council.

[249] Majone, 2003.

[250] See, e.g., 114(4), (5), (10) TFEU.

[251] Commission staff working document accompanying the Communication from the Commission – Inquiry pursuant to Article 17 of Regulation (EC) No. 1/2003 into the European gas and electricity sectors (Final Report) {COM (2006) 851 final}.

[252] Talus and Hunt, 2011, p. 35.

[253] Haase, 2008.

has been established, the foundations of Member States' gas markets have originated prior to and in the absence of EU energy policy goals. Such foundations have been largely determined by Member States' specific energy mixes, access to raw resources, relationships with external suppliers, political ideology, etc. The liberalization waves have shaped the development of these foundations through the imposition of minimum regulatory market harmonization standards. In the Third Energy Package, Member States' discretion has been significantly reduced with respect to a number of key issues regarding the EU gas market.[254] At the same time, several supranational bodies have been put in place to coordinate cross-border gas market restructuring measures. Despite Member States' response to the different stages of the EU gas market reform, such developments signal EU's intention to achieve deeper gas market integration through regulatory means.

5.2.2.2 Enforcement of gas sector regulation

The rationale for continuing the restructuring of energy markets under the Third Energy Package relied on the findings of the Energy Sector Inquiry, which showed that one of the main drawbacks of energy sectors has been the maintenance of vertically integrated networks. The findings of the Sector Inquiry were backed up by evidence collected by the Commission during dawn raids at gas and electricity undertakings' premises during 2005–2007.[255] One of the main conclusions was that former unbundling measures did not achieve their purpose owing to remaining incentives for preferential treatment within vertically integrated operators. It has been argued that ineffective unbundling favours discrimination against third-party access, allows information leakage between supply and network affiliates and contributes to distorted investment incentives, as a result of allowing network operators only a low degree of autonomy over investment decisions.[256] In contrast to this, ownership unbundling has been advocated as the best guarantee for full network opening and free energy supply competition.[257] The argument is based on the assumption that ownership unbundling could solve the

[254] Including consumer protection and mechanisms for ensuring the security of energy supply.

[255] European Commission, 2006: Competition: Commission confirms inspections in the energy sector.

[256] Kroes, 2007, pp. 1406–1409.

[257] Proposal for a Directive of the European Parliament and of the Council amending Directives 96/92/EC and 98/30/EC, /*COM/2001/125 final – 2001/0077/COD*/.

dilemma of incumbent gas TSOs. These face the decision whether to grant or not network access for third parties who compete with the supply business to which the said TSOs have been traditionally affiliated. Under unbundling conditions, it is the infrastructure manager (TSO, ISO, ITO) – more or less regulated – who makes the investment decisions.[258] Hence, by severing the common ownership of supply activities and network assets, ownership unbundling increases the independence of TSOs over network third-party access and investment decision-making, which should contribute to the levelling of competitors in the EU gas market. If indeed ownership unbundling is able to enhance competitiveness, this should allow the entrance of more gas undertakings in the market. Complementary to this, the cost of energy should drop and the ability of consumers to choose from more suppliers should increase.

5.2.2.2.1 Ownership unbundling The first question that should be looked at is to what extent ownership unbundling satisfies or impedes the satisfaction of EU gas market goals. In order to answer this question three elements are relevant: the enforceability of ownership unbundling provisions at Member State level, the capacity of ownership unbundling to meet EU gas market goals and the possible limitations of these.

Ample discussions have been carried out on the legal, political and economic effects and limitations of employing ownership unbundling to restructure the energy sector.[259] From a legal and possibly also political point of view, ownership unbundling has been contested at the level of national jurisdiction owing to its invasive approach to property rights and further constitutional constraints regarding the provision of public service obligations.[260] In the general European constitutional framework, property rights are protected from expropriation in the absence of demonstrated public interest and adequate compensation. Among EU Member States, the arguments for and against ownership unbundling have oscillated between deeming it unconstitutional – owing to its imposition of expropriation without compensation, and considering it legal – as long as it did not entail the selling of assets at less than their full market value.[261] This oscillation has been reflected in the final text of Directive 2009/73/EC, which acknowledged the right of Member States to decide whether

[258] Glachant, 2009.

[259] Pollitt, 2008a; Kroes, 2007, pp. 1387–1441; Alesina *et al.*, 2005, pp. 791–825; Schmitt and Rammerstorfer, 2010; Pielow and Ehlers, 2008.

[260] Pielow and Ehlers, 2008.

[261] See Gromitsaris, 2011; del Guayo *et al.*, 2010, pp. 326–359.

or not to adopt ownership unbundling and to choose among three unbundling models suggested by the legal text.[262]

Stricto sensu, limitations to the enforceability of ownership unbundling derive from article 345 TFEU. This article prohibits the EU from interfering with Member States' manner of governing property ownership. The CJEU has however narrowed down the application of the article by interpreting it in the way that national ownership concepts are subject to basic freedoms and to the principle of prohibition against discrimination,[263] and that EU's intervention in national property structure should be restricted to reaching EU's goals.[264] In the light of these, *a priori*, the EU seems to be entitled to demand effective unbundling in the Member States, but cannot impose a harmonized implementation of ownership unbundling, as long as it cannot demonstrate that it is the only way to reach EU energy market goals. The protest of several Member States against ownership unbundling[265] has exerted a considerable impact on its implementation at national level and has resulted in the compromise unbundling provisions of Directive 2009/73/EC. As a consequence, Member States continue to enjoy a considerable degree of regulatory autonomy in relation to unbundling choices. Furthermore, the choice of unbundling regime allows the coexistence of various gas market models across the EU, which may contribute to cross-border investment foreclosure in situations where an ownership unbundled TSO on one side of the border faces a less unbundled TSO on the other side of the border. The 2009 gas Directive contains also important derogations from unbundling measures with regard to emergent and isolated markets. It is difficult to see how this will benefit EU gas market integration and competitiveness, especially in light of the fact that the addressees of such derogations are the Member States most dependent on foreign gas supply and most prone to gas market failures.

---

262 On condition that the gas transmission system belonged to a vertically integrated undertaking on 3 September 2009. See recitals 11, 14 and 18 of Directive 2009/73/EC.

263 *Fearon v Irish Land Commission*, Case 182/83, para. 7, *Klaus Konle v Republic of Austria*, Case C-302/97.

264 *Daniele Annibaldi v Sindaco de Commune di Guidonia and Prezidente Regione Lazio*, Case C-309/96; see also del Guayo *et al.*, 2010, pp. 326–359.

265 Représentation permanente de la France auprès de l'Union européenne, 2008: Effective and Efficient Unbundling of Transmission System Operators … .

Notwithstanding the limited enforceability of ownership unbundling on European gas markets, several Member States have adopted this unbundling model[266] for its ability to fulfil EU energy policy objectives.

The effect of ownership unbundling on market competitiveness has generally been a positive one. Nondiscriminatory access to gas infrastructure is essential for the development and enhancement of competition.[267] Joskow and Tirole maintain that, if generators possess physical transmission rights, this increases the ability of the vertically integrated undertaking to withhold transmission capacity.[268] In return, in an unbundled regime transmission system operators are released from the vertical control of traditional market incumbents, which results in a level playing field for all market players.

There have been examples where Member States practising ownership unbundling have scored low concerning market competitiveness.[269] However, this might not necessarily mean that ownership unbundling is unable to enhance the competition on a given market, but that several other factors have to be taken into consideration, such as a Member State's external dependence on gas imports, its attachment to national gas champions, national consumers' switching practices and the evolution of tariff deficit.[270]

In relation to energy security of supply, the Commission's Impact Assessment on the Third Energy Package observes that ownership unbundling seems to have exerted a positive effect on the increase in investment in the gas network and infrastructure.[271] As production declines, it is important that investments in storage are enhanced in order to mitigate the potential negative effects of import dependence on a stable supply to gas customers. In correlation to this, investment in new gas infrastructure may contribute to the diversification of supply routes, while investment in existent infrastructure might reduce the risk of accidents.

---

266 UK, Spain, Romania, Netherlands, Denmark, Portugal.

267 Pollitt, 2008a.

268 Joskow and Tirole, 2000, pp. 450–487.

269 Pielow and Ehlers, 2008.

270 Shortfall between regulated revenues and costs.

271 Commission Staff Working Document: Accompanying the legislative package on the internal market for electricity and gas – Impact Assessment … , /*SEC/2007/1179*/. However, the Commission admits important constraints to its observation in terms of the difficulty of comparing the absolute investment levels across the EU owing to country-specific characteristics, multiple factors influencing the data regarding interconnector investment and the limited availability of public data on network investment in the case of vertically integrated TSOs.

Investments in gas infrastructure have been characterized as irreversible (sunken) and indivisible (lumpy). They are irreversible because gas infrastructure comprises high-cost extraction rigs, high-pressure natural distribution and storage pipelines, and lower-pressure regional and local distribution networks. The installation is extremely costly owing to a high specificity of assets. Investments in gas infrastructure are indivisible because they are largely responsible for the economies of scale that make some network industries natural monopolies.[272] Owing to gas investments' specificity and the lack of coordination, command and control in liberalized sectors, it is considered that the task of incentivizing investments in this area should be entrusted to regulation.[273]

A 2005 study on regulation and investment observed that regulatory reform of product markets exerts a positive effect on investments and investment decisions and that investments in energy sectors increase according to higher levels of unbundling.[274] The observations are verified by the example of Spain, where investments seem to have blossomed between 1998 and 2008, most likely owing to the introduction of ownership unbundling in 2002.[275]

Pollitt[276] seems to believe that ownership unbundling increases the incentives to invest in the gas sector owing to the fact that it offers extensive opportunities for market expansion and deepening. However, he also emphasizes the fact that enhanced regulatory oversight might be required in order to prevent potential excessive expansions in transmission assets. In this sense, energy investors in Romania in the period 2003–2006 seem to have been influenced to search for new investment opportunities by potential ownership unbundling requirements and pressure to divest in order to comply with third-party access needs in their home countries.[277] In this particular example, however, the results of the investments with respect to market competitiveness are questionable. This is because they represent downstream acquisition strategies of traditional incumbents and new vertically integrated national champions,[278] who have attempted to gain a significant market position outside their

---

272 Bergman *et al.*, 1998.

273 Jamasb and Pollitt, 2008, pp. 4584–4589.

274 Alesina *et al.*, 2005, pp. 791–825.

275 Schmitt and Rammerstorfer, 2010.

276 Pollitt, 2008a.

277 Haar and Marinescu, 2011, pp. 2245–2255.

278 ENEL SpA and E.ON Energie AG are the traditional major incumbents of Italy and Germany; CEZ is an integrated energy utility created in 2003 controlled by the Czech state with 65 per cent of shares.

countries in order to compensate for the loss of market share owing to liberalization.

With regard to sustainability, under Directive 2009/73/EC, each transmission, storage and LNG system operator is supposed to pay due regard to environment when performing its duties. However, the Directive does not distinguish between the three unbundling models. The Commission's Impact Assessment[279] takes into consideration the situation where unbundling would lead to a decrease in consumption prices, which would result in a possible increase in demand with environmental effects – owing to an increase in CO_2 and nitrogen oxide emissions. At the same time however, the hypothetical decrease in natural gas prices owing to unbundling would strengthen the substitutability of gas in opposition to more polluting fuels, which would benefit the environment. This remains, however, a discussion of principle, since for the time being a direct link between ownership unbundling and a decrease in gas prices has not been demonstrated.

Indeed, in an empirical study conducted in 2008, Growitsch and Stronzik concluded that no significant effect for the reduction of consumer end-prices could be observed from the application of ownership unbundling.[280] It could be assumed that a correct application of sector-specific rules – which demand a fair reflection of the cost of energy in the final price – should determine a drop in the gas prices. This is of direct interest for energy consumers, who at least in the short term are motivated by low and affordable energy prices.[281] However, this has not been the case in reality. As derived from Eurostat's observations, the level of gas prices in the first half of 2011 was the same for households and slightly different for the industry in UK – which is one of the Member States with the oldest tradition of ownership unbundling – and Estonia – which is one of the six beneficiaries of the highest derogations from unbundling provisions under TEP.[282] Their score in the Eurostat final gas price ranking could lead to the assumption that ownership unbundling does not influence final consumer prices.[283]

---

[279] Commission Staff Working Document: Accompanying the legislative package on the internal market for electricity and gas – Impact Assessment …, /*SEC/2007/1179*/.

[280] Growitsch and Stronzik, 2008.

[281] Allegedly, in the long-run a steady and secure supply of energy becomes more important.

[282] European Commission, Eurostat, 2011: File: Half-year electricity and gas prices, first half of year, 2009–2011 … .

[283] See Growitsch and Stronzik, 2009; Schmitt and Rammerstorfer, 2010.

A final point with respect to ownership unbundling and gas market goals concerns the case of public service obligations. France, which has a strong culture of PSOs and a strong tradition of state involvement and control in the governance of utilities to ensure the respect of PSOs, has been one of the major opponents of the ownership unbundling of energy networks. France's concerns, however, seem to have rather been a possible expropriation of the state from its participation in the governance of EdF and GdF, the main electricity and gas French incumbents, than the inability of European regulation to ensure the provision of public service obligations.[284] It could be added that, as long as ownership unbundling is able to ensure enhanced market competition, powerful incentives for investments and sustainability, it will positively contribute to the fulfilment of public service obligations in the Member States practising it.

5.2.2.2.2 Third-party access Access-based policies are ruled by a double dilemma. On the one hand, the granting of access might lead to an enhancement of the short-term welfare of consumers, at the expense of long-term market dynamic efficiency, through a decrease in investment incentives as a result of lower investment returns.[285] On the other hand, the refusal of access might temporarily stimulate investments, which would lead to a better satisfaction of security of supply and consumer protection requirements, while potentially exerting a detrimental effect on market competitiveness.

The role of specific-sector regulation is to secure and manage open access to the network, through the monitoring of supply markets, generation of network codes, analysis of third-party access exemptions requests and elaboration of guidelines for market conduct. Third-party access (TPA) in its negotiated or regulated form has been a part of gas restructuring reforms since the very beginning.[286] Directive 2009/73/EC has preserved most of the TPA provisions of Directive 2003/55/EC. The principle of granting nondiscriminatory access for third parties to the gas network and infrastructure facilities is based on article 18 TFEU, which codifies the principle of equality or nondiscrimination. It has been already shown that the structure and nature of the gas grid is rigid and that the existent infrastructure remains scarce. In recent years, several

284 Pielow and Ehlers, 2008.
285 de Hauteclocque *et al.*, 2011.
286 Article 3(2) of Directive 1998/30/EC.

pipeline development projects have been launched.[287] However, instead of enhancing the gas transmission network and ensuring market interconnection, these projects have been conducted in an incoherent and largely divergent and antagonistic manner by their initiators. This has occurred because EU network development strategies have not always matched the priorities of main non-EU gas suppliers and producers. As a result, the EU gas market is still largely dependent on traditional pipelines and pipeline networks, which augments the importance of third-party access for the development of a competitive gas market and the multiplication of consumer choices.

As seen above, there is a trade-off that market regulation must moderate between investments in new infrastructure and the development of competition in individual markets. Transmission system operators managing the gas infrastructure are obliged to provide effective third-party access in accordance with objective and transparent criteria and on the basis of tariffs or methodologies adopted for their calculation, which must be approved by the national regulatory authorities before their entry into force.[288] However, at the same time, they must guarantee that necessary investments to enhance and maintain the gas infrastructure are made by market participants.[289] Article 22 of Directive 2009/73/EC could be seen as an innovation in this respect. On the one hand, the article gives TSOs powers to make network development plans and take investment decisions. On the other hand, it ensures that NRAs are confronted annually with a 10-year network development plan, indicating the necessary investments.

Under article 36 of Directive 2009/73/EC, exemptions from TPA rules can be granted in the case of new major infrastructure projects or significant increases in capacity in existing interconnectors. Exemption decisions may be granted on a temporary basis and may cover the entire infrastructure or only parts of it. There are six conditions that an investment must fulfil in order to benefit from an exemption to third-party access:

---

[287] For example, the Nord Stream pipeline project launched by Gazprom under the Baltic Sea, the Southstream project initiated by ENI (Italy) and Gazprom under the Black Sea, the Nabucco pipeline project envisaged by an Austrian–Turkish–Hungarian–Romanian trust, and the Trans-Adriatic-Pipeline project aiming at linking Greece to Italy through Albania.

[288] Articles 32, 33 and 34 of Directive 2009/73/EC.

[289] Articles 14, 17, 22 and 23 of Directive 2009/73/EC.

- it must constitute major gas infrastructure;
- it must enhance competition and security of gas supply;
- the level of risk attached to it should be such that the investment would not take place unless an exemption was granted;
- the infrastructure must be owned by a natural or legal person separated at least legally from directly concerned system operators;
- charges must be levied on users of that infrastructure;
- the exemption must not be detrimental to competition or the effective functioning of the market.

Directive 2009/73/EC does not precisely define the meaning of "major" in the concept of "major new infrastructure". However, based on the formulation of article 22 of Directive 2003/55/EC it can be concluded that major infrastructure refers to interconnection between Member States, LNG and storage facilities. The understanding of the term is confirmed by the list of EU exemptions for new gas infrastructure between 2004 and 2011.[290]

The second exemption condition regards the simultaneous fulfilment of two gas market goals, which have been at odds in certain cases. An investment that is able to enhance the security of gas supply might impede the competitiveness goal only in the short term. However, in the longer term, the same investment could prove positive for competition by increasing allocative efficiency.[291] The role of regulation in this case is to establish a balance between the two and observe whether the proposed investment respects that balance or not.

The risk condition describes the uncertainty of the investor concerning the profitability of the investment. From the point of view of the gas investor, three major elements might exert a deterring effect on the investment. These are the irreversibility of investment costs, the uncertainty of the gas demand on a specific gas infrastructure segment and the submission of new infrastructure to regulated access.[292] It has been maintained that the fears of the investor draw upon the inability of the market regulator to credibly commit *ex-ante* to mitigate the *ex-post* risks and costs of the investment.[293] It is correct that an NRA is unable to guarantee the return of an investment in the liberalized gas market. Even if there is long-term planning of necessary investment in the gas

290 European Commission, 2013: Exemptions for new infrastructures, http://ec.europa.eu/energy/infrastructure/exemptions/doc/exemption_decisions.pdf.

291 Szydło, 2009, p. 255.

292 *Ibid.*, p. 255.

293 Spanjer, Aldo, 2007.

infrastructure, the NRA cannot offer anticipatory reward to investors interested in taking up projects on the investment plan. In this sense, it has been wrongly maintained that the granting of an exemption is a kind of remedy helping the investor to overcome the risk.[294] However article 36 of the 2009 gas Directive represents a form of *ex-ante* gratification of the investor, by temporarily providing it with better conditions than its competitors on the same market.

In order to benefit from an exemption under article 36 of Directive 2009/73/EC, the infrastructure that should be exempted must be legally unbundled from the system operators in whose systems that infrastructure should be built. The logic of this condition stems from the necessity to avoid cross-subsidization between regulated and nonregulated parts of the infrastructure and to prevent the situation where the exempted interconnector, which also benefits from price regulation exemptions, does not have any of its costs reduced through regulated transmission charges.[295]

The total amount of the charges that must be levied on users of gas infrastructure is not mentioned in the legal text. The provision, however, entails an obligation for the operator of the exempt investment to collect the said charges. The Commission regards the nondiscriminatory application of the same tariff method to all system users as fulfilling the criterion.[296]

The final condition of article 36 of Directive 2009/73/EC concerns the lack of detrimental effects to competition or the effective functioning of the market. This is a different condition than that of enhancing market competition, while it entails a negative obligation on the investment not to bring any harm to market competition. In opposition, the second condition discussed above refers to a positive obligation of the investment of adding up to the existent level of competition. NRAs enjoy a large margin of discretion when assessing the impact of an investment on the effective functioning of the market.

The six conditions commented above are cumulative. Should an investment fail to accomplish one of the exemptions, the decision might be granted provided that additional conditions are fulfilled or might be revoked. In relation to the first case, in the Commission's decision[297] regarding an exemption on the Austrian section of the Nabucco pipeline,

[294] See Szydło, 2009, pp. 254–269.

[295] Note of DG Energy and Transport on Directives 2003/54–55 and Regulation 1228/03 in the Electricity and Gas Internal Market.

[296] Exemption decision on the Austrian section of the Nabucco pipeline. Brussels, 8 February 2008, CAB D (2008) 142.

[297] *Ibid.*

the European institution observed that several of the markets that were envisaged to receive gas through the Nabucco pipeline were at the moment of the decision dominated by a single market player. Hence, it appeared to the Commission that an exemption of the proposed investment could strengthen one or more dominant positions in the market, or could lead to a situation where the dominant incumbents could have the power to adopt a common market policy and that cooperation agreements between the two could negatively impact on the fair auction of the Nabucco shares. At the same time, it was observed that the investment enhanced the security of supply by adding new gas routes from the East to the EU, by improving market interconnection and by attracting and connecting new gas suppliers. It was also noticed that the level of risk was such that the project would have not taken place without the exemptions and that the segments of the networks are legally unbundled. No objections were brought with regard to Nabucco's charging system and tariff methodologies. Additionally, it was concluded that the investment was supposed to improve the functioning of the internal market by increasing its liquidity owing to additional gas imports and some short-term capacity. Therefore, the only concerns of the Commission remained in relation to the capacity of the investment to enhance market competition. The Commission decided nevertheless to grant the exemption under certain conditions, which limited the capacity cap[298] to 50 per cent. However, in a case regarding underground gas storage in the Czech Republic, the Commission decided to withdraw a previous exemption decision owing to the absence of sufficient evidence that the second condition of Article 36 of the 2009 gas Directive was satisfied.[299]

5.3 SOME CONCLUSIONS AND CRITIQUES TO THE COHERENCE OF THE MARKET DESIGN

The pursuit of EU gas market goals lacks coherence both at the level of policy formulation and with regard to the manner in which competition law and sector regulation are enforced. The ambiguous formulation of gas policy goals at EU level correlated with the fuzzy pursuit of EU gas

---

[298] Share of annual capacity that a dominant undertaking can book at the total of all Austrian exit points of a pipeline.

[299] Commission Decision of XXX on the exemption of an Underground Gas Storage Facility in Damborice, Czech Republic, from the internal market rules on third party access, http://ec.europa.eu/energy/infrastructure/exemptions/doc/doc/gas/2011_damborice_decision_en.pdf

market integration has allowed and contributed to the preservation of divergent national gas market designs. It has been shown that sector regulation is able to set a balance between market goals, but fails to foster harmonization of regulatory regimes across Member States. At the same time, it has been observed that some of the traditional constraints of applying competition tools on the European gas markets are gradually disappearing, but new limitations are appearing. These derive from the reduced scope of the Commission to use competition law so as to alter the gas market structure and the questionable extent to which far-reaching, infrastructure-divesting commitments could be achieved in alleged competition-distortion cases involving non-EU gas incumbents. In several instances, the application of the same legal instrument issued from either competition law or market regulation has led to the satisfaction of one of the market goals, while causing detrimental effects on another.

Unlike the US Supreme Court in *Trinko*, which concluded that the existence of a regulatory structure aimed to correct the effects of anticompetitive behaviour precludes the application of antitrust measures, the European practice has been to allow overlap between sector-specific rules and competition principles.

It has been shown above that competition law and sector regulation in the EU gas market have had in general a complementary application. Traditionally, competition law has acted as a market corrector, through the imposition of case-by-case sanctions. In parallel, sector-specific regulation has been aimed at curtailing the ability and incentives of incumbents of engaging in discriminatory, restrictive and abusive market behaviour through changes in the market structure. The use of competition law in energy matters has been favoured by the initiation of regulatory reforms in the gas and electricity sectors. Prior to the adoption of the first Directives there was little use of competition law for the regulation of energy markets.[300] A more pro-active enforcement of competition law in the gas market is a more recent practice, dating from the time of the Energy Sector Inquiry. As a consequence, the outcome of more effective competition law enforcement in recent years has been reflected in enhanced support to gas sector regulation. The enforcement of competition law rules represents a guarantee that barriers to fair competition removed by sector-specific legislation will not be curtailed by anticompetitive, abusive market behaviour from market actors, which attempt to maintain or regain their prior market position. One of the most

300 Cameron, 2007, p. 279.

obvious examples is provided by the Commission's commitment decisions with major EU gas incumbents, which target market restructuring through ownership unbundling.[301] The Commission's commitments practice appears to pursue an increase in the role of competition law in the gas market. However, this has not necessarily exerted a positive impact on the satisfaction of all of the gas market goals. This has been observed particularly with respect to consumer protection and enhancing security of supply. Specifically concerning the second aspect, the scope of applicability of the commitments practice to the regulation of gas relationships with third-country suppliers remains questionable.

Despite the generally positive coexistence of competition law and regulation, their combined application in the gas market could also lead to potential normative tensions owing to their sometimes divergent goals. Competition law and sector regulation have a cumulative application in the EU gas market notwithstanding that the former's application is superior to the latter's. However, this is an important aspect of their relationship, which has an influence on the achievement of the gas market goals. Competition law is part of the Treaty, while sector regulation is not. According to the hierarchy of EU norms, the application of competition law should preclude the application of sector regulation. However, with regard to the regulation of utilities sectors, both the Commission and the European Courts seem to have opted for a complementarity approach between the Commission and the NRAs.[302] This has not meant in principle that the application of specific sector regulation could exonerate an undertaking from observing antitrust rules or that the existence of a specific sector regulation could exclude more general antitrust provisions.[303] However, the Lisbon Treaty has codified for the first time all of the policy goals applicable to the gas market, placing security of energy supply and energy sustainability on the same level as market competitiveness and the functioning of the internal energy market.

To a certain extent, competition law and gas legislation envisage different aims. The former is focused on the promotion of economic efficiency by improving market competitiveness, even if that occurs at the expense of the rest of the gas market goals. The latter takes into consideration market goals deriving from public welfare and envisages

---

[301] For example, *RWE gas foreclosure*, Case COMP/39.402.

[302] *T-Mobile Deutschland/O2 Germany*, Case COMP/38.369; *Deutsche Telekom AG*, Case COMP/C-1/37.451, 37.578, 37.579; *Deutsche Telekom AG*, Case T-271/03.

[303] Nyssens and Schnichels, 2007.

the gradual correction of market imperfections. A costs-and-benefits type of analysis of the concurrent application of competition law and gas legislation in view of achieving the gas market goals is not entirely able to offer an accurate solution. Competition law suffers from a series of internal limitations that prevent it from being able to forge an integrated gas market while exerted alone. At the same time, sector-specific legislation allows for a significant margin of interpretation specifically with respect to the enforceability of regulatory obligations.[304] Hence, there might be a need to reinterpret the relationship between competition law and sector regulation in the EU gas market, in the context of allowing regulation to fulfil its balancing role among gas policy goals.

---

[304] For example, unbundling, transparency, nondiscrimination, consumer protection, public service obligations, etc.

6. Integration of the EU gas market through administrative bodies

The implementation of the rules governing the EU gas market relies on specific institutional arrangements between European energy regulators, NRAs, NCAs, Member States' governmental services and the European Commission. As shown in the following, these institutional arrangements entail both top-down and bottom-up features of administrative integration and vertical and horizontal patterns of cooperation. Their setup is influenced by the application of both competition law and regulatory tools to satisfy EU gas market goals, as well as by the concurrent competences of different authorities mandated by one of the two distinct sets of legal tools ruling a particular aspect of the gas market.

6.1 THE MODEL OF ADMINISTRATIVE IMPLEMENTATION OF THE RULES GOVERNING THE EU GAS MARKET

Competences in the energy field are shared between the EU and its Member States. This makes the EU's action subsidiary to that of the Member States and maintains the principle of autonomy of national administrative systems. Specifically, paragraphs (2) and (3) of article 194 TFEU curtail significantly the EU's competence in energy matters. In this sense, the paragraphs provide that measures in the field of energy taxation are subject to unanimity. Also subject to unanimity are Member States' rights to decide upon their energy mixes, conditions for exploiting their energy resources and the general structure of national energy supply. Additionally, Member States are allowed to retain the right to conduct bilateral energy relations with non-EU countries as they see fit.[1]

A priori the EU lacks a legitimate monopoly to bring Member States into compliance with EU law. However, the supranationalization of energy policy allows the possibility of hierarchical coordination. This

[1] Braun, 2011.

type of coordination has been defined as taking the form of authoritative decisions that actors must obey, owing to their legal obligations towards legitimate institutions.[2] In this sense, in the EU gas market Member States must act in the spirit of loyal cooperation[3] and in observance of the pre-emption doctrine,[4] despite their large margin of appreciation with regard to national energy mixes and relationships with third-country suppliers. In this market, coordination is achieved through a combination of hierarchal-ordered conduct and voluntary compliance. The latter is necessary owing to the very nature of EU rules, which require Member State-specific application.[5] Glachant and Lévêque[6] appear to acknowledge this assumption, but they underline that, despite their variety, national gas market designs must ensure the respect of an absolute minimum of competitiveness and openness. It has been shown above that the first two EU legislative packages addressing the liberalization of the EU gas market demanded a minimal level of harmonization between Member States' market designs, while allowing the latter enough discretion in implementing the European liberalization provisions at a national level. The TEP gives a further push for harmonization by curtailing the autonomy of Member States with respect to the organizational features of the regulatory authorities and their way of functioning.

The model of institutional arrangement in the EU gas market can be developed to a certain extent on the basis of a top-down mechanism of

---

2 Börzel, 2010.

3 The duty of loyal cooperation set in article 24(3) TEU states that domestic administrations are expected to ensure the full implementation of Union provisions regarding energy policy applicable to the gas market. Such an obligation derives from the more general principle that national administrations are supposed to abstain from any measures that could jeopardize the attainment of the Treaty's objectives. Moreover they should respect the integrity and independence of the Union's legal order and guarantee through their employed measures the scope and effectiveness of the Union's law.

4 The pre-emption doctrine prevents the national legislator from adopting national rules that are contrary to Union laws governing the same matters. This means when the Union law covers the area exclusively, when the national law would be in breach of Union law or when the latter would constitute an obstacle to the smooth functioning of the common market.

5 For example, affordability of energy prices due to public service obligations. See Craig, 2009.

6 Glachant and Lévêque, 2006, p. 4.

administrative integration elaborated by Chiti.[7] The characteristics of the original model encompass:

- the establishment of European common systems composed of national independent authorities, a European independent authority in embryo and the Commission;
- EU direct involvement in the implementation process of national independent regulators;
- the integration of the national independent regulators in a unitary administration; and
- the establishment of a collegiate body at Union level composed of representatives of the national regulators exercising a number of relevant tasks, which does not have the attributes of a genuine European regulatory authority.

Looking at the EU gas market institutional arrangements one can observe a mixture between Commission's direct administrative responsibilities, elements of bottom-up and top-down delegation to ACER and implementation carried through national authorities, stemming from both competition law and sector regulation. Collaboration between institutions and bodies governing the EU gas market occurs through forums and transnational regulatory networks.

The EU's involvement in the implementation process of national independent regulators is direct, because the Commission participates in the monitoring of Member States' gas markets and it is responsible for opening infringement procedures when a violation is detected.[8] However, both national and European energy regulators are independent and are expected not to take instructions from political power at national and European level.[9]

The relationship between the Commission and ACER entails both vertical and horizontal elements. Concerning the former, ACER is accountable to the Commission[10] and the achievement of part of its tasks

---

[7] Chiti, 2009, pp. 19–22.

[8] Commission Staff Working Document: 2009–2010 Report on progress in creating the internal gas and electricity market, http://ec.europa.eu/energy/gas_electricity/legislation/doc/20100609_internal_market_report_2009_2010.pdf; article 29 of Regulation (EC) 715/2009.

[9] See Directive 2009/73/EC, Regulation (EC) 713/2009; Hofer, 2011.

[10] Article 6(6) of Regulation (EC) 713/2009: the Agency is expected to report to the Commission on the implementation of the network codes and Guidelines adopted by the Commission; article 7(5) of Regulation (EC) 713/

depends on the participation of the Commission.[11] The Commission is also involved in coordinating through guidelines the situations where the Agency is competent to decide upon the terms and conditions of access to infrastructure.[12] Elements of horizontal cooperation between ACER and the Commission are visible in the monitoring of the internal markets in energy.[13]

The creation of ACER has resulted in what Chiti identifies as a European independent authority in embryo.[14] The newly established ACER is a body with legal personality and a Board of Regulators, which has the "last say" on regulatory decisions. Unlike ERGEG, ACER has not been established by a Commission decision and its tasks go beyond the facilitation of consultation, coordination and cooperation between the regulatory bodies in Member States, and between those bodies and the Commission.[15] However, the implementing powers of the Agency are weak and limited to contributing to the implementation of guidelines on trans-European energy networks.[16]

Apart from the arrangements introduced above the EU gas market institutional framework entails elements of direct and indirect execution, as well as of bottom-up administrative integration.[17] An example of direct execution is the straightforward application of competition law to the regulation of gas supply relationships in the EU. This mandates the Commission to conduct investigations in cases of competition law infringement caused by Member States or by private actors, to impose sanctions and to take legal recourse to CJEU.

2009: the Agency must inform the Commission of the noncompliance of a national regulatory authority with the Agency's opinion; recital 24 of the Regulation (EC) 713/2009: the Agency should be accountable to the European Parliament, the Council and the Commission, where appropriate.

[11] Recital 9, articles 4(e) and 6(4) of Regulation (EC) 713/2009: adopt the nonbinding framework guidelines elaborated by the Agency; article 6(11) of Regulation (EC) 715/2009: the Commission is empowered to adopt one or more network codes in the areas listed in article 8(6) of the Regulation, on its own initiative, or upon the failure of ENTSOG or ACER to develop a network code, or upon recommendation of the Agency.

[12] Article 8(4) of Regulation (EC) 713/2009.

[13] Article 11 of Regulation (EC) 713/2009: monitoring is shared between ACER, the Commission, the Member States and relevant national authorities.

[14] Chiti, 2009, pp. 19–22.

[15] Article 1(2) of Commission Decision 2003/796/EC.

[16] Recital 14 of Regulation (EC) 713/2009.

[17] See more in Chiti, 2009, pp. 19–22.

The influence of EU sector regulation on the organization and functioning of national administrations[18] represents an element of indirect execution. On the EU gas market the pressure of EU regulation takes the shape of a top-down mechanism of administrative coordination and integration. EU sector regulation influences the manner in which NRAs and other competent authorities interact with each other, by setting up platforms of cooperation such as ACER and ENTSO for gas and by requiring the internal harmonization of NRAs' powers.[19] Furthermore, Member States' discretion with regard to the creation and identification of NRAs is limited. In this sense, notwithstanding national specificities, NRAs have to be independent from national governments and the gas industry.[20]

The memberships of ENTSO and of the Madrid forum are examples of bottom-up administrative integration. In this sense, the organization and activities of both ENTSO and the Madrid forum reflect the European and the national levels of administration taken together. ENTSO for gas is composed of 39 TSOs belonging to 22 Member States and two associate partners from Latvia and Lithuania.[21] At the same time, the composition of the Madrid Forum groups NRAs, Member State governments, the European Commission, TSOs, gas suppliers and traders, consumers, network users and gas exchanges.[22]

The proportion between elements of direct and indirect execution, as well as bottom-up and top-down administrative integration is not static. Rather, as Chiti observes, it is dependent on the on-going evolution of the global legal space.[23] Moreover, EU administrative governance structures vary and evolve according to the specific needs of the policy area in which they operate.[24] In the case of the gas market, an example of this is provided by the increase in the role of information sharing and transparency in the exchanges between European and national regulatory

---

[18] For example by obliging the Member States to designate independent national regulatory authorities responsible for the effective competition and the efficient functioning of the market.

[19] Recital 34 of Directive 2009/73/EC.

[20] Recital 29 and article 39 of Directive 2009/73/EC.

[21] Estonia, Malta and Cyprus are not represented. See http://www.entsog.eu/aboutus_members.html

[22] Gas Regulatory Forum Madrid, http://ec.europa.eu/energy/gas_electricity/gas/forum_gas_madrid_en.htm

[23] Chiti, 2009, pp. 19–22.

[24] Hofmann and Türk, 2006, p. 584.

levels, as a result of the enhancement of consumer protection requirements among gas policy priorities. Additionally, informal institutions such as Member States' attitudes towards liberalization and European energy policies exert a direct influence upon the shape of the gas market's institutional arrangements.[25] This assumption is validated when looking at the heterogeneity of NRAs throughout the EU and the creation of regulatory bodies with coordinating roles, albeit with weak decision-making powers. It could be further assumed, drawing on the findings in Chapter 3, that non-EU gas suppliers may also exert an indirect pressure on EU gas market governance, to the extent to which they are able to influence Member States' governments approach to the gas market restructuring reform.[26]

6.2 OPPORTUNITIES AND CONSTRAINTS TO GAS MARKET INTEGRATION DERIVING FROM THE MULTIPLE ROLE OF THE COMMISSION

The Commission's powers and actions in the EU gas market have a triple basis. They stem from the Treaty, the specific gas sector legislation and competition law. Article 17(1) TFEU contains an implicit mandate for the Commission to ensure the external representation of the Union on energy matters, which generally falls under the scope of operation of DG ENER.[27] Article 106(3) TFEU has provided the Commission with specific powers to deal with State measures concerning legal monopolies and to adopt decisions and directives in this respect, outside the procedures laid down in articles 289 and 297 TFEU. The logic of using article 106(3) TFEU has been to force Member States to act where they proved unwilling to negotiate a subsequent liberalization stage. Notwithstanding the success of using this article to forge market opening in other sectors, in energy markets recourse to article 106(3) TFEU occurred only as late

---

[25] Haase, 2008, p. 130.

[26] See for example the discussion in Chapter 3 on the scrapping of the "reciprocity clause" from the initial proposal on a third Directive addressing the EU gas market.

[27] Notwithstanding the mandate of the Commission to represent the EU in external energy matters, there is not one but multiple configurations between the directorate-generals of the Commission in view of ensuring the external representation on projects of "European" interest, the furthering of the cooperation and dialogue with non-EU countries and regions, climate change and energy sustainability, etc.

as 2003 and solely in individual cases.[28] Article 291(2) TFEU provides for the conferral of clear, limited and nondiscretionary powers on the Commission. Read in conjunction with Council Regulation (EC) No. 58/2003, this article allows the Commission to establish executive agencies in the EU gas market.

One of the most important provisions in the Treaty allowing the Commission to bring about a harmonization of the Member States' gas market regimes derives from article 258 TFEU. The interpretation of the article has been that the Commission as the "guardian of the Treaties" is responsible for ensuring that EU law is correctly applied and that Member States' actions do not prejudice EU's objectives. The Commission's decision to begin infringement proceedings based on article 258 TFEU may take place whenever the Commission considers that an infringement of EU law has occurred. Owing to their decisiveness for the success of the gas Directives, actions taken by the Member States to properly implement the energy legislative packages beyond the mere transposition are of the utmost interest for the Commission. National implementation, which represents the transposition of Union law into national law by the national legislature, provides the guarantee that Union law is properly applied and should a violation occur a sanction may follow if required. In a 2008 case,[29] the Court found that EU Directives must be implemented by appropriate measures taken by the Member States in conformity with article 288 TFEU. The fact that certain provisions of the Directive in question were directly applicable in the national legal system did not relieve the Member States of their obligations to transpose EU law.[30]

The first infringement procedure round for noncompliance with EU energy regulation started as late as 2006.[31] This has been followed by seven subsequent infringement rounds. Regarding the implementation of the TEP, on 1 June 2011 – three months after the expiration of the transposition deadline – no Member State had yet notified transposition measures to the Commission. A Commission's report concluded that "overall the state of implementation of internal market legislation at national level is disappointing; with still many open infringements under

---

28 Nyssens and Schnichels, 2007.

29 *European Commission v Belgium*, Case C-475/08.

30 *Ibid.*, para. 44.

31 European Commission, 2006: Infringement procedures opened in the gas and electricity market sector, by Member State, http://europa.eu/rapid/press ReleasesAction.do?reference=MEMO/06/152&language=en

the second package".[32] Following the Commission's report, at the beginning of 2012, legislation in eight of the EU Member States was still not in line with EU rules deriving from the TEP.

Examples of recent infringements of the third gas Directive regard the misapplication of the TEP provisions, specifically with respect to market interconnection and adoption of national transposition measures.[33] Such infringements are likely to perturb the advancement of the internal gas market project. The Commission seems to have developed its infringement proceedings, complementary to the strengthening of the competition law enforcement in the EU gas markets. The purpose of infringement proceedings is to stop the violation. However the recourse to formal legal action cannot occur before attempting to put a stop to the infringement through negotiation.[34] In parallel to infringement proceedings initiated by the Commission, the Court has developed its own jurisprudence, which until now seems to concur with the Commission's findings.[35] This seems to imply a certain degree of effectiveness of the infringement practice in determining the Member States to comply with secondary energy legislation.[36]

The role of the Commission as "guardian of the Treaty" is recognized in Recital 19 of Directive 2009/73/EC, which underlines that, notwithstanding the market design, the Commission should be consulted on the compatibility between the measures taken by Member States to ensure a level playing field on the one hand and the respect of the Treaty and EU law on the other. In line with the Court's jurisprudence, this provision instils an obligation for Member States to communicate to the Commission their envisaged transposition measures.

Under the TEP, the Commission has been able to develop administrative rule-making through guidelines. Administrative guidelines allow the Commission to structure its margin of discretion and to regulate the private and public activity through publication of standard review.[37] The use of Commission guidelines in sector regulation appears to follow

---

[32] Commission Staff Working Document, 2009-2010 Report on progress in creating the internal gas and electricity market.

[33] IP/12/52 Commission refers Ireland and the United Kingdom to Court to ensure European law is properly implemented; IP/11/1437 Commission refers Bulgaria and Romania to Court to ensure European law is properly implemented.

[34] Craig and de Búrca, 2007.

[35] *European Commission v Belgium*, Case C-475/08.

[36] See Internal Market Scoreboard 25, http://ec.europa.eu/internal_market/score/docs/score25_en.pdf

[37] Hofmann, 2006, p. 177.

the aim of enhancing harmonization between EU Member States' market designs. The integrated gas market aim has been pursued at different speeds during the market restructuring process. This has happened nonetheless owing to the fact that, in the early years of the gas market reform, the European Commission, as the main agent in charge of the completion of the single market programme, advocated the idea of a liberalized market integration necessitating a certain degree of regulatory harmonization at the national level, without prescribing a coherent best-practice model.[38] The lack of such a model concurred with Member States' allegiance to their different regulatory traditions and did not impair the pursuit of a European integrated gas market, although arguably it slowed it down. As a result, the harmonization of the regulatory regimes has been hindered to some extent.

Under Directive 2009/73/EC, the Commission is entitled to adopt guidelines to provide the minimum degree of harmonization required to achieve the aim of the Directive.[39] Following the same logic, article 23 of Regulation (EC) 715/2009 delegates to the Commission the adoption of guidelines for providing the minimum degree of harmonization required to achieve the aims of the Regulation, with regard to third-party access services, capacity-allocation mechanisms and congestion-management procedures and tariff methodology related to cross-border trade.

Furthermore, under the provisions of Directive 2009/73/EC, the Commission is entitled to adopt legally binding guidelines to achieve the necessary degree of harmonization.[40] Such a degree is not defined in the legal texts, which may lead to the assumption that the Commission benefits from a significant margin of discretion. This discretion is however curtailed in article 52(1)(g) of Directive 2009/73EC. In this sense, the article clearly states that, if the Commission identifies the need for possible harmonization requirements, which are not linked to the provisions of the Directive, it must signal this fact to the Parliament and the Council. The text does not entitle the Commission to adopt guidelines on its own in such a situation.

Some of the most interesting competences of the Commission stemming from the TEP concern its relationship with third-country suppliers. Specifically, the accent has been placed on involving the Commission in the process of granting non-EU incumbents access to the EU gas market and in managing a possible future gas crisis. According to article 11(1) of

---

38 Haase, 2009, p. 3.
39 Recital 63 of Directive 2009/73/EC.
40 Recital 61 of Directive 2009/73/EC.

Directive 2009/73/EC, the Commission must be notified when certification is requested by a third-country TSO. The NRA is also required to inform the Commission of any certification decision.[41] The Commission is not entitled to take the final certification decision, which stays with the concerned Member State. However, the NRAs must take the utmost account of the Commission's opinion.[42] This latter statement implies an element of vertical hierarchy in the relationship between the Commission and NRAs.

Under Regulation (EU) 994/2010 the Commission's competences with respect to the coordination of actions regarding third countries include arrangements for handling crisis situations and the adoption of measures to guarantee a stable gas flow to the Union.[43] The Commission is further entitled to deploy a task force to monitor gas flows into the Union in crisis situations in consultation with the third countries involved,[44] to assume a mediation and facilitation role where a crisis arises owing to difficulties in a third-country,[45] to coordinate the action of competent authorities on its own or through the Gas Coordination Group[46] and to declare a Union emergency or a regional emergency for a specifically affected geographical region.[47] In close connection to this, ensuring the security of supply is considered a shared responsibility of the Commission, natural gas undertakings and Member States or respective national competent authorities.[48]

The role of the Commission in relationships with third-country suppliers is in line with the provisions of articles 17(1) and 194(1) TFEU. The first one calls for external representation of the EU in energy matters through the Commission, while the latter implies that the competences of the EU to adopt measures on matters of energy must obey the principle of subsidiarity.

Since the beginning of the liberalization process, the Commission has, to a great extent, based its intervention on the legitimacy provided to it

---

41 Article 11(4) of Directive 2009/73/EC.

42 Article 11(8) of Directive 2009/73/EC.

43 Recital 44 of Regulation (EU) 994/2010.

44 Recital 44 and article 11(7) of Regulation (EU) 994/2010.

45 Recital 44 of Regulation (EU) 994/2010.

46 Article 3(5) of Regulation (EU) 994/2010.

47 Article 11(1) Regulation (EU) 994/2010. The Commission may declare (discretion) a Union emergency or a regional emergency for a specifically affected geographical region at the request of a Competent Authority, but it must declare (obligation) a Union emergency at the request of at least two Competent Authorities.

48 Article 3(1) of Regulation (EU) 994/2010.

by EU competition law. In regards to the enforcement of competition law, the European Commission has direct administrative responsibilities.[49] Recent years have witnessed a reinforcement of its role in the EU gas market as a result of a more focused and outcome-oriented competition law enforcement. It has been noticed above that, in relation to energy matters, the choice between remedies, punitive measures – for example, fines – and enforcement of competition rules through commitments given willingly by the concerned parties seems to favour the latter. From the point of view of the Commission this has often meant a more expeditious process and additional means for the institution to intervene in the restructuring of the market beyond the scope of article 106(3) TFEU.[50] The logic followed by the Commission appears to have been that of determining a restructuring of the EU gas markets that escapes regulation and regulators, which cannot overcome Member States' opposition on controversial political issues.[51] The Commission's preference for the commitments practice, combined with the results achieved and the critiques that it has attracted, suggests that a re-assessment of the institution's role in the EU gas market is required.

It has been maintained that, in relation to energy, the Commission has made use of its antitrust enforcement powers in a regulatory manner, rather than a corrective one, in an attempt to substitute or regulate the regulator.[52] In the aftermath of the *E.ON* and *RWE* cases,[53] the Commission's commitment practice based on article 9 of Regulation (EC) No. 1/2003 has attracted both political and academic critique.[54] Both are of interest for the present discussion.

---

[49] These are shared with national competition authorities. This subsection is concerned with the Commission's enforcement practice and how this influences its position on the EU gas market. For a more in-depth approach to the practice of NCAs in the recent years see Scholtz and Purps, 2010.

[50] Having regard to the limitation in article 106(3) TFEU, it is difficult to imagine that the Commission might reach the same effects by turning to the procedure encompassed in article 7 of Regulation 1/2003. At the same time, although the fines are consistent in energy cases and might act as a deterrent for a specific type of behaviour of market players, they are applied only on a case-by-case basis and cannot significantly affect the entire structure of the market.

[51] For example, ownership unbundling.

[52] Tapia and Mantzari, 2012.

[53] *E.ON AG*, Case COMP/39.388; *RWE gas foreclosure*, Case COMP/39.402.

[54] del Guayo *et al.*, 2010.

The most advocated objections to the Commission's commitments practice regard the proportionality of the commitments and the use of competition law to force deeper regulatory harmonization at the level of the Member States. The first claim, that such a practice does not respect the principle of proportionality, has been based on the observation that the Commission would not have been able to achieve similar results by imposing behavioural and structural remedies in view of article 7 of the Regulation (EC) No. 1/2003.[55] The second claim in its turn draws on the fact that the most far-reaching commitments achieved by the Commission concern the German major gas and electricity incumbents, whose national government vehemently opposes the restructuring of its market through ownership unbundling. The two objections are different, but they both found their counter-arguments in the same feature of the commitments procedure: the voluntary proposal of commitments by the concerned parties.

This particularity of article 9 of Regulation (EC) 1/2003 has two important consequences for the critiques brought to the commitments practice. First, the undertakings concerned might be tempted to offer extensive (disproportionate) commitments instead of risking the incurrence of a high level fine.[56] Second, the fact that commitments are formulated and voluntarily provided by the concerned undertakings without direct instruction from the Commission might void the "market restructuring" claim of its content.

Notwithstanding the fact that commitments are offered on a voluntary basis, questions have been formulated as to the standard of proportionality applicable to the Commission's acceptance of commitments.[57] In the landmark *Alrosa* case, the General Court found that the standard for remedies under article 9 of Regulation (EC) No. 1/2003 was the same as under article 7 of the same Regulation.[58] This meant that commitments accepted by the Commission should not exceed what was appropriate and necessary for terminating the infringement of competition law. The basis for the General Court's finding was the fact that the commitments offered by the parties did not have a legally binding effect. On the contrary, the acceptance of commitments by the Commission had such an effect.[59]

---

[55] Monti, Giorgio, 2008.

[56] Scholtz and Purps, 2010.

[57] Koch and Gauer, 2011.

[58] *Alrosa Company Ltd v Commission of the European Communities*, Case T-170/06.

[59] *Ibid.*, para. 86.

Concluding, the General Court stressed that the Commission was obliged to apply a proportionality test in commitment decisions.[60]

Contrary to the position of the General Court, the CJEU in the same case overruled the General Court's judgment and projected a new light on the interpretation of proportionality under article 9.[61] Thus, the CJEU maintained that article 9 did not require the Commission to find an infringement, but to examine and possibly accept commitments formulated by the parties. Hence, the principle of proportionality should be restricted to verifying that the commitments in question address the competition concerns formulated by the Commission and that the undertakings concerned have not previously formulated less onerous commitments satisfying the concerns.[62]

Turning to the "market restructuring" argument, it could be maintained that, since the Commission has not orchestrated the divestiture commitments undertaken by the undertakings concerned in the contested cases,[63] it cannot be accused of attempting a deliberate restructuring of the gas market outside the harmonization competences conferred upon it by the Treaty and the specific sector regulation. However, as observed by the General Court in the *Alrosa* judgment,[64] the acceptance of commitments by the Commission makes them legally binding and able to create effects outside the scope of the particular competition case in which they have been adopted. The Commission becomes thus the sole author of the commitments from the moment when it accepts them and accordingly assumes sole responsibility for them.[65]

It was shown in Chapter 5 that the market restructuring function of the commitment practice raises several problems in terms of achieving the gas market goals and could possibly lead to even more disparities between the Member States. Hence, it may be observed that, although *a priori* the Commission's intervention in the EU gas market has been positive for the advancement of its harmonization and subsequent integration, certain clashes may occur owing to the double – competition law/regulation – mandate that the Commission has for regulating the market. This double mandate complicates the model of administrative

---

60 *Ibid.*, para. 99.

61 *Commission v Alrosa*, Case C-441/07 P [2010].

62 *Ibid.*, para. 41.

63 *E.ON AG*, Case COMP/39.388; *RWE gas foreclosure*, Case COMP/39.402.

64 *Alrosa Company Ltd v Commission of the European Communities*, Case T-170/06.

65 *Ibid.*, para. 88.

implementation. In this respect, when performing competition law enforcement the Commission retains an almost absolute position to instruct cases, take decisions on the validity of the conduct in question and decide upon penalties that should be imposed on the parties that infringed competition law. At the same time, in the light of sector-regulation the involvement of the Commission in the gas market must remain moderate. Indeed, as identified in the Commission's Impact Assessment, the European institution does not have enough autonomous decision powers and network leverage in the regulation of energy markets.[66]

It should be observed that the role of the Commission as the "guardian of the Treaty" allows it to decide whether or not to intervene in those cases in which gas-specific rules already offer appropriate solutions to competition-related issues.[67] While performing this role, the Commission must judge whether its intervention by means of enforcing competition law, where gas sector rules have already provided solutions, is necessary. The elements that are taken into consideration are the proved efficiency of the regulator's activity and the possibility that a subsequent action based on competition law would lead to a contradictory decision in the specific case under consideration. In general, the expertise of the sector regulator, as well as its advantageous position with regard to the market behaviour that should be regulated, renders the Commission's intervention obsolete. However, in the specific case where the regulatory authority would prove to be inefficient, competition law is to be applied. As shown below, owing to the particularities of the regulatory framework of the EU gas market, such arrangements are advisable, but do not constitute the rule.

6.3 THE ROLE OF REGULATORY AUTHORITIES IN THE EU GAS MARKET

6.3.1 Principal-agent Theory Applied to the EU Gas Market

The principal-agent theory generally refers to contractual agreements between two parties, where one of them – the principal – delegates to the other – the agent – the responsibility for carrying out functions or tasks

---

[66] Commission Staff Working Document Accompanying the legislative package on the internal market for electricity and gas – Impact Assessment ..., /*SEC/2007/1179*/.

[67] Geradin, 2005.

on his behalf. The theory applied to the study of the EU has aimed at understanding the complexities of institutional settings and alignments in the Union and at explaining the reasoning behind delegated governance.[68] Principal-agent theory is based on three behavioural assumptions: that actors are rational utility maximizers; that principals and agents may develop different preferences; and that between principals and agents there is an informational asymmetry.[69]

On the EU gas market, the demarcation lines between principals and agents are often blurred. The three main principals – European authorities, national governments and national regulatory authorities – wear a double hat.[70] This renders the definition and separation of their roles ambiguous, owing to the fact that the institutional designs that each of them shapes as principal – to determine the power and statutory discretion delegated to their respective agents – overlap.

Thus, the European Commission is both a principal for the national governments and NRAs and an agent of national governments, with regard to the formulation of the EU gas policy. Delegation in this context takes place in both a top-down – from the supranational to the national level – and a bottom-up manner – from the national to the supranational level. ACER acts as an agent of the European Commission and as a principal of NRAs. Furthermore, NRAs are agents of the national governments and the European authorities and principals of the national gas firms and the manager of the network infrastructure. The double delegation towards NRAs represents a relevant example of overlapping institutional design. NRAs have a dual allegiance to national and supranational levels, owing to the fact that they are structurally part of their respective national administrations, while at the same time functionally belonging to the EU administration.[71] From the government's point of view, their existence is legitimized by the need to increase efficiency in complex and technical domains,[72] as well as to enhance credible commitments in the gas sector in order to exert a higher appeal for investors.[73] From the EU's point of view, the delegation of tasks and powers to NRAs is justified by the need to ensure uniform implementation and enforcement of the liberalization package across Member States. The double delegation leads to an overlapping of allegiances of

---

68 Kassim and Menon, 2003.
69 Héritier, 2005.
70 *Ibid.*
71 Nizzo, 1999.
72 Majone, 2001, pp. 103–122.
73 Levy and Spiller, 1996.

NRAs towards the national and supranational levels, which might impact upon the effectiveness of NRAs in the liberalization process. This happens owing to the conflicting purposes of the delegating principals. Thus, the European principal requires NRAs to implement and interpret EU legislation in a manner increasing regulatory homogeneity across the Union. At the same time, Member States expect NRAs to interpret the EU sector legislation according to their specific national context, which allows Member States to maintain discretion on a number of key features of the gas sector liberalization process.

It may be also argued that national governments have to a certain extent and in very specific cases acted as agents of national gas incumbents, specifically in situations where they have protected the interests of such incumbents against market openness requirements formulated by the specific-sector legislator. This has been observed for example in relation to Spain and Norway and their policy to back up the bargaining power of European gas incumbents.[74] Such examples should, however, be dealt with cautiously, as it is difficult to demonstrate the existence and potential scope of incentive and influencing mechanisms through which gas incumbents as hypothetical principals could ensure the compensation of the state as agent.[75] At the same time, if established, such an agency relationship would be immediately denounced by the Commission, owing to its hindering effects on the development of a liberalized gas market.[76]

It has been observed that, as a result of the enhancement of the technical and institutional expertise of agents vis-à-vis their principals, the information asymmetry between the two grows deeper, allowing agents to pursue their market and policy objectives in an increasingly autonomous manner.[77] In the EU gas market such an asymmetry might occur in information exchanges between regulatory authorities and transmission system owners, owing to the latter's right to not disclose certain commercially sensitive information obtained in the course of carrying out their activities.[78] The increased autonomy of agents is met by attempts of the principal to limit the autonomy of the agent efficiently – for example,

---

[74] In 2001 the Norwegian government claimed that European competition law could not be applied to the GFU case, while the selling of Norwegian gas through the GFU scheme has been ordered by the government itself. See also *Gas Natural + Endesa*, COMP/37.542.

[75] See more on "agency theory and the state" in Rowe, 2002, pp. 287–316.

[76] *Commission v Spain*, Case C-196/07.

[77] Majone, 1996, p. 72.

[78] Article 16 of Directive 2009/73/EC.

through *ex-ante* codes of conduct and rules and *ex-post* monitoring of task fulfilment and imposition of sanctions in the event of failure.[79] *Ex-ante* codes of conduct may consist of the description of the tasks of national regulatory agencies in gas Directives and Regulations. An example of *ex-post* monitoring of task fulfilment and imposition of penalties is provided by the Commission's ability to sanction the infringement of competition and sector-regulation rules by Member States and gas undertakings.

Relationships between principals and agents in the EU gas market entail the act of delegation, which generally presumes the conferral of a certain degree of discretion upon the agents. Under EU law such a discretion is submitted to important legal and doctrinal constraints. In this respect articles 290 and 291 TFEU set barriers to the delegation of power to the Commission by stipulating that the legislature may delegate to the Commission the power to supplement or amend "nonessential elements of the legislative act" and that "[t]he essential elements of an area shall be reserved for the legislative act and accordingly shall not be the subject of a delegation of power".[80] Article 290 TFEU in particular has been interpreted as maintaining the distinction between essential elements, which are "provisions [...] intended to give concrete shape to the fundamental guidelines of Community policy"[81] and which must be reserved to the Council's power and an act of a merely implementing nature which may be delegated to the Commission.[82]

The jurisprudence of the European Courts[83] shows that a delegation of wide discretionary powers is clearly unlawful under present primary law and that the creation of regulatory agencies which typically would dispose of such powers is violating the principle of institutional balance, and is therefore not permissible.[84] Any delegation of power to agencies must be limited to implementing powers that are precisely defined and

---

[79] Haase, 2009.

[80] Article 290(1) TFEU.

[81] *Germany v Commission*, Case C-240/90.

[82] Council of the European Union, 2011: Opinion of the Legal Service Application of articles 290 (delegated acts) and 291 (implementing acts) TFEU.

[83] *Meroni & Co, Industrie Metallurgiche SpA v High authority*, Case 9/56; *Draft Agreement establishing the European laying-up fund for inland vessels*, Opinion 1/76; *Giuseppe Romano v Institut national d'assurance maladie-invalidité*, Case 98/80; *The Queen, on the application of Alliance for Natural Health and Nutri-link Ltd v Secretary of State for Health*, Cases C-154 and 155/04; *Tralli v ECB*, Case C-301/02 P.

[84] Griller and Orator, 2004, p. 8.

can be rigorously monitored.[85] In the Court's reasoning, the Commission may delegate certain tasks to administrative agencies, provided that the delegation abides by five strict conditions:

- a delegation may regard only powers that the Commission itself possesses;
- power may be delegated in view of the preparation and performance of executive acts alone;
- independent bodies may not be afforded any discretionary powers;
- the oversight of the delegated powers has to remain with the Commission which is responsible for the delegated act; and
- the delegation should not disturb the "balance of powers" among European institutions.[86]

The lawfulness of delegation depends on the degree of supervision and control that the delegating institution exercises upon the delegated body. Two preliminary principles have to be respected. First, a conferral of powers cannot be presumed, but the delegating authority has to take an express decision transferring them. Second, European institutions cannot entrust agencies with powers they do not themselves enjoy. Hence, a lack of autonomous powers of the Commission in the energy sector would justify the failure of establishing a powerful energy regulator with a wide scope of tasks and decision-making powers. This has been one of the most advocated explanations for the adoption of ACER, regarding the fact that, at the time of the elaboration of the TEP, the Treaty of Lisbon and its Energy title had not yet entered into force.

6.3.2 Energy Regulators

6.3.2.1 The national level

National regulatory authorities play a significant role in both gas policy-making and gas policy enforcement. Regulation through NRAs has become one of the most important modes of regulation and implementation of the EU gas market liberalization project at the national level.[87] Their participation in the development of the internal gas market takes place through cooperation with the Commission in view of building a comprehensive and transparent regulatory framework for the internal gas

85 Curtin, 2006, p. 92.
86 Majone, 2005, pp. 88–89.
87 Majone, 1996, p. 47.

market and to ensure competition law enforcement in the gas sector.[88] As the main implementing bodies of EU gas legislation, NRAs are expected to find a balance between the satisfaction of public good, the protection of consumers, the interest of stakeholders, market competitiveness, energy efficiency, security of supply, the needs of a specific national context and the European drive for harmonization. In performing this balancing task the regulators must act objectively, through the interpretation of available data. There are three important obstacles, however, that NRAs must overcome in order to perform their task efficiently. These refer to the tendencies of principals to minimize "agency loss", the "risk of regulatory capture" and information asymmetry.

"Agency loss" refers to the fact that, when delegating, principals will act to minimize the possibility of agents acting against their preferences and they will try to design weaker bodies.[89] It is the principals who decide what powers should be delegated to their agents and who maintain control over the performance of the delegated tasks. In this sense, article 11 of Directive 2009/73/EC grants the NRAs the power to decide upon the certification of TSOs. However, notwithstanding NRAs' opinion on the matter, it is the Member States that have the power of "last say" in the certification process. Furthermore, NRAs are entitled by article 36 of the Directive to decide whether or not to grant an exemption to third-party access to infrastructure in the case of new investments. Their decision-making power is, however, curtailed by the Commission, which may decide to revoke the granted exemption.[90]

Based on the above examples, it can be observed that agency loss acts as a deterrent for granting autonomous regulatory authorities wide discretionary powers, which could prevent the NRA from acquiring a "detailed knowledge of, and intimate involvement with, the regulated activity".[91]

"Regulatory capture" refers to the situation where NRAs established to regulate a specific sector or industry end up being influenced by the undertakings they were supposed to control or by their respective Member States.[92] It has been shown in the preceding chapters that,

---

88 Vasconcelos, 2005.

89 Coen and Thatcher, 2006.

90 Commission Decision of XXX on the exemption of an Underground Gas Storage Facility in Damborice, Czech Republic, from the internal market rules on third party access, http://ec.europa.eu/energy/infrastructure/exemptions/doc/doc/gas/2011_damborice_decision_en.pdf

91 Majone, 1994, p. 81.

92 Hofmann and Morini, 2012.

during the EU gas market liberalization process, not only national governments, but also the industry has exerted pressure upon the NRAs. In response to this, Directive 2003/55/EC increased significantly the legal status of NRAs.[93] That has been achieved through two sets of measures. The first concerned the obligation of Member States to charge one or more competent bodies with the function of regulatory authorities. The second referred to bestowing upon such bodies a minimum set of functions and competences, pre-established by the text of Directive 2003/55/EC in the interest of harmonization. Despite the positive advancements brought by the second gas Directive, Member States continued to make use of their discretion in the setup of national energy regulatory authorities. Such a tendency raised important questions with regard to the NRAs' independence,[94] their access to information and their conformity with European standards.[95] In its 2007 Energy Sector Inquiry the Competition DG acknowledged the need for a consistent strengthening of the powers of NRAs and their enhanced coordination at the European level.[96] Under Directive 2009/73/EC this has been translated into a focalization of the attention upon cross-border issues, previously discussed only in the informal environment provided by the Madrid Forum and ERGEG.

The creation of independent NRAs is legitimated by the need to prevent the interested parties – national governments, gas market stakeholders – from exerting pressure upon the market regulator.[97] In the gas sector, which is dependent on constant and reliable investment sources, an independent regulator provides the potential investors with enough positive signals regarding the stability of the regulatory framework encompassing that sector. Moreover, such independence is essential to

---

[93] Cameron, 2005a, pp. 19–20.

[94] For example, the Romanian government has obliged the National Authority for Energy Regulation to respect the law of unitary payment, which subordinates to a certain extent the NAER to the government, whereas its financing is dependent on the budget of the Government's General Secretary.

[95] An example of controversial NRA setup is provided by Germany, which despite the European legislative requirements, has opted not to create a specific regulatory authority for gas, its role being entrusted to the Bundesnetzagentur, an agency that ensures the supervision of the German electricity, gas, telecommunications, postal and railway markets.

[96] DG Competition Report on Energy Sector Inquiry, /*SEC/2006/1724/, p. 15.

[97] Vasconcelos, 2005.

enable the NRAs to perform their tasks and duties under article 41 of Directive 2009/73/EC.[98]

Under the TEP the regulators' independence is strengthened away from any public or private entity, and specifically from their national governments from which they are expected not to take any instruction.[99] This development is in line with the role of energy regulators, which is to protect the development of the liberalized gas market from various interest groups, "each possessing a substantial private power of information and significant social legitimacy".[100] NRAs see their regulatory oversight increased in order to be able to ensure a balanced allocation of resources, as well as more transparency in the entire gas supply process. Transparency is necessary in order to diminish the risk of agency capture of regulatory authorities by the undertakings that the former are regulating. Article 41 of Directive 2009/73/EC provides an extension of the powers of NRAs in order to guarantee their independence from national governments and their capacity to enforce gas market opening principles. Specifically, in reference to enforcing consumer protection, paragraph 4 of the article sanctions the power of regulators to issue binding decisions in relation to natural gas undertakings, as well as impose penalties on the latter, should they fail to comply with their obligations. Additionally, the regulatory authorities can suggest a competent court to propose such penalties.

An additional important limitation of regulation is the information asymmetry that occurs between the regulator and other market agents and the incomplete framework of information on the market where those agents make decisions.[101] Access to information determines the ability of the national regulator to regulate. Asymmetric information creates several regulatory problems with respect to actions that interest groups might take in order to influence preferential decisions within the NRA. Further regulatory difficulties may occur in respect to the optimal equilibrium of market prices and the efficiency of the activities of a regulated firm that is placed in an advantageous informational position.[102] In order to gain access to such information, the regulator may be motivated to foster a

---

[98] Such tasks include the capacity to review investment plans, monitor compliance with TPA and impose sanctions where the compliance has been ignored, as well as decide upon exemptions of third-party access to network infrastructure in case of new investments.

[99] Article 39 of Directive 2009/73/EC.

[100] Glachant, 2009.

[101] *Ibid.*

[102] Spanjer, 2008, p. 85.

closer relationship with the undertakings it regulates. However, such a relationship could lead to an increased possibility of "regulatory capture".[103]

Under the TEP, NRAs' access to information is enhanced both with regard to information obtained from other regulatory authorities active on the market and from national gas undertakings.[104] Such information might prove essential for the enhancement of market interconnection. In this respect, article 42 of Directive 2009/73/EC provides the duty of NRAs to cooperate with each other and exchange the necessary information to solve cross-border issues.

6.3.2.2 Agency for the Cooperation of Energy Regulators

The Agency for the Cooperation of Energy Regulators was established as a regulatory agency in conformity with article 114 TFEU. According to article 2(2) of Regulation (EC) 713/2009, it enjoys legal personality. Its independence is guaranteed by the fact that its Director, Administrative Board and Board of Regulators are expected to act objectively in the public interest and not to take instructions from any government, the Commission or other public or private entity.[105] According to article 12 of the Regulation, the Administrative Board of the Agency is composed of nine members (each having an alternate) appointed by the Council, the European Parliament and the Commission. It is interesting to observe that the composition of the Board reflects perfectly the institutional balance in the EU energy market, following the adoption of the Lisbon Treaty. In this sense, the Council has the right to appoint the majority of the members of the Board – five out of nine. This fact confirms the assumption that the Member States continue to preserve the greatest leverage over national energy policies. At the same time, the enhancement of powers of the European Parliament is reflected in its right to adopt two members of the Board. The new powers of the Parliament curtail to a certain extent the position of the Commission vis-à-vis ACER, while they limit the Commission's appointment margin to two of the Administrative Board's members.

The Board of Regulators is composed of senior representatives of NRAs, one alternate per Member State from the current senior staff of those authorities and one nonvoting representative of the Commission.[106] It has been observed that the composition of the Board of Regulators

---

[103] Glachant, 2009.

[104] Articles 41(2), 41(4) and 48 of Directive 2009/73/EC.

[105] Articles 12(7) and 16(1) of Regulation (EC) 713/2009.

[106] Article 14 of Regulation (EC) 713/2009.

reflects an attempt to mix the independence of a regulatory network with the inclusion of Member States' specific interests.[107] This is true to the extent that NRAs continue to approach the implementation of EU energy legislation in a manner corresponding to a specific national context.

Article 18 of Regulation (EC) 713/2009 establishes a Board of Appeal comprising of six members and six alternates selected from current or former senior staff of NRA's, NCA or other national or EU institution with relevant experience in the energy sector. The participation of both NRAs and NCAs in ACER's Board of Appeal reflects the enhancement of the use of competition to regulate European energy markets.

The establishment of the Agency for the Cooperation of Energy Regulators adds to the complexity of the institutional framework characterizing the EU gas market, by introducing an intermediary level between the national and the supranational ones. This intermediary level should be able to foster an integrated implementation of EU gas policy across Member States and ultimately an integrated European energy administration.[108] In the specific case of ACER, however, the limited scope of action of the Agency makes the achievement of this task questionable.

ACER's tasks are mainly technical and include the harmonization of grid codes and developing new market infrastructure. In general, its competences are reduced to issuing opinions and recommendations for TSOs, NRAs, the European Parliament, the European Council and the Commission.[109] This sets important limitations to ACER's ability to make effective use of the decision-making power bestowed upon it regarding regulatory issues of cross-border infrastructure and exemption decisions for new interconnectors.[110] Let us take the example of third-party access exemptions for new infrastructure as provided in article 36 of Directive 2009/73/EC. According to the article, if the investment under discussion concerns cross-border infrastructure, ACER has the right to participate in the final exemption decision. A first limitation of ACER's decision-making powers derives from linking this power to an action or inaction of the concerned NRAs. In this sense, the Agency is entitled to adopt an exemption decision on the cross-border investment only as a last resort and only if the concerned NRAs could not reach an agreement within a period of six months or if they made a joint request to the Agency to take

---

107 Hofmann and Morini, 2012.

108 *Ibid.*

109 Article 6 of Regulation (EU) 994/2010.

110 Article 8 of Regulation (EC) 713/2009; article 36 of Regulation (EC) 715/2009.

the decision.[111] An additional limitation stems from the fact that ACER's exemption decision might be discarded by the Commission through an *ex-post* control.[112]

ACER's limited decision-making power is visible also in relation to the development of network codes. According to article 6 of Regulation (EC) 715/2009, ACER is entrusted with the development of framework guidelines for network codes. However, in order for such guidelines to become legally binding, they have to be adopted through the comitology procedure.[113] The priorities of network guidelines are established by the Commission, which is also the body that initiates the guidelines development process and which has the last say in establishing whether the developed guidelines contribute to nondiscrimination, effective competition and the efficient functioning of the market. Nevertheless, article 6(9) of Regulation (EC) 715/2009 establishes a limit to the Commission's discretion. Thus, in cases where the Commission is satisfied with the framework guidelines and the Agency submits a network code recommending the latter to adopt it through the comitology procedure, any inaction of the Commission in this regard must be duly justified.

In light of the above it could be maintained that ACER is an example of the EU's intention to push forward the harmonized integration of the EU gas market, combined with doctrinal obstacles, Member State reluctance and the Commission's attempt to minimize "agency loss".

In this sense, ACER's envisaged purpose is the establishment of a framework for cooperation between national regulatory authorities.[114] Such a cooperation is justified because it is expected that it will improve the performance of individual regulators and the advancement of the internal gas market. The framework provided by ACER is expected to lead in time to the coordination and harmonization of NRAs' actions across EU Member States. Within the framework, NRAs should coordinate horizontally with each other and vertically with the European Agency by sharing information on cross-border issues. As seen in the preceding section, a decrease in information asymmetry is expected to positively contribute to an increase of market competitiveness.

Cooperation between energy regulators is closely related to the strengthening of regulatory oversight within the Member States. During the Energy Sector Inquiry it was observed that the internal gas market

[111] Article 8(1) of Regulation (EC) 713/2009; article 9(1) of Regulation (EC) 713/2009; article 36(4) of Regulation (EC) 715/2009.

[112] See Section 5.2.2.2.2.

[113] Article 6(11) of Regulation (EC) 715/2009.

[114] Recitals 5, 10, 20 and 29 and article 24 of Regulation (EC) 713/2009.

suffered from both a lack of liquidity of resources and a lack of transparency. Such market failures have and continue to hinder the efficient allocation of resources, as well as consumers' trust in the market. More regulatory oversight has been judged necessary to render the EU gas market more transparent, to facilitate the access to gas resources and to ensure an adequate consumer protection. Accordingly, ACER has been designed to cover the regulatory gaps deriving from the establishment of unequal regulatory authorities across the Member States and from the differences in decision-making priorities at NRAs and Commission levels. An example of this derives from recitals 7 and 11 of Regulation (EC) 713/2009, which provide that ACER is entrusted with regulatory oversight of the cooperation between transmission system operators.[115]

Both the Commission and the Council of European Energy Regulators considered that the platform of regulatory cooperation offered by ERGEG was insufficient to cover the regulatory gap, owing to ERGEG's lack of binding decision-making powers, specifically in relations to cross-border issues involving NRAs and relevant market players. In contrast to this, ACER has been enabled to handle specific issues regarding cross-border activities, such as exemption requests for interconnectors.[116] The new body is also expected to provide advice to the Commission on cross-border matters. In this way, ACER appears to help reduce decision-making costs through the expertise of the executive branch, as well as enhance the credibility of long-term policy commitments.[117]

Furthermore, ACER's Board of Regulators has the "last say" on regulatory decisions. Its tasks go beyond the facilitation of consultation, coordination and cooperation between the regulatory bodies in Member States, and between those bodies and the Commission.[118]

Notwithstanding such developments, the Commission's stance regarding the adoption of ACER reflects its tendency to adopt restrictive doctrine interpretations, when sensing that its competences might be reduced, in contrast with its traditional avant-gardism in promoting innovative readings of the Treaty.[119] Discussions anticipating the establishment of ACER envisaged modifications to the EU energy institutional setting, including the alternative between the creation of an energy

---

115 Recitals 7 and 11 of Regulation (EC) 713/2009.
116 Articles 8 and 9(1) of Regulation (EC) 713/2009.
117 The reasons behind delegation of powers. See, Majone, 2005, p 64.
118 Article 1(2) of Commission Decision 2003/796/EC.
119 Geradin and Petit, 2004.

regulator, the extension of the tasks and powers of ERGEG and the Commission taking up the pursuit of new tasks, emulating the competition regulatory model.[120] Neither of these alternatives was considered in the final proposal for an Energy Agency. The main arguments formulated by the Commission against the adoption of a strong European energy regulator with autonomous decision-making powers and legally binding effects have been the lack of a Treaty basis legitimizing such an agency and limitations deriving from the so-called *Meroni* doctrine.[121] It seems, however, highly unlikely that the Commission would have suggested a more powerful European energy body should the Treaty have provided a basis for it. This observation is based on the initial proposal of an Energy Agency formulated by the Commission. Despite acknowledging the limits of the powers of ERGEG and the lack of efficiency of the gradual convergence method regarding the adoption of harmonized network grids, the Commission's first suggestion for an Energy Agency restricted its role to monitoring the execution of the ENTSOs' tasks and the provision of nonbinding opinions.[122] It is difficult to imagine how such an agency could have helped close the regulatory gap between the decision-making by the Commission and the decision-making by NRAs and put forward the cooperation mechanisms envisaged by ERGEG.

It appears thus that the Commission's position towards the independence and increased powers of energy regulators entails a certain duality. At a national level, the Commission has been one of the most active promoters of the re-definition of the roles and tasks of NRAs and the conferral of enhanced powers upon them so as to allow them to foster an integrated gas market. Concerning the European regulator, it appears that the Commission attempts to retain an authoritative definition of the EU energy policy and a quasi-monopoly upon the supranational decision-making process in this area, enhanced by the Commission's new competences in energy as a result of the Lisbon Treaty. Taking into consideration this dual position of the Commission, one may ask if this does not identify it as the main culprit for the lack of a real European energy regulator, which ultimately contributes to a poorly integrated EU gas market.

---

[120] Hofer, 2011.

[121] Deriving from *Meroni & Co, Industrie Metallurgiche SpA v High authority*, Case 9/56.

[122] Hofer, 2011.

6.4 COLLABORATIVE GOVERNANCE APPLIED TO THE EU GAS MARKET

Resulting from the above, in the TEP the accent is placed on cooperation between energy regulators.[123] These are supposed to cooperate with each other and with different regulators operating on both national and European levels in order to gain a correct overview of the markets concerned. Patterns of cooperation in the EU gas market entail both hierarchical and negotiation-based frameworks and have been developed in both a formal and an informal way. An example of a formal and hierarchical framework for cooperation is provided by ACER, which codifies the horizontal cooperation between NRAs and places constraints on the autonomy of subordinate actors. Diametrically opposed to this, the Forum process appears as an example of an informal, negotiation-based framework, which serves as a venue for the meeting of all the levels of the regulatory structure on an equal footage, having the aim to facilitate the development of an efficient internal gas market. At the crossroads of these two frameworks, network cooperation occurs in a formal, but largely negotiated manner.

The second and the third patterns of cooperation appear to be focused on joint problem-solving by all stakeholders, through collaborative governance.[124] While the first pattern of cooperation was dealt with above, the next subsection will concentrate on gas market governance through forums and networks of regulators.

6.4.1 The Forum Process

The Forum process in European energy markets has been perceived as an informal venue for cooperation between energy regulators and stakeholders, necessary to cover the regulatory gap created by the First Energy Package.[125] The 1999 creation of the Madrid Forum for gas coincided with the Commission's proposal for an "open method of coordination" through an informal horizontal network of regulators.[126] The institutional setting of the Madrid gas Forum brings together representatives delegated by the Commission, the Council of European Energy Regulators, NRAs,

---

[123] The establishment of ACER and the European Network of Transmission System Operators (ENTSO) for gas has been aimed at fostering frameworks of cooperation between NRAs and TSOs across EU Member States.

[124] Koch, 2009.

[125] Vasconcelos, 2005; Eberlein, 2005.

[126] Coen and Thatcher, 2006.

ACER and European Network of Transmission System Operators for Gas (ENTSOG). Its biannual meetings are also attended by the International Energy Agency and major interest associations[127] – including those representing consumers' interests – as well as additional speakers, who owing to their expertise might be invited to attend a specific meeting.

The purpose of the Forum is two-fold: to allow stakeholders to participate in shaping various "soft law" instruments required to facilitate action in a specific area of energy policy and to foster transnational communication between NRAs, TSOs, consumers and traders, the relevant government departments and the Commission. It has been observed that, in the context of a large, nonhomogenous market – such as the EU gas market – harmonization is determined by the similarity of needs and preferences in the Member States and not always by top-down pressure.[128] In this respect, the expected success of the Forum has aimed at the building of voluntary consensus and delivery of agreements between participants in order to facilitate the development of the internal gas market.

The limitations of the forum process are ample and have been widely discussed.[129] The Madrid Forum has no decision-making structure, it entails lengthy procedures to reach consensus between various participants and it lacks the ability to coerce its participants to comply with the codes of practice and guidelines achieved within the Forum's platform.[130] Hence, it would appear that the enforceability of the voluntary agreements achieved within the Forum is very restrained. It has been observed in this sense that the ineffective unbundling of national transmission networks has contributed to rendering difficult the implementation and monitoring of voluntary agreements.[131] The enforceability of the Forum's outcomes seems, however, to have increased since 2005 nonetheless

---

[127] Eurogas, the Electricity Association Eurelectric, the European Federation of Energy Traders, the International Federation of Industrial Energy Consumers, the International Association of Oil and Gas Producers, the European Chemical Industry Council, the Gas Transmission Europe, the Association of European Energy Exchanges, the International Oil and Gas Producers Association, the Gas Infrastructure Europe, GEODE, CEDEC, the Technical Association of the European Natural Gas Industry (MARCOGAZ) and the European Free Trade Association, EASEE.

[128] Majone, 2003.

[129] For example, Cameron, 2005a; Eberlein, 2005.

[130] For example, very low compliance with the first set of Guidelines for good practice regarding TPA services, tariff structures and balancing adopted in February 2002.

[131] Vasconcelos, 2005.

owing to the formalizing of voluntary agreements in guidelines annexed to Regulations.[132] Such guidelines are binding on the parties to whom they are addressed and the addressees are obliged to implement them. It has been observed that the "soft law" effect of guidelines is their ability to steer the behaviour of Member States policies or stakeholders in a direction that prevents their actions from being regarded as a violation of EU law.[133] Article 23 of Regulation (EU) 715/2009 gives an almost absolute right to the Commission to adopt and amend guidelines in order to ensure a minimum degree of harmonization on third-party access services, principles underlying capacity-allocation mechanisms, the provision of information necessary for network users to gain effective access to the system, tariff methodology related to the cross-border trade of natural gas and network codes. Before adopting or amending the guidelines, the Commission must consult ACER and ENTSOG. However the legal text does not condition the decision of the Commission on the result of the consultation. Since the Commission's power to adopt Guidelines is complemented by the possibility of making recourse to the comitology procedure[134] in order to amend the text of the guidelines in a manner fostering a homogeneous implementation in the Member States, the willingness of the concerned parties to cooperate appears to be not essential.

In the Forum process the Commission has played an important role from the very beginning through its monopoly on the accreditation and acceptance of participants, the setting of the agenda and the "last say" on allowing revisions to the agenda suggested by participants.[135] Additionally, it has also acted as a moderator, guide and driving force behind the process. It is interesting to see that, although the Commission lacks formal powers in the Forum's institutional setting, it has been able to steer the voluntary agreements of parties towards rendering the market more competitive through its informal agenda setting powers.[136] Its role seems to have increased since 2005, complementary to the initiation of

---

132 The first Guideline for good practice adopted in 2002 was the first one to be included in a Regulation (Regulation (EC) No. 1775/2005).

133 Hofmann, 2006, pp. 153–178.

134 In exercising its implementing powers, the Commission is assisted by representatives of Member States through committees, in accordance with the "comitology" procedure. There are four types of possible committees: advisory, management, regulatory and regulatory with scrutiny.

135 Haase, 2008.

136 Eberlein, 2005.

the Commission's competition review of energy sectors and the strengthening of the enforcement of EU competition law on the gas and electricity markets.

6.4.2 Trans-national Regulatory Networks

The concept of network entails both a reference to "types of interactions" and an indication of "more permanent structures".[137] There are many explanations for the emergence of transnational regulatory networks in the EU. It has been maintained that networks can be explained as an increased pressure upon state actors, for example, national regulatory agencies, to enter a horizontal form of cooperation owing to their dependency on other actors.[138] Also, the creation of transnational regulatory networks has been linked to the need for covering the regulatory gap between the Commission's vision of the development of an EU policy and the uncoordinated national implementation of such a policy through NRAs.[139] In this sense, the legitimacy of networks would derive from their ability to foster information exchange, administrative support and joint planning.[140] Furthermore, it has been observed that the Commission's limited oversight of NRAs coupled with its fear of losing decision-making powers through delegation to regulatory authorities has contributed to the emergence of the need of a stronger EU-wide coordination through formal and informal networks of regulators.[141] The participation of the Commission in such governance networks has been welcomed as an increase of its capacity to supervise and monitor the implementation and application of EU law.[142]

Traditionally, transnational regulatory energy networks have been perceived as "children of multiple parents",[143] owing to the fact that the delegation of tasks and powers towards the networks comes from the EU, the NRAs and national governments. This multiple parenthood ensures a balance between the preservation of the autonomy of national administrations and the proper and uniform implementation of Union's rules.

---

[137] Scharpf, 1993, pp. 125–165.
[138] Bevir, 2012, p. 49.
[139] Kourtikakis, 2005.
[140] Hofmann *et al.*, 2011, p. 307.
[141] Coen and Thatcher, 2006.
[142] Weiss, 2009, pp. 45–70.
[143] The concept has been coined by Coen and Thatcher. See Coen and Thatcher, 2006.

In the EU, governance through networks occurs in the form of negotiation between equal actors with similar bargaining powers and social coordination based on mutual agreement.[144] An example of such a network is generally provided by the European Competition Network, whose members' equality is formally emphasized in the EU afferent legislation.[145] However, it has been observed that the participation of supranational institutions in networks submits their operation to hierarchical constraints,[146] which exert pressure upon the achievement of agreements. Furthermore, the role of private actors is limited, which makes participation in network governance an exclusive, rather than inclusive, practice.[147] It also puts constraints on the ability of the participating regulators to receive feedback from stakeholders and consumer groups and to further project these in the implementation of the EU legislation.

Contrary to this model, collaborative governance through networks of regulators entails participation of industry and nonindustry stakeholders in joint action in order to facilitate efficient markets and reliability.[148] As in the case of traditional EU network governance, collaborative governance is centred on information sharing, cooperation, dialogue, consensus-oriented decision making and the corrections of the failure of downstream implementation and of the high cost and politicization of regulation.[149] However, unlike in the case of traditional networks, collaborative governance has not been designed as a second-best option to the delegation of more powers from the Member States to the supranational level.[150] Instead, it has been pursued as a problem-solving oriented institutional setting, where stakeholders are directly involved in the decision-making process.

6.4.2.1 Cooperation of national authorities in the EU gas market

The European Competition Network (ECN) represents an example of negotiated cooperation in the shadow of hierarchy in the EU gas market. Not only the Commission, but also national competition authorities have seen an enhancement of their enforcement role in the energy sector in the aftermath of the Energy Sector Inquiry. Starting in 2010, one of the

---

144 Scharpf, 1993, pp. 125–165.

145 Joint Statement of the Council and the Commission on the Functioning of the Network of Competition Authorities.

146 Börzel, 2010, pp. 191–219.

147 Weiss, 2009, pp. 45–70.

148 Ansell and Gash, 2007, pp. 543–571.

149 Koch, 2009, pp. 543–571.

150 Weiss, 2009, pp. 45–70.

identified trends of competition law enforcement in the energy sector has been the gradual transfer of enforcement practices from the Commission to the NCAs.[151] This trend continued in 2011, when the Commission appears to have been rather focused on the implementation of gas sector regulation at the level of Member States. In their turn, NCAs seem to have become very active in this sector, through sector inquiries[152] and taking a position against abusive behaviour and collusions in gas.[153]

The mandate of NCAs to apply competition on national gas markets derives from Regulation (EC) No. 1/2003. Their cross-border cooperation has been facilitated by the establishment of an ECN,[154] which is dedicated to the effective enforcement of European competition law throughout the EU. In performing their function, the NCAs are able to make use of discretionary powers. Furthermore, they are expected to carry out extensive economic assessments and to balance different conflicting public policy interests before deciding to enforce the EU competition law.[155]

Within ECN, cooperation between NCAs and the Commission is expected to occur on the basis of equality, respect and solidarity.[156] The equality between the two levels of competition law enforcement is, however, qualified in Regulation (EC) No. 1/2003. According to article 11(3) and (4) of the Regulation, NCAs are obliged to inform the Commission of proceedings that have been initiated in the Member States. Furthermore, according to the same article, NCAs are obliged to inform the Commission before adopting a decision on the infringement of articles 101 and 102 TFEU or before accepting commitments offered by the national undertakings concerned. Furthermore, NCAs are "relieved of their competence" to apply articles 101 and 102 TFEU, if the Commission initiates proceedings for the adoption of a decision.[157] Last but not least, NCAs cannot apply articles 101 and 102 TFEU to adopt a counter-decision to the one already adopted by the Commission on the

---

151 Scholtz and Purps, 2010, pp. 37–51.

152 Bundeskartellamt, 2009: Sektoruntersuchung Kapazitätssituation in den deutschen Gasfernleitungsnetzen … .

153 Conseil de la Concurrence, 2011.

154 Commission Notice on cooperation within the Network of Competition Authorities; Joint Statement of the Council and the Commission on the Functioning of the Network of Competition Authorities.

155 Weiss, 2009, pp. 45–70.

156 Joint Statement of the Council and the Commission, *supra* note 154.

157 Article 11(6) of Regulation 1/2003.

same matter.[158] In light of these provisions it may be concluded that the ECN provides a framework for the networked enforcement of competition law, while granting leadership to the Commission. However, as seen before, the establishment of networks aims at creating an institutionalized platform for information-sharing between national regulatory authorities, which ultimately should increase the expertise of the Commission. Hence, in order to allow ECN to fulfil its purpose, it has been observed that, despite the possible tensions that might occur between the NCAs and the Commission in specific enforcement cases, the latter has abstained from diminishing the powers of NCAs in order to preserve their incentive to cooperate.[159]

A similar network of energy regulators does not exist in the EU gas market for the enforcement of the TEP. However, in the institutional arrangements provided by gas sector regulation, NCAs and NRAs are brought together in several instances in a manner similar to that of a network of regulators. Article 10(1) of Regulation (EC) 713/2009 emphasizes that ACER must consult competition authorities where relevant in an open and transparent manner. Furthermore, NRAs and NCAs are expected to cooperate in the Board of Appeal of ACER and exchange their relevant experience in the energy sector.[160] In article 14(6) of Directive 2009/73/EC monitoring activities of the compliance of TSOs with their obligation are divided between regulatory and competition authorities, which are expected to cooperate with each other. According to article 41(4)(b) of Directive 2009/73/EC, the carrying out of investigations in the gas sector is submitted to cooperation between regulatory and competition authorities.

Notwithstanding the cooperation patterns between NCAs and NRAs under the rules of sector regulation, certain situations may lead to tensions between the two sets of authorities owing to overlapping competences. An example of this is the application of fines in cases of abusive behaviour on the market by NCAs and the application of fines for noncompliance with gas market legislation by NRAs.[161] Furthermore, additional tensions might occur owing to constraints placed upon the role of NRAs in the gas market by concurrent competences of NCAs. However, in such cases regulation gives priority to the competences of

158 Article 16(2) of Regulation (EC) 1/2003.
159 See also Salerno, 2008.
160 Article 18(1) of Regulation (EC) 713/2009.
161 Article 41(3)(a) of Directive 2009/73/EC.

NCAs,[162] as a reflection of the supremacy of competition law and afferent institutions above regulatory tools and regulators. This is especially visible in article 40 of Directive 2009/73/EC, which requires NRAs in the fulfilment of their tasks to closely consult with the relevant NCAs and avoid bringing prejudices to the latter's competencies in the gas market.

6.4.2.2 The European Network of Transmission System Operators for Gas

Under the Third Energy Package, networks of transmission system operators have been created to ensure an optimal management of the gas transmission network in the EU.[163] Article 4 of Regulation (EC) 715/2009 requires all TSOs to cooperate at EU level through ENTSOG to promote the functioning and completion of the internal market, cross-border trade in gas and the optimal gas network management. The working method of ENTSOG should ensure efficiency, transparency and the representative nature of gas networks and its tasks should be carried out in compliance with EU competition rules.[164] One of its main roles in accordance with recital 15 of Regulation (EU) 994/2010 is to contribute to the strengthening of the security of gas supply at EU level through their participation in regulating investments in cross-border infrastructure.

Article 17(2)(b) of Directive 2009/73/EC states that the independent transmission operator shall ensure the representation of the respective TSO within ENTSOG. Furthermore, article 12(1) of Regulation (EC) 715/2009 requires TSOs to establish regional cooperation within the ENTSOG to contribute to the realization of their tasks. Accordingly, the membership of ENTSOG gathers 39 TSO Members and two associate partners from 24 European countries.[165] Although the participation of stakeholders and consumer representatives is not emphasized in ENTSOG's membership, article 10 of Regulation (EC) 715/2009 represents a proof of the inclusive nature of ENTSOG. The article provides for an extensive consultation process in an open and transparent manner initiated by ENTSOG and including all relevant market participants. These encompass organizations representing all stakeholders, NRAs, other national authorities, supply and production undertakings, network users including customers, DSOs, relevant industry associations, technical bodies and stakeholder platforms. In order to serve the aim of a stronger

[162] Recital 8 and article 11(1) of Regulation (EC) 713/2009; article 41(1)(j)(k)(p) of Directive 2009/73/EC.

[163] Recital 16 of Regulation (EC) 715/2009.

[164] Recital 16 of Regulation 715/2009.

[165] ENTSO for gas Members, http://www.entsog.eu/aboutus_members.html

EU-wide coordination, ENTSOG is expected to identify in the consultation process all the views and proposals of the relevant parties during the decision-making process. The effectiveness of the coordination exchange between stakeholders through the framework provided by ENTSOG lies in the obligation of ENTSOG to indicate how the observations received during the consultation have been taken into consideration before adopting the annual work programme and the network codes.

It should, however, be observed that, unlike the theoretical model of collaborative governance briefly described at the beginning of this subsection, the ENTSOG model does not presume the participation of flexible, engaged governmental institutions. Instead, the performance of its tasks requires the acceptance of the Commission. ENTSOG's tasks are detailed in articles 6 and 8 of Regulation (EC) 715/2009. According to article 8 of the Regulation, ENTSOG is entrusted with:

- the elaboration of network codes;
- the adoption of common network operation tools to ensure coordination of network operation in normal and emergency conditions;
- a nonbinding EU-wide 10-year network development plan;
- recommendations relating to the coordination of technical cooperation between EU and third-country TSOs;
- an annual work programme;
- an annual report and annual summer and winter supply outlooks.

According to article 6(1) of the Regulation, ENTSOG is expected to play a consultancy role in the establishment of network codes and the drafting of framework guidelines by the Agency. ENTSOG may also submit network codes[166] for the Commission, if so requested by it or propose amendments for the network codes.[167] However, none of its proposals has a binding effect, whereas it is the Commission that has the final say in the adoption of network codes and framework guidelines. In this sense, ENTSOG is very similar to a traditional EU network of regulators, which despite having been given wide-ranging tasks and membership, continues to be highly dependent on the Commission and enjoys few formal powers or resources.[168]

It further results from the text of article 6 of Regulation (EC) 715/2009 that the power of ENTSOG is limited to formulating recommendations.

---

166 Articles 6(6) and 8(1) of Regulation (EC) 715/2009.

167 Articles 6(8) and 7(1) of Regulation (EC) 715/2009.

168 Coen and Thatcher, 2006.

The network has no real decision-making powers. By way of contrast, collaborative governance settings require stakeholders to participate directly in the decision-making process, even if ultimate authority remains with the public authority – for example, the Commission in the case of ENTSO for gas. The main difference between network and collaborative governance is thus the fact that, while the former model entails only the consultation of ENTSO by the public authority, the latter implies a collective decision-making process, with a two-way communication and influence between agencies and stakeholders in a deliberative and multilateral process.[169]

Unlike in the case of ACER, where third-country participation is limited by the conclusion of prior agreements between the Commission and the interested third country, participation in ENTSOG does not seem to be confined to European TSOs. Articles 12 and 17 of Regulation (EC) 715/2009 indicate only that TSOs are required to collaborate and that the relevant ISO shall ensure their participation in ENTSOG. The condition of TSO participation in ENTSOG appears to be contained in articles 10 and 11 of the Regulation, which establish the rules for the certification of both EU and non-EU TSOs. In contrast to this, article 31 of Regulation (EC) 713/2009 stipulates that agreements between the Commission and third parties leading to participation in ACER may be concluded only if the interested third country already applies the EU's energy *acquis* and in particular the TEP.

Recital 17 of Regulation (EC) 715/2009 states that market participants should be offered the possibility to establish effective consultation with ENTSO in the frameworks of the European Association for the Streamlining of Energy Exchange (EASEE), national regulators or ACER. The new rules on unbundling and third-party access in the sector-specific Directive and Regulation make it highly unlikely that the major EU gas suppliers will participate in ACER in the near future. In contrast to this, EASEE counts among its members representatives of Russian and Norwegian gas undertakings. It would appear thus that, within the platform offered by EASEE, ENTSOG has a possibility to be exposed to the feedback of non-EU stakeholders.

169 Ansell and Gash, 2007, pp. 543–571.

6.5 POSSIBILITIES AND LIMITATIONS TO THE ADOPTION OF A EUROPEAN ENERGY REGULATOR

Pollitt argues that the necessity of independent sectoral regulatory agencies might be questionable to a certain extent in the case of energy markets.[170] This happens because in general regulatory agencies are legitimized by the need of immature markets to receive a special regulatory attention that cannot be provided by general competition authorities. He maintains, however, that this is not the case for EU energy markets where competition authorities have proved to be very active and the Commission has demonstrated a powerful forcing effect in bringing about structural change.[171]

Notwithstanding the ambiguous outcome of the Commission's new activism when enforcing competition law on energy markets in relation to the achievement of the gas market policy goals, regulation through competition is not enough for the EU gas market.[172] Instead a powerful regulator is required – among other things – to mitigate the effects of the gas policy of third-country suppliers on the market's harmonization. In this respect, it has been shown in the previous subsections that the adoption of ACER has questionable benefits and important limitations for the furthering of the gas market integration.

Hence, the present subsection does not question the need for a European energy regulator having a wide scope of tasks and decision-making powers, but the legal, doctrinal and institutional possibilities and limitations to the adoption of such a regulator. In this sense, having due consideration of the constraints in the current legal framework for delegation of executive powers in the EU, it looks at three models of delegation that might be employed in the EU gas market.

Hofmann and Morini identify several models of delegation of executive powers to European agencies.[173] Out of these, three might be employed in the EU gas market with interesting effects on the opportunity for adopting a European energy regulator. These are the subdelegation

---

[170] Pollitt, 2008b.

[171] *Ibid.*

[172] This book does not, however, consider the specific case of EU electricity markets, which to a certain extent might require a different balance of competition law and regulation for the advancement of the electricity single market project.

[173] Hofmann and Morini, 2012.

model, the delegation-for-harmonization model and the direct delegation by Member States model.

The subdelegation model derives from the economics of article 291(2) TFEU, which states that "[w]here uniform conditions for implementing legally binding Union acts are needed, those acts shall confer implementing powers on the Commission". Although the subdelegation model does not require the existence of an explicit Treaty basis for the adoption of an energy regulator,[174] it is probably the most restrictive delegation model that could be applied in the gas market, notwithstanding the recent accrual of Commission competences in the energy sector.[175] This model has in general led to the adoption of executive agencies under Council Regulation (EC) No. 58/2003, which gives the Commission the possibility to set up agencies through a decision and to delegate tasks of implementing EU programmes to such agencies. These agencies do not enjoy any discretionary powers and they are under the direct control of the Commission, despite having their own legal personality.[176]

The delegation-for-harmonization model draws on article 114(1) TFEU, which entrusts the European Parliament and the Council with the adoption of measures for the approximation of laws, regulations and administrative action in the Member States in view of advancing the internal market. This model has been already used for the establishment of the European Food Safety Authority (EFSA),[177] the European Supervisory Authorities (ESA)[178] and ACER.[179] Powers delegated to agencies through this model refer to the adoption of detailed technical rules (e.g. ACER), ensuring scientific advice for the Commission (e.g. EFSA) and quite uncommonly the drafting of regulatory and implementing standards (e.g. ESA).

The jurisprudence of the Court has been to uphold article 114 TFEU as a basis for delegating power to a European institution or body, in view of legislative harmonization across Member States.[180] In the *Product Safety*

---

174 *Meroni*, *supra* note 83.

175 Braun, 2011.

176 Articles 1 and 6(1) of Regulation (EC) No. 58/2003.

177 Regulation (EC) No. 178/2002.

178 See Regulation (EU) 1092/2010, Regulation (EU) 1093/2010, Regulation (EU) 1094/2010 and Regulation (EU) 1095/2010.

179 Other agencies based on article 114 TFEU are, for example, the Community Plant Variety Office and the European Aviation Safety Agency.

180 *Federal Republic of Germany v Council (Product Safety)*, Case C-359/92; *UK v Parliament and Council (Smoke Flavourings)*, Case C-66/04; *UK v Parliament and Council (ENISA)*, Case C-217/04.

decision,[181] the Court found that the Commission's powers to take decisions replacing those of national authorities, in order to ensure the compliance of national laws with Directive 92/59 on general product safety, were within the scope of article 114 TFEU, owing to the fact that the partially harmonizing provisions of the Directive could give rise to the necessity of temporary emergency intervention. In the *Smoke Flavourings* case, the Court held that the expression "measures for approximation" in article 114 TFEU conferred on the Commission discretion with regard to the "harmonization technique most appropriate for achieving the desired result, in particular in fields which are characterized by complex technical features".[182]

The Court, however, also established two conditions curtailing the discretion of the European legislature. The first concerned the fact that EU's legislature was obliged to determine in the basic act the essential elements of the harmonizing measure concerned and that the mechanisms for implementing those essential elements was designed in a way leading to harmonization in the sense of article 114 TFEU.[183] In this specific case law, however, the Court also referred to the already established delegation doctrine, by further circumscribing the delegation of power from the EU legislature to the Commission to detailed rules for making decisions at each stage of the authorization procedure, which determine and outline precisely the powers of the Commission as the body that has to take the final decision.[184]

Of most relevance for the adoption of a European energy regulator is the decision of the Court in *ENISA*,[185] which regarded the establishment of the European Network and Information Security Agency based on article 114 TFEU. Specifically, paragraph 44 of the decision states that nothing in the wording of article 114 TFEU excludes the delegation of powers to an EU body responsible for contributing to the implementation of a process of harmonization. There are two additional conditions to the discretion of conferring harmonizing measures on an EU body. These refer to the existence of a close link between the tasks conferred on such a body and the subject matter of the approximating acts[186] and the appropriateness of creating the body in order to prevent "the emergence

---

181 *Product Safety*, *supra* note 180.
182 *Smoke Flavourings*, *supra* note 180, para. 45.
183 *Ibid.*, paras 48, 49.
184 *Smoke Flavourings*, *supra* note 180, para. 49.
185 *ENISA*, *supra* note 180.
186 *ENISA*, *supra* note 180, paras 45, 47.

of disparities likely to create obstacles to the smooth functioning of the internal market in the area".[187]

It has been shown in the previous subsections that the level of gas market harmonization achieved through Directive 2009/73/EC is only partial, owing to the coexistence of three unbundling models, the persistence of derogations and exemptions with regard to emergent and isolated gas markets and the maintenance of Member States' discretion with regard to a variety of gas market aspects. Hence, while looking at the decision in *Product Safety*,[188] it could be assumed that the objective of the Directive, which is the achievement of integrated EU gas markets, could be fulfilled by mandating ACER to take autonomous decisions, which could overrule national decisions.

However, it has been maintained that the Treaty does not offer a basis for such a delegation of power.[189] Indeed, in the EU system of delegating power, no explicit reference is made to agencies, as bodies on which such power might be conferred. In this context, it is important to observe that the *Product Safety* case concerned the delegation of harmonizing powers to the Commission and not to a European agency. Regarding the decision in *ENISA*, this did not concern the delegation of powers to an agency to take externally binding decisions, but resumed to address only limited delegations of preparatory or coordinating roles of agencies.[190] In a more recent case, however, the Courts accepted that an agency may be entrusted with the power of taking externally binding decisions under EU law, which entails the exercise of administrative discretion.[191] This goes in line with article 263(1) and (5) TFEU, which acknowledges the fact that agencies of the EU might adopt acts intended to produce legal effects vis-à-vis third-parties.

The ESAs in the banking, securities and insurance sectors[192] represent one of the most advanced examples of agency rule-making stemming from this second model of delegation and it would be interesting to look at as a possible model for a future Energy Regulator. The Authorities are

---

187 *Ibid.*, paras 62, 64.

188 *Federal Republic of Germany v Council (Product Safety)*, Case C-359/92.

189 Hofer, 2011.

190 Hofmann and Morini, 2012.

191 *Ralf Schräder v Community Plant Variety Office* (CPVO), Case C-38/09. See also *Schräder v OCVV – Hansson (Lemon Symphony)*, joined cases T-133/08, T-134/08, T-177/08, T-242/09.

192 European Banking Authority, European Insurance and Occupational Pensions Authority and European Securities and Markets Authority.

entitled to adopt draft regulatory technical standards[193] and implement technical standards,[194] which must be adopted by the Commission in conformity with articles 290 and 291 TFEU. The discretion of the Commission in adopting the regulatory and implementing standards drafted by ESAs is clearly qualified in the text of the Regulations. According to article 15(1) of each of the Regulations, the Commission must explain to the concerned ESA its decision to not endorse or to endorse with amendments a draft implementing technical standard. Furthermore, the article provides that the Commission may not bring any change to the draft without prior consultation with the concerned ESA. The same conditions are set out in article 10 of the Regulation, regarding the adoption of draft regulatory technical standards. In regards to the draft regulatory technical standards, the preambles of the regulations circumscribe their amendment by the Commission to restricted and extraordinary circumstances, such as: if they are incompatible with Union law; did not respect the principle of proportionality; or ran counter to the fundamental principles of the internal market for financial services as reflected in the EU's financial services *acquis*. In the rest of the cases priority must be given to the choice of the Authority owing to its expertise and close involvement with the daily functioning of financial markets.

Article 18 of the Regulations gives ESAs the power to coordinate the actions undertaken by the relevant national competent supervisory authorities in emergency situations. The preamble of the Regulations affirms the ability of ESAs to temporarily ban certain financial activities. In cases where such a ban entails a cross-sectoral impact, coordination between the three authorities is demanded. However, the Commission is not included in this scheme.

In cases regarding a breach of Union law, ESAs may formulate legally binding decisions directly applicable to financial market participants.[195] Such a possibility must occur, however, as a last resort, where a competent authority has failed to comply with a formal opinion of ESAs requesting it to act in line with EU law. Furthermore, according to article 19(3) of the Regulations, where the competent authorities concerned fail to reach an agreement in the case of a cross-border conflict situation, ESAs are entitled to adopt legally binding decisions, which shall ensure

193 Article 10 of Regulations (EU) No. 1094/2010, No. 1095/2010 and No. 1093/2010.

194 Article 15 of Regulations (EU) No. 1093/2010, No. 1094/2010 and No. 1095/2010.

195 Article 17(6) of Regulations (EU) 1093/2010, 1094/2010 and 1095/2010.

compliance of the concerned national authorities with EU law. Such decision-making powers have been perceived as similar to the discretion enjoyed by the Commission in areas of competition law and state aid.[196]

The third model of delegation that could be applicable in the EU gas market concerns the direct delegation by Member States. In this sense, the granting of implementation powers to agencies does not amount to a delegation of power from an EU institution to a regulatory authority, but rather to the "Europeanization" of powers traditionally belonging to national administrators.[197] As seen above, the principal and agent roles in the EU gas market entail a high degree of complexity, which includes the fact that European authorities – particularly the Commission and ACER – act as both principals and agents of national governments. For this specific model of delegation, of interest is the direct conferral of powers from the Member States on ACER or another potential European energy regulator. Article 291(1) TFEU maintains that implementing powers for EU policies generally reside with EU Member States. Such a provision has constituted the grounds for the initial delegation of powers from the original Member States to the European Commission, as a supranational actor[198] and *a priori* makes possible the delegation of powers from the Member States to European agencies.

However, such a model of delegation, despite its obvious advantages, might be at odds with the EU gas market where the lack of a real European energy regulator can be explained to a certain extent by "the determination of the Member States to resist the emergence of a single European energy regulatory authority".[199] The new regulatory framework certainly imposes higher standards of harmonization of the national levels of the European gas market than any of its predecessors. One of its important focuses is the increased homogeneity of NRAs. This might lead in the future to more similarities among Member States' preferences for gas regulatory models, which could result in enhanced opportunities for adopting a European energy regulator through direct delegation of powers from the Member States. However, at this moment and in the absence of any evidence confirming a convergence of Member States gas market models,[200] such a conclusion appears to be rather speculative.

---

[196] Hofmann and Morini, 2012.

[197] Geradin and Petit, 2004.

[198] Pollack, 2003.

[199] Cameron, 2005b.

[200] For example, only seven Member States have adopted ownership unbundling of the transmission network.

6.6 SOME CONCLUSIONS

The attenuation of market forces' influences in the gas sector has facilitated the appearance of different institutional arrangements across Member States, as a consequence of their specific regulatory cultures. At the level of the EU a coordination and cooperation of national and European gas administrative bodies has been attempted in the form of a new Agency and the parallel development and advancement of the Forum process and governance through networks. Such patterns of cooperation entail hierarchical and negotiation-based frameworks and have been developed in either a formal or an informal way.

The creation and adoption of ACER as opposed to a more powerful Energy Regulator has been justified by the Commission through the persistence of a strict approach at EU level to the doctrine of delegation and the absence of a Treaty basis for the delegation of extensive decision-making powers to such a regulator.[201] However, the recent adoption of ESAs in the banking, securities, pension funds and insurance sectors demonstrates that the delegation doctrine may be curtailed by turning to article 114 TFEU, whose interpretation legitimizes the creation of agencies for achieving harmonization in the internal market.[202] The example of ESAs heralds the emergence of real regulatory agencies at the EU level with extended decision-making and rule-making attributions. The path of article 114 TFEU has been followed also in the adoption of ACER. Nevertheless, the outcome in the latter case has been the creation of an Agency entrusted mainly with technical harmonization, the requirement to create a platform for the cooperation of energy regulators and highly limited decision-making powers. Notwithstanding the many structural differences between the sectors regulated by ACER and ESAs, a relevant institutional difference that could explain the success of one and the arguable failure of the other is the role played by the Commission in their administrative arrangements. It appears at a first comparative reading of their foundational Regulations that the Commission preserves a more comprehensive control in the case of ACER. This is reflected, for example, in the Commission's development of the practice of adopting and amending guidelines for the harmonization of EU gas market rules across Member States.

Network governance is expected to provide a platform for the equal and open participation of gas market stakeholders, EU and Member

201 Hofer, 2011.
202 *ENISA*, *supra* note 180.

States institutions and other gas administrative bodies. The creation of the European Network of Transmission System Operators for gas represents to a certain extent an advancement from the traditional model of network governance in the EU, where the participation of stakeholders is limited, towards collaborative governance, which is based on an enhanced participation in joint action of industry and nonindustry stakeholders, including consumers and various governments.[203] However, except for its inclusionary nature, ENTSO does not bring significant modifications to the EU network governance model. Its decision-making powers are limited, despite its extensive tasks, and the success of implementing the agreements among its participants depends highly on the Commission. The Commission, although not a member of ENTSO for gas, plays a key role in the achievement of ENTSO's tasks, while it continues to retain a monopoly on rendering legally binding the recommendations, guidelines and network codes emanating from the network of regulators. The Commission also retains a rather hierarchical approach to ACER, despite the Agency's independence and autonomy from the European institution.

Both ACER and ENTSOG are children of multiple parents,[204] as a result of receiving delegated powers and tasks from both the EU and the Member States. However, in the fulfilment of their role, which is essentially covering the regulatory gap between the development of the EU gas policy at a supranational level and its implementation at Member State level, the Commission appears to exert a much closer and more intense control than the Member States. This is due to two complementary factors: the Commission's attempt to preserve a quasi-monopoly on the development of an EU gas policy and the reluctance of the Member States to delegate more powers towards the supranational level.

It has been shown that the Commission represents the engine of institutional arrangements in the gas market, where its control levels vary, but where its presence remains constant. The Commission is also the initiator of the internal gas market project. However, its preference for a strict interpretation of the delegation doctrine in the discussions leading to the adoption of ACER could suggest that the Commission intends to take advantage of the new EU energy competences in the Lisbon Treaty and enhance its own oversight of the advancement of the EU gas market in accordance with EU-elaborated rules. This ultimately should lead to a harmonious implementation of the EU gas policy across Member States. Nevertheless, having regard to the limited resources and expertise of the

---

203 Koch, 2009, pp. 543–571.
204 Coen and Thatcher, 2006.

Commission, it is questionable how and when such a harmonious implementation will occur.

At the moment, the failure to adopt a powerful Energy Regulator appears to be shared between "the determination of the Member States to resist the emergence of a single European energy regulatory authority"[205] and the Commission's ambition to maintain a quasi-monopolistic approach to the formulation of the directions taken by the EU energy policy. Since an Energy Regulator could potentially foster a more appropriate venue for achieving enhanced gas market integration than the recently adopted Agency, it appears that not only the practices of the Member States, but also those of the Commission, have contributed to the hindering of the ultimate gas market goal.

[205] Cameron, 2005b.

7. Conclusion

It has been shown in the above how the development of the European legislation addressing the single EU gas market is influenced by the constantly evolving geopolitical environment, which serves as its background. Here the regulatory preferences of third-country suppliers, Member States' different levels of gas import dependencies and attachment to traditional gas market relationships, and the rigidity and scarcity of the gas pipeline grids interact with a weak energy regulator, with questionable abilities to foster market interconnection and integration, and a European Commission with redefined gas market competences, whose ambition for retaining an overall control of the market appears to have contributed to the undermining of its internal gas market project.

The coexistence of law and politics in the EU gas market acts as the main cause and facilitating factor for gas market regulation choices and influences the adoption and evolution of the afferent gas legislative framework and institutional arrangements. This is primarily reflected in the current EU gas market structure and EU gas market design.

7.1 EVALUATION OF THE CURRENT GAS MARKET STRUCTURE

The EU gas market structure continues to entail both a national and a European dimension. It has been shown that the weak interconnection among Member States' gas markets and the lack of an energy external representation of the EU render the bidimensional European gas market problematic. Hence, in time it is advisable that the EU reinforces its ability to speak with one voice with regard to external energy matters. The effectiveness of the new basis for common action in the field of energy provided by article 17(1) TFEU is still to be tested, particularly owing to the provisions of article 194(2) TFEU. These continue to allow Member States to preserve their right to conduct bilateral energy relationships with non-EU suppliers as they see fit.[1] The coexistence of a

[1] Braun, 2011.

two-level political and institutional framework in the EU gas market does not constitute *per se* a barrier to the achievement of market integration. However, the sometimes divergent interests at EU and Member State levels impact upon the uniform implementation of the EU legislation across the Member States and the existing market arrangements.

EU energy competences are shared with the Member States and are submitted to the principle of subsidiarity.[2] However, the EU has embarked upon a more profound harmonization of Member States' natural gas regimes, which contradicts to a certain extent the general rule that the Union legal system should not interfere in the institutional and administrative organizations of the Member States.[3] Even before the adoption of the TEP, the Commission affirmed the necessity of bringing Member States and their energy policies together in a coherent approach, in order to enhance the progress of the internal energy market and ensure a further EU external representation.[4] However, the response across Member States has not been uniform.[5] This situation has attracted a series of infringement proceedings carried out until the present, initiated by the Commission against several Member States that did not respect their Treaty obligations. It has been observed that Member States continue to enjoy high levels of discretion upon unbundling models, the insurance of security of supply, public service obligations and consumer protection. This should, however, be linked not only to their resistance to a convergence of their regulatory regimes towards an EU model, but also to the weakness of such a model in terms of gas policy definition and regulatory constraints. An example of this is provided by the definition of "solidarity" in primary and secondary legislation and the persistence of importance exemptions which, to a certain extent, defeat the purposes of solidarity action. Thus, the solidarity spirit between the Member States, recommended when aiming at the establishment and functioning of the internal market,[6] has visible limitations owing to its rather vague formulation and the lack of accompanying sanctions, should the Member States decide to travel in different directions.

Taking a step further from the Treaty provisions, the Regulation (EU) 994/2010 underlines the preventive aspect of the required solidarity and

---

2 Article 194 TFEU.
3 Nizzo, 1999.
4 Green Paper /*COM/ 2006/ 105 final*/.
5 European Commission, 2011b.
6 Article 194 TFEU.

coordination among Member States in response to supply crises.[7] Certainly a positive reaction of the Member States to the demand to adopt a unified stance in the face of security of supply threats would benefit the formation of an integrated gas market. However, such a reaction may be severely impaired by current energy exemptions and derogations from improved competition conditions,[8] allowed for in national markets, which necessitate a boost in investment and connection to the rest of the EU. Other gaps in the current legislative framework, such as those favouring bad balancing rules, contribute to the maintenance of poorly interconnected national gas markets.

Another element of the gas market structure is provided by the current position of European gas incumbents. It has been shown during the Energy Sector Inquiry that European gas incumbents continue to remain strong owing to the backup they receive from their governments and recent market strategies focused on cross-sector mergers and acquisitions allowing them to strengthen their position in the gas market.[9] Evidence from the industry demonstrates that some of the Member States have reacted to the European energy liberalization agenda by creating or reinforcing national champions. Gas incumbents in their turn have played an important role in influencing their governments regarding the application of the European liberalization agenda.

Currently, despite the significant bargaining power and position on their respective national markets of traditional gas incumbents, new unbundling measures under the TEP and the strengthened enforcement of competition law by both the Commission and by NCAs could be regarded as more powerful deterrents than their predecessors against the proliferation of gas market abuse by traditional incumbents. However, the combined application of such legal tools has not necessarily led to an increase in third-party access to the gas market. On the contrary, the strengthening of competition law enforcement has put a strain on investment in the new infrastructure, which might potentially affect third-party access, security of gas supply and ultimately the end consumers. Furthermore, both the competitive and regulatory measures have been contested by the industry and academia as representing an intrusive approach to national property rights and a misuse of competition tools to enhance gas market restructuring.

---

7 Regulation (EU) No. 994/2010.

8 Articles 36 and 49 of Directive 2009/73/EC.

9 Haar and Marinescu, 2011.

Third-country gas producers and suppliers continue to exert an indirect pressure upon the evolution of EU gas policy, while attempting to preserve their position in the EU gas market. Such pressure entails both political and economic components and takes place either through informal venues hinting at the response of national governments to EU proposals[10] or through the formalized context of long-term contracts with EU gas undertakings.[11] Their leverage on the internal market process is maintained by the specificities of natural gas – which entail a highly regional and rigid trade and lack of flexibility of network infrastructure – and is influenced by different national gas policy preferences and gas import dependencies. Moreover, dominant third-country gas producers continue to manifest no actual rivalry between each other, while concentrating on launching parallel infrastructure expansion and diversification plans to those originated in the EU. These sometimes take the form of bilateral agreements that favour the maintenance of the functional segmentation of the EU gas market.

7.2 EVALUATION OF THE GAS MARKET DESIGN

The pursuit of EU gas market goals lacks coherence both at the level of policy formulation and with regard to the enforcement of the available tools. The ambiguous formulation of gas policy goals at an EU level correlated with the fuzzy pursuit of EU gas market integration has allowed and contributed to the preservation of divergent national gas market designs.

7.2.1 The Formulation of the Policy at EU Level

Without being mutually exclusive, the three affirmed gas market goals of competitiveness, security of supply and sustainability determine a specific direction for the EU gas policy. The clashes between the three main policy goals – both in pairs and, even more complicated, all three together – are mostly the outcome of a lack of prioritization among them in the EU energy policy.[12] Hence, as seen above, it has often been the role of sector regulation to determine which goal will prevail in situations

[10] For example, this has been visible in the reaction of some EU governments regarding the introduction of the so-called "Gazprom clause" in the Third Energy Package. See also Grätz, 2009, pp. 66–80.

[11] *Gazprom*, Case COMP/39.816.

[12] Haase and Bressers, 2008, pp. 15–17.

where the pursuit of one contradicts the achievement of another. Furthermore, when assessing the coherence of the European approach to the current gas market design it has to be remembered that energy goals pursued on the market co-exist with the goal of market integration.

For the purpose of this book, gas market integration has been defined as the harmonized enforcement of gas regulation in the Member States, reflected in the convergence of policy objectives among regulatory authorities and cooperation among relevant stakeholders, in view of ensuring the territorial interconnection of European gas markets. In the specific case of natural gas, the book maintains that the accomplishment of a fully liberalized market has not necessarily led to market integration. This has happened owing to the fact that the asymmetrical implementation of the sector specific legislation has created different market conditions across EU Member States, affecting both energy consumers and companies.[13] Furthermore it has also led to the persistence of regulatory gaps at European level and the attachment to sovereignty over national network assets and energy policy at the level of Member States. As a result of this, integration misfits continue to exist. Despite the claims of the EU legislator that the Third Energy Package will foster deeper EU gas market integration, recital 19 of Directive 2009/73/EC limits the potential effects of supranational harmonization by allowing different types of natural gas market organizations to coexist. This explains the legitimacy of the persistence of Member States' discretion with regard to the implementation of several key issues of the liberalization directives.[14] There is a trade-off between Member States' right to be solely responsible for their energy mixes and the harmonious coexistence of the three policy goals of competitiveness, security of gas supply and sustainability. Certainly, markets of different maturity, nature and risk profile might require different governance structures, rendering obsolete the quest for a regulatory approach tailored to fit all.[15] However, the large margin of discretion granted by the EU legislator to the Member States with respect to the implementation of the gas market regulatory reform (in terms of choice of competent authorities, unbundling regime, network access exemptions, etc.) has allowed them to grant preference to some

---

[13] Commission Staff Working Paper First benchmarking report on the implementation of the internal electricity and gas market, /*SEC/2001/1957*/.

[14] Choice of unbundling model, definition of public service obligations, choice between regulated and negotiated access to storage, exemptions from infrastructure standards, choice of regulatory supervision in each Member State, etc.

[15] van der Linde *et al.*, 2006, p. 11.

market goals above others and has ultimately led to the coexistence of divergent market designs among them.

7.2.2 Competitive and Regulatory Choices

As observed above, sector-specific rules and competition principles have generally had a complementary application on the EU gas market. In the past, their overlapping has been contained by the fact that competition law, whose main role is to correct market failures, has generally addressed the behaviour of gas undertakings through *ex-post* control, while regulation has served as an *ex-ante* incentive for good behaviour. In the aftermath of the Energy Sector Regulation, the risk of an overlapped application of competitive and regulatory tools has increased owing to the enforcement of competition law in an increasingly regulatory manner. This has not necessarily produced positive effects for the achievement of the gas market goals.

In several instances, the application of the same legal instrument issued from either competition law or market regulation has led to the satisfaction of one of the market goals, while causing detrimental effects on another. In the case of regulation enforcement, this has occurred as a result of the non-negligible trade-off between Member States' right to maintain considerable discretion with regard to the implementation of several key issues of the liberalization directives[16] and the harmonious coexistence of the three policy goals. In the case of competition law, inconsistencies in the achievement of gas policy goals are no longer restricted to a dilemma of principle between achieving economic efficiency in the market, as opposed to public welfare. Instead, they derive from the legal uncertainty surrounding the commitment practice of the Commission and the limited scope for the private enforcement of commitment decisions.

It has been observed that some of the traditional constraints of applying competition tools in the European gas markets are gradually disappearing. One of the most important critiques of the past regarding the use of competition in energy markets has drawn upon the assumption that the effect of competition tools is rather market failure corrective than market failure preventive. As a result, competition law has been thought to be poorly equipped to liberalize the formerly monopolistic energy

[16] Choice of unbundling model, definition of public service obligations, choice between regulated and negotiated access to storage, exemptions from infrastructure standards, choice of regulatory supervision in each Member State, heterogeneous pursuit of the competitiveness goal among Member States, etc.

markets. From 2007, however, the European gas markets have been exposed to full liberalization. In such markets competition has been correctly used not to open the markets but to restore and maintain market competition. Thus, the principally *ex-post* enforcement of competition rules is finally serving its market corrective purpose.

It has been shown above that one of the particularities of applying competition law in the gas market is to restore competition, not punish anticompetitive behaviour. This has been reflected in the Commission's preference for commitment decisions. As argued before, such a preference has somewhat curtailed the effective application of competition tools in the gas market. Two important limitations have been identified. At the level of the internal gas market the Commission has a limited scope for using competition law so as to alter the gas market structure. This has been and should continue to be achieved through gas market regulation. At an international level, it is questionable to what extent far-reaching, infrastructure-divesting commitments could be achieved in alleged competition-distortion cases involving non-EU gas incumbents. The fining practice of the Commission is almost nonexistent in the EU gas market. In contrast to this, regulatory developments regarding third-country TSO certification could be applicable in such cases to prevent the market failures.

A further advantage of using sector regulation to achieve the EU gas market goals is its ability to foster their pursuit in a balanced manner. Regulation takes into consideration market goals deriving from public welfare and envisages the gradual correction of market imperfections in accordance with its balancing role for the satisfaction of gas market goals.[17] It has been shown that regulation fares better in satisfying the public interest in the gas market in relation to the security of supply and consumer empowerment. This can be observed in the case of third-party access exemptions generated by new infrastructure where regulation plays a mediation role between short-term gas market competitiveness and the long-term security of gas supply and consumer satisfaction.

The biggest failure of sector regulation remains, however, its limited scope to foster the integration of the Member State gas markets. The inability of sector regulation to ensure a harmonized regulatory approach in all Member States should not *a priori* constitute a problem.[18]

---

17 Dreyer, Dominique, 2010.

18 For example, Majone questions the ability of supranational harmonization to ensure welfare-enhancement in an environment where citizens' preferences, economic development and social organization differ significantly across jurisdictions. See Majone, 2005, p. 161.

However, the asymmetrical implementation of the sector specific legislation has created different market conditions across EU Member States, affecting both energy consumers and companies.[19] Differences between Member States' gas regulatory regimes may lead to the maintenance of barriers to trade. This could further result in the preservation of the market segmentation. As an example of this, emergent Member States creating the so-called gas islands are still allowed extensive derogations from gas market unbundling, third-party access and competition rules in the TEP. Such measures do not achieve their purpose of gradually integrating the gas islands with the rest of the EU gas market. Instead their effect is the maintenance and encouragement of gas market segmentation, poor interconnectivity, lower security of supply, lower consumer protection and decreased competitiveness. Added to this, in these emergent markets, competition in the gas sector is reduced or absent. This decreases the scope of application of competition in such markets, despite the manifested support that the General Court and the CJEU have shown for the Commission forging the openness of markets previously not opened to competition.

A costs-and-benefits type of analysis of the concurrent application of competition law and gas legislation in view of achieving the gas market goals is not entirely able to offer an accurate solution. As seen above, competition law suffers from a series of internal limitations that do not allow it to achieve an integrated gas market alone. At the same time, the current form of the sector-specific regulation allows for a significant margin of interpretation, specifically with respect to the enforceability of regulatory obligations.[20]

According to the findings of this book a further reinterpretation of sector regulation in the EU gas market is necessary in order to foster gas market integration. The coexistence of significant diversity among EU Member States with efforts towards market integration might not necessarily disturb the primary goal of competition, which is the attainment of competitive gas markets. As Glachant and Lévêque have observed, there are different acceptable ways to accomplish a competitive energy market.[21] However, the rest of the goals of the gas market appear to require deeper cross-border integration. This could be achieved through an enhanced cooperative regulatory oversight among various accountability

[19] Commission Staff Working Paper First benchmarking report on the implementation of the internal electricity and gas market, /*SEC/2001/1957*/.

[20] For example, unbundling, transparency, non-discrimination, consumer protection, public service obligations, etc.

[21] Glachant and Lévêque, 2006, p. 4.

instances. The process is clearly on the right track, as demonstrated by the enhanced transfer in TEP of decision-making powers in the gas markets from national governmental structures to independent national regulatory authorities. The new institutional settings occur in a multilevel system in which market players have become accountable to a complex arrangement of European and national agencies and institutions. The implementation of the rules governing the current EU gas market relies on specific institutional arrangements comprising mechanisms of delegation, distribution and sharing of executive power between European energy regulators, national regulatory authorities, national competition authorities, Member States' governmental services and the European Commission. The integration of these administrative bodies occurs both in a hierarchical, formalized framework and through horizontal patterns of cooperation. Their setup is influenced by the combined application of competition law and specific sector regulation to achieve the EU gas market goals, as well as by the concurrent competences of the competition and regulatory authorities over the same aspect of EU gas market regulation.

It has been shown that the Commission represents the engine of institutional arrangements in the gas market, where its control levels vary, but where its presence remains a constant. In relation to the new activism of the Commission on the EU gas market it has been argued – while looking at the example of the Nabucco pipeline project and the role played by the Commission in it – that the traditional role of the Member States in supporting their energy companies has not disappeared but instead has been absorbed by the European Commission.[22] The Commission is also the initiator of the internal gas market project. However, its preference for a strict interpretation of the delegation doctrine in the discussions leading to the adoption of ACER could suggest that the Commission intends to take advantage of the new EU energy competences in the Lisbon Treaty and enhance its own oversight of the advancement of the EU gas market in accordance with EU elaborated rules. This is also confirmed by the Commission's recent practice to use its far-reaching, formal powers under competition law in order to achieve results in the restructuring of the European gas sector.

Resulting from the three delegation models explored in the sixth chapter, article 114 TFEU provides enough grounds for the adoption of a powerful EU Agency, with regulatory and decision-making powers. This has been demonstrated while looking at the example of ESAs. It should

---

[22] Talus, 2010.

nevertheless be underlined that an important difference between ESAs and ACER lies in the role played by the Commission in their respective markets and in relation to the fulfilment of their tasks and roles. In this sense, the Commission's position on the gas market is much more pronounced and the dependency of ACER on the Commission's action in order to achieve its tasks is much more developed. As shown above, it appears that the Commission has developed an ambition to maintain a quasi-monopolistic approach to the formulation of the directions taken by the EU energy policy. This has led to a hybrid form of governance of the EU gas market, where regulatory cooperation between energy regulators and market stakeholders occurs in the shadow of hierarchy.[23] Hence, the ultimate goal of achieving a harmonious implementation of the EU gas policy across Member States in order to foster deeper EU gas market integration might be undermined both by the Member States and the main herald of gas market liberalization – the Commission.

7.3 FINAL REMARKS

"All good things come in threes", or so the saying goes. The adoption of the Third Energy Package in 2009 has been heralded as a breakthrough advancement of EU energy market regulation towards closing the regulatory gap, making the liberalized energy market fully effective and creating a single European market in gas and electricity.[24] Notwithstanding the claims and promises of the European legislator, the findings of this book indicate that the measures undertook in the latest gas market reform are not sufficient to attain their purposes. The failure of the package lies not only with the Member States, whose attachment for sovereignty over own energy resources and own energy regulatory regimes have been identified as the source of the regulatory gap between EU and national levels of implementation of the EU gas legislation and as a great obstacle to achieving an integrated gas market. Nor can it be justified as mainly the outcome of the residual resistance of traditional gas market incumbents to gas market liberalization, translated into new

---

23 Eberlein, 2005.

24 European Commission, 2011: Questions and Answers on the third legislative package for an internal EU gas and electricity market; SPEECH/11/144 Günther Oettinger EU Commissioner for Energy: The Importance of ACER for the Internal Energy Market Speech of Commissioner Oettinger at the Opening ceremony of the Agency for the Cooperation of Energy Regulators (ACER) Ljubljana, 3 March 2011.

strategies for market dominance and market foreclosure. Such elements have certainly played a key role in the current structure and design of the EU gas market. However they are no longer the principal weak shackles of the chain.

Instead, as seen above, a significant amount of responsibility for the questionable results of TEP could be attributed to the ambiguous and nonhierarchical formulation of gas policy goals at EU level, combined with the different pursuit of such goals through the available legal tools. The recent strengthening of competition law enforcement on the EU gas market has been welcomed by many as an important step in the gradual replacement of sector-specific regulation.[25] However, this does not concur with the findings of this book. Certainly, a more effective enforcement of competition policy is welcomed in order to allow competition to serve its supporting role for sector regulation. Nevertheless, gas sector regulation should continue to be applied and further developed in the gas market, owing to the importance of its role in balancing the pursuit of the gas market goals and the structural and functional specificities of the natural gas, which render its market rather unique in comparison with other EU utilities' markets.

---

[25] United Nations Economic Commission for Europe Committee on Sustainable Energy Working Party on Gas Study: The Impact of Liberalization of Natural Gas Markets in the UNECE region – Efficiency and Security, 17 March 2012, http://www.unece.org/fileadmin/DAM/energy/se/pdfs/wpgas/pub/ImpactLibNGM_UNECE_EffSec.pdf

Bibliography

Aalto, Pami, 2009: Policy Platforms for regionalizing EU–Russia energy relations. Expert article 359 Baltic Rim Economies, Bimonthly Review, March 2009.

About the WTO – A statement by the Director-General. http://www.wto.org/english/thewto_e/whatis_e/wto_dg_stat_e.htm.

Agency for the Cooperation of Energy Regulators, 2011: Work Programme of the Agency for the Cooperation of Energy Regulators.

Alesina, Alberto; Ardagna, Silvia; Nicoletti, Giuseppe; Schiantarelli, Fabio, 2005: Regulation and investment. Journal of the European Economic Association, Vol. 3, No. 4, pp. 791–825.

Almunía, Joaquin, 2011: Public enforcement and private damages actions in antitrust European Parliament, ECON Committee Brussels, 22 September 2011. SPEECH/11/598. http://europa.eu/rapid/pressReleasesAction.do?reference=SPEECH/11/598&format=HTML&aged=0&language =EN&guiLanguage=en.

Andersen, Mikael Skou, 2010: Europe's experience with carbon energy taxation. SAPIENS Vol. 3, No. 2. http://sapiens.revues.org/1072.

Andersen, Svein S., 2000: EU Energy Policy: Interest Interaction and Supranational Authority. Arena Working Papers, WP. 00/5. Arena Centre for European Studies.

Andersen, Svein S.; Sitter, Nick, 2007: Re-politicising regulation: Politics: Regulatory variation and fuzzy liberalisation in the Single European Energy Market. EUSA Conference, May 2007. http://www.unc.edu/euce/eusa2007/papers/sitter-n-03g.pdf.

Ansell, Chris; Gash, Alison, 2007: Collaborative governance in theory and practice. Journal of Public Administration Research and Theory, Vol. 18, No. 4, pp. 543–571.

Ascari, Sergio, 2011: An American model for the EU gas market? EUI Working Papers. Robert Schuman Centre for Advanced Studies. Florence School of Regulation.

Asmus, Ronald D.; Dimitrov, Konstantin; Forbrig, Joerg (eds, 2004): A New Euro-Atlantic Strategy for the Black Sea Region. The German Marshall Fund of the United States.

Auer, Josef, 2003: Liberalization of the German gas industry – under pressure to increase competition. http://www.dbresearch.de/PROD/DBR_INTERNET_EN-PROD/PROD0000000000066473.pdf.

Azaria, Danae, 2009: Energy transit under the Energy Charter Treaty and the General Agreement on Tariffs and Trade. Journal of Energy and Natural Resources Law. Vol. 27, No. 4, 2009, pp. 559–596.

Bachour, Kirsten; Conte, Giuseppe; Eberl, Peter; Martini, Clémentine; Paolicchi, Alessandro; Redondo, Philippe; van Haasteren, Augustijn; Wils, Geert, 2007: Gaz de France; Suez: Keeping energy markets in Belgium and France open and contestable through far-reaching remedies. Competition Policy Newsletter No 1, Spring.

Baldwin, Robert; Scott, Colin; Hood, Christopher, 1998: A Reader on Regulation, p. 3. Oxford University Press, Oxford.

Bartl, Marija, 2010: The affordability of energy: How much protection for the vulnerable consumers? Journal of Consumer Policy, Vol. 33, pp. 225–245.

Barysch, Katinka, 2008: Introduction, p. 1. In Baev, Pavel et al., Pipeline, Politics and Power: The Future of EU–Russia Energy Relations. Centre for European Reform, London.

Baur, Jürgen F.; Pritzsche, Kai; Pooschke, Sebastian, 2008: "Ownership Unbundling" von Energienetzen und der europäische Schutz des Eigentums. Deutsches Verwaltungsblatt, Vol. 8, pp. 483–492.

Beckman, Karel, 2012: The birth of a European gas infrastructure market. European Energy Review. http://www.europeanenergyreview.eu/site/pagina.php?email=gring@ulb.ac.be&id_mailing=313&toegang=158f3069a435b314a80bdcb024f8e422&id=3870.

Belgium – Energy mix fact sheet. http://ec.europa.eu/energy/energy_policy/doc/factsheets/mix/mix_be_en.pdf.

Belkin, Paul, 2008: The European Union's Energy Security Challenges. CRS Report for Congress. http://www.fas.org/sgp/crs/row/RL33636.pdf.

Bellantuono, Giuseppe, 2010: The Limits of Contract Law in the Regulatory State. European Review of Contract Law, Vol. 6, No. 2, pp. 115–142.

Belton, Catherine, 2009: Gazprom Neft grabs slice of Sibir. Financial Times 24 April, p. 21.

Belyi, Andrei V., 2009: A Russian perspective on the Energy Charter Treaty. http://www.realinstitutoelcano.org/wps/portal/rielcano_eng/Content?WCM_GLOBAL_CONTEXT=/elcano/elcano_in/zonas_in/europe/ari98-2009.

Belyi, Andrei V., 2011: Trends of Russia's gas sector regulation. Paper presented at the Fourth Annual Conference on Competition and Regulation in Network Industries.

Bergman, Lars; Doyle, Chris; Gual, Jordi; Hultkrantz, Lars; Neven, Damien; Röller, Lars-Hendrik; Waverman, Leonard, 1998: Europe's Network Industries: Conflicting Priorities, 22. Centre for Economic Policy Research, London.

Bevir, Mark, 2012: Democratic Governance. Princeton University Press, Princeton, NJ.

Blas, Javier, 2009: Market cautious ahead of OPEC meeting. Financial Times, 26 May, p. 4.

Blas, Javier; Hoyos, Carola; Dombey, Daniel, 2009: Chinese companies supply Iran with petrol. Financial Times, 23 September, p. 1.

Boltz, Walter, 2007: A regulatory framework for the electricity and gas industry. European Review of Energy Markets, No. 4. http://www.eeinstitute.org/european-review-of-energy-market/eei-review-issue-4.

Borisocheva, Ksenia, 2007: Analysis of the oil- and gas-pipeline links between EU and Russia: An account of intrinsic interests. CERE Publications.

Börzel, Tanja, 2010: European governance: Negotiation and competition in the shadow of hierarchy. Journal of Common Market Studies, Vol. 48, No. 2, pp. 191–219.

Bosmans, Willy, 2007: A contribution to the current EU debate on energy – Focus on the European gas market. European Review of Energy Markets, No. 4. http://www.eeinstitute.org/european-review-of-energy-market/eei-review-issue-4.

Boudet, Jennifer, 2009: Le contrôle des concentrations dans le secteur de l'énergie: un outil au service du processus de libéralisation? http://www.mayerbrown.com/public_docs/Thesis_Boudet_Jennifer.pdf.

Bourguignon, Didier, 2008: Quel Rapport Entre Libéralisation du Marché Gazier et Sécurité Energétique? Comment Bruxelles tente d'assurer son approvisionnement en gaz naturel sur fond de tensions avec Moscou. Revue du Marché commun et de l'Union européenne, No. 516, March.

Boussena, Sadek; Locatelli, Catherine, 2011: Gas market developments and their effect on relations between Russia and the EU. OPEC Energy Review, Vol. 35, No. 1, pp. 27–46.

Brau, Rinaldo; Doronzo, Raffaele; Fiorio, Carlo V.; Florio, Massimo, 2010: EU gas industry and consumers' prices. The Energy Journal, Vol. 31, No. 4, pp. 167–182.

Braun, Jan Frederik, 2011: EU energy policy under the Treaty of Lisbon Rules Between a new policy and business as usual. Working Paper No. 31/February. http://www.ceps.eu/system/files/book/2011/02/EPIN%20WP31%20Braun%20on%20EU%20Energy%20Policy%20under%20Lisbon.pdf.

Brinkley, Andrew; McIlven, Robert, 2010: Running on empty: Coming to terms with UK gas dependence. http://www.policyexchange.org.uk/images/publications/pdfs/Running_on_Empty_-_Jan__10.pdf.

Brownlie, Ian, 2008: Principles of Public International Law, 7th edn, pp. 309–311. Oxford University Press, Oxford.

Bryant, Chris; Tait, Nikki, 2009: Berlin set to cut industry power bills. Financial Times 26 May, p. 4.

Buigues, Pierre-André, 2006: Competition policy versus sector-specific regulation in network industries – the EU experience. UNCTAD's Seventh Session of the Intergovernmental Group of Experts on Competition Law and Policy, Geneva, 30 October to 2 November 2006. http://archive.unctad.org/sections/wcmu/docs/c2clp_ige7p14_en.pdf.

Bundeskartellamt, 2009a: Summary. http://www.bundeskartellamt.de/wEnglisch/download/pdf/10-001_Zusammenfassung_Sektorberichtes_Gas-E.pdf.

Bundeskartellamt, 2009b: Sektoruntersuchung Kapazitätssituation in den deutschen Gasfernleitungsnetzen. Abschulssbericht gemäaa § 32e ABS. 3 GWB. December. http://www.bundeskartellamt.de/wDeutsch/download/pdf/Stellungnahmen/0912_Abschlussbericht_SU_Gasfernleitungsnetze.pdf.

Burgis, Tom, 2009: Chinese seek huge stake in Nigeria oil. Financial Times, 29 September, pp. 1–3.

Busa, Lucrezia; Zaera Cuadrado, Elisa, 2008: Application of Article 21 of the Merger Regulation in the E.ON; Endesa case. Competition Policy Newsletter 2008-2.

Cameron, Peter D., 2005a: Legal Aspects of EU Energy Regulation: Implementing the New Directives of Electricity and Gas across Europe. Oxford University Press.

Cameron, Peter D., 2005b: Creating an internal market in energy: How can the tools be more effective. Policy Papers. Florence School of Regulation. Robert Schuman Centre for Advanced Studies. European University Institute.

Cameron, Peter D., 2007: Competition in Energy Markets Law and Regulation in the European Union, 2nd edn. Oxford University Press, Oxford.

Cameron, Peter D., 2011: The EU and energy security: A critical review of the legal issues. In Antoniadis, Antonis; Schütze, Robert; Spaventa, Eleanor, 2011: The European Union and Global Emergencies: A Law and Policy Analysis. Hart, Oxford, pp. 125–140.

CEER, 2011: Draft Vision for a European Gas Target Model: A CEER Public Consultation Paper. http://www.energy-regulators.eu/portal/page/portal/EER_HOME/EER_CONSULT/CLOSED%20PUBLIC%20

CONSULTATIONS/GAS/Gas_Target_Model/CD/C11-GWG-77-03%20GTM%20PC_5-July-2011.pdf.

CEER: Gas Regional Initiative (GRI). http://www.energy-regulators.eu/portal/page/portal/EER_HOME/EER_ACTIVITIES/EER_INITIATIVES/GRI.

Centre for European Reform, 2008: Pipelines, politics and power: The future of EU–Russia energy relations. The German Marshall Fund of the United States. http://www.cer.org.uk/pdf/rp_851.pdf.

Chauve, Philippe; Godfried, Martin; Kovács, Kristóf; Langus, Gregor; Nagy, Károly; Siebert, Stefan, 2009: The E.ON electricity cases: an antitrust decision with structural remedies. Competition Policy Newsletter 2009-1.

Chiti, Edoardo, 2009: The administrative implementation of European Union law: A taxonomy and its implications, pp. 19–22. In Hofmann, Herwig C. H.; Türk, Alexander H. (eds), Legal Challenges in EU Administrative Law Towards an Integrated Administration. Edward Elgar, Cheltenham.

Clancy, Joy S.; Skutsch, Margaret; Batchelor, Simon, 2003: The gender–energy–poverty nexus: Finding the energy to address gender concerns in development. http://www.sarpn.org/documents/d0000378/P342_Skutsch_Batchelor.pdf.

Clingendael International Energy Programme, 2008: The geopolitics of EU gas supply: The role of LNG in the EU gas market. Part II. A/DMV.

Clover, Charles, 2009: Russia asks EU to pay Ukraine gas bill. Financial Times, 23 May, p. 4.

CMS, 2008: Fortress Europe? The Risk of Rising Protectionism in Europe, p. 7. Oxford Analytica, Oxford.

Coen, David; Thatcher, Mark, 2006: Network governance and delegation: European networks of regulatory agencies. Paper presented at "Towards a European Administrative Space" Conference, 16–18 November, London.

Conclusions of the first meeting of the EU–Russia Gas Advisory Council Vienna, 17 October 2011. http://ec.europa.eu/energy/international/russia/dialogue/doc/20111017_gac_conclusions.pdf.

Conil-Lacoste, Jean-François, 2007: It is time for Germany and France to clearly join forces. European Review of Energy Markets, No. 4. http://www.eeinstitute.org/european-review-of-energy-market/eei-review-issue-4.

Conseil de la Concurrence, 2011: L'Auditorat auprès du Conseil de la concurrence a retiré sa demande de renvoi du projet d'acquisition d'International Power par GDF Suez. http://economie.fgov.be/fr/binaries/Communiqu%C3%A9%20de%20presse26012011_tcm326-116024.pdf.

Conway, Gerard, 2010: Conflicts of competence norms in EU law and the legal reasoning of the ECJ, p. 969. German Law Journal, Vol. 11, No. 9. pp. 966–1005.

Coppens, François; Vivet, David, 2004: Liberalization of network industries: Is Electricity an exception to the rule? NBB Working Paper No. 59.

Correljé, A. F., 1997: "Energiebeleid" in Aanpassing onder druk: Nederland en de gevolgen van de internationalisering. In Hout,W.; Sie Dhian Ho, M. (eds), Aanpassing onder druk: Nederland en de gevolgen van de internationalisering, pp. 165–178. Koninklijke Van Gorcum BV, Assen.

Cottier, Thomas; Matteotti-Berkutova, Sofya; Nartova, Olga, 2011: Third country relations in EU unbundling of natural gas markets: The "Gazprom Clause" of Directive 2009/73 and WTO law. In Buschle, Dirk; Hirsbrunner, Simon; Kaddous, Christine (eds), European Energy Law Dossier de droit européen No. 22. Helbing Lichtenbahn, Bâle.

Craig, Paul, 2009: Shared administration, disbursement of community funds and the regulatory state. In Hofmann, Herwig; Türk, Alexander (eds), Legal Challenges in EU Administrative Law: Towards an Integrated Administration. Edward Elgar, Cheltenham.

Craig, Paul; de Búrca, Gráinne, 2007: EU Law Text, Cases and Materials, 4th edn. Oxford University Press, Oxford.

Craig, Paul; de Búrca, Gráinne, 2011: EU Law: Text, Cases and Materials, 5th edn. Oxford University Press, Oxford.

Cronshaw, Ian; Marstrand, Jacob; Pirovska, Margarita; Simmons, Daniel; Wempe, Joost, 2008: Development of competitive gas trading in continental Europe: How to achieve workable competition in European gas markets? IEA Information Paper. International Energy Agency, Paris.

Crooks, Ed, 2009a: Back to petroleum. Financial Times, 8 July, p. 8.

Crooks, Ed, 2009b: Climate of opinion. The future of energy. Financial Times, 4 November.

Crooks, Ed, 2009c: EU energy curb strategy hailed. Financial Times, 24 April, p. 4.

Crooks, Ed, 2009d: Lack of support threatens future of UK nuclear power, EDF warns. Financial Times, 26 May, p. 13.

Cseres, Katalin J., 2008: What has competition done for consumers in liberalized markets? Competition Law Review, Vol. 4, No. 2, pp. 77–121.

Curtin, Deirdre, 2006: Delegation to EU non-majoritarian agencies and emerging practices of public accountability. In Geradin, Damien; Muñoz, Rodolphe; Petit, Nicolas, 2005: Regulation through Agencies in the EU: A New Paradigm of European Governance. Edward Elgar, Cheltenham.

Curtin, Deirdre, 2007: Holding (quasi-) autonomous EU administrative actors to public account. European Law Journal, Vol. 13, No. 4, pp. 523–541.

Ćwiek-Karpowicz, Jarosław; Gawlikowska-Fyk, Aleksandra, 2012: European Commission's antitrust inquiry of Gazprom: Consequences for the gas market in Europe. In Zaborowski, Marcin; Staniewska, Katarzyna (eds), Bulletin, No. 84 (417), 13 September. Polski Instytut Spraw Miedynarodowych.

Daultrey, Sally, 2009: Russia, energy and the rule of law. Centre for European Studies, Cambridge Judge Business School.

de Hauteclocque, Adrien, 2009: Legal uncertainty and competition policy in European deregulated electricity markets: the case of long-term exclusive supply contracts. World Competition: Law and Economics Review, March.

de Hauteclocque, Adrien; Marty, Frédéric; Pillot, Julien, 2011: The essential facilities doctrine in European competition policy: the case of the energy sector. In Glachant, Jean-Michel; Finon, Dominique; de Hauteclocque, Adrien (eds), Competition, Contracts and Electricity Markets: A New Perspective. Edward Elgar, Cheltenham, pp. 259–293.

Dehousse, Renaud, 2002: Misfits: EU law and the transformation of European governance. The Jean Monnet Center for International and Regional Economic Law and Justice.

de Jong, Hanneke, 2008: Regulatory mode decision-making in the European process of electricity market integration. Network Industries Quarterly, Vol. 10, No. 3, pp. 11–14.

del Guayo, Iñigo; Kühne, Gunther; Roggenkamp, Martha, 2010: Ownership unbundling and property rights in the EU Energy Sector. In McHarg, Aileen; Barton, Barry; Bradbrook, Adrian; Godden, Lee, 2010: Property and the Law in Energy and Natural Resources. Oxford University Press, Oxford.

Delvaux, Bram; Guimaraes-Purokoski, Alice, 2008: Vertical division of competences between the European Community and its Member States in the energy field – some remarks on the evolution of community energy law and policy, 10. In Delvaux, Bram; Hunt, Michael; Talus,

Kim, EU Energy Law and Policy Issues, 1st edn. ELRF Collection. Euroconfidentiel.

de Parfouru, Anatole Abaquesne, 2008: Research notes: International relations theories, Article 95 EC, Legal basis and the delimitation of competences between the EU and the Member States. Journal of Contemporary European Research, Vol. 4, No. 3, pp. 248–252.

Derdevet, Michel, 2008: Entre service public et concurrence européenne – Quinze ans de valse-hésitation énergetique française … et d'incompréhension mutuelle. Revue du Marché commun de l'Union européenne, No. 516, March.

Diathesopoulos, Michael D., 2010: From energy sector inquiry to recent antitrust decisions in European energy markets: Competition law as a means to implement sector regulation. http://papers.ssrn.com/sol3/papers.cfm?abstract_id=1639883.

di Jole, Bertone, 2008: The use of remedies under the EC merger regulation and the liberalization of the energy sector. In Ghidini, Gustavo; Libonati, Berardino; Marchetti, Piergaetano, Concorrenza e Mercato. Rassegna degli orientamenti dell'Autorità Garante della Concorrenza e del Mercato. Giuffrè Editore 16/2008.

Dinmore, Guy, 2009: Italian solar energy rush risks overheating. Financial Times, 19 May, p. 18.

Dorigoni, Susanna; Portatadino, Sergio, 2008: LNG development across Europe: Infrastructural and regulatory analysis. Energy Policy, Vol. 36, No. 9, pp. 3366–3373.

Dreyer, Dominique, 2010: L'accès non discriminatoire aux réseaux européens de l'électricité et du gaz. In Buschle, Dirk; Hirsbrunner, Simon; Kaddous, Christine (eds), 2010. Dossier de droit européen, No. 22.

Dreyer, Iana, 2009: EU should use international law to tackle Russian energy relations. Business New Europe. http://www.ecipe.org/archived-articles/eu-should-use-international-law-to-tackle-russian-energy-relations/File.

Dreyer, Iana; Erixon, Fredrik; Winkler, Robin, 2010: The quest for gas market competition: Fighting Europe's dependency on Russian gas more effectively. ECIPE Occasional Paper No. 1/2010.

Eberlein, Burkard, 2005: Regulation by cooperation: The "third way". In Cameron, Peter D., Legal Aspects of EU Energy Regulation: Implementing the New Directives of Electricity and Gas across Europe, pp. 58–88. Oxford University Press, Oxford.

EBN, 2010: Focus on Dutch Gas. http://www.ebn.nl/files/ebn_focus_on_dutch_gas_2010.pdf.

EBRD: Russia Country Profile. http://www.ebrd.com/downloads/legal/irc/countries/russia.pdf.

Ecritt, Thomas, 2009: Freeze risks energy future of Europe. Financial Times, 18 June 2009, p. 6.

Ecritt, Thomas; Wagstyl, Stefan, 2009: Energy sale raises Hungarian ire. Financial Times, 2 April 2009, p. 6.

Egeberg, Morten; Martens, Maria; Trondal, Jarle, 2009: Building executive power at the European level: On the role of EU-level agencies. Arena Center for European Studies, Oslo. http://www.arena.uio.no/publications/working-papers2009/papers/WP10_09.pdf.

Egenhofer, Christian, 1997: Understanding the politics of European Energy Policy: The driving and stopping forces, the politics of European energy, the energy of European politics and Maastricht II. The Journal (online Journal of the Centre for Energy, Petroleum, Mineral Law and Policy at the University of Dundee), Vol. 2, No. 9. http://www.dundee.ac.uk/cepmlp/journal/html/article 2-9.html.

Eikeland, Per Ove, 2004: The long and winding road to the internal energy market – Consistencies and inconsistencies in EU policy. FNI Report 8/2004. http://www.fni.no/doc&pdf/FNI-R0804.pdf.

Energy Charter Secretariat, 2002: The Energy Charter Treaty: A reader's guide. http://www.encharter.org/fileadmin/user_upload/document/ECT_Guide_ ENG. pdf.

Energy Charter Secretariat, 2009: Fostering LNG trade: Developments in LNG trade and pricing, p. 8.

Energy Delta Institute: Portugal. http://www.energydelta.org/mainmenu/edi-intelligence-2/our-services/Country-gas-profiles/country-gas-profile-portugal.

Energy Viewpoints, 2007: Market liberalization: Conditions for effective unbundling. http://www.moffatt-associates.com/energy_services/forecasting_market_trends/energy_viewpoints/documents/12/market_liberalisation_conditions_for_effective_unbundling.pdf.

Engerer, Hella, 2009: Erdgas für Europa: Die Importe steigen deutlich. Wochenbericht des DIW Berlin, No. 17, pp. 274–282. http://www.diw.de/documents/publikationen/73/97193/09-17-1.pdf.

ERGEG, 2007: ERGEG's response to the European Commission's Communication "An Energy Policy for Europe". http://www.ergeg.org/portal/page/portal/ERGEG_HOME/ERGEG_DOCS/ERGEG_DOCUMENTS_NEW/Energy%20documents/C06-BM-09-05_RespToSER_2007-02-06.pdf.

ERGEG, 2009: Status review of the definitions of vulnerable customer, default supplier and supplier of last resort. Ref: E09-CEM-26-04, 16 July.

ERGEG, 2010: Success in regional and inter-regional energy market integration. Press Release, Brussels, 10 November.

Essex, Ben, 2010: Gas and electricity prices: how are they calculated? http://www.bbc.co.uk/news/business-11767544.

Estrada, J; Bergesen, H. O.; Moe, A.; Sydnes, A. K., 1988: Natural Gas in Europe. Markets, Organisation and Politics. Pinter, London.

EU Energy Policy Monitoring, 2009: Erdgas- und Elektrizitätbinnen markt Entwicklungsstand. http://energy.iep-berlin.de/php/1_binnenmarkt_analyse.php.

European Commission, 2000: Commission closes investigation on Spanish company Gas Natural. http://europa.eu/rapid/pressReleasesAction.do?reference=IP/00/297&format=HTML&aged=1&language=EN&guiLanguage=en.

European Commission, 2005: Energy: Member States must do more to open markets, competition inquiry identifies serious malfunctions. http://europa.eu/rapid/pressReleasesAction.do?reference=IP/05/1421&format=HTML&aged=0&language=EN&guiLanguage=en.

European Commission, 2006a: Competition: Commission confirms inspections in the energy sector. http://europa.eu/rapid/press-release_MEMO-06-220_en.htm.

European Commission, 2006b: Infringement procedures opened in gas and electricity market sector, by Member State. http://europa.eu/rapid/pressReleasesAction.do?reference=MEMO/06/152&language=en.

European Commission, 2011a: Questions and answers on the third legislative package for an internal EU gas and electricity market. http://europa.eu/rapid/press-release_MEMO-11-125_en.htm?locale=en

European Commission, 2011b: The internal energy market – Time to switch into higher gear. http://ec.europa.eu/energy/gas_electricity/legislation/doc/20110224_non_paper_internal_nergy_market.pdf.

European Commission, 2012a: Internal energy market: national legislation in 8 Member States still not in line with EU rules. http://europa.eu/rapid/pressReleasesAction.do?reference=IP/12/181&format=HTML&aged=0&language=en&guiLanguage=fr.

European Commission, 2012b: Commission refers Ireland and the United Kingdom to Court to ensure European law is properly implemented. http://europa.eu/rapid/press-release_IP-12-52_en.htm.

European Commission, 2012c: Antitrust: Commission opens proceedings against Gazprom. http://europa.eu/rapid/press-release_IP-12-937_en.htm.

European Commission, 2013: Exemptions for new infrastructures. http://ec.europa.eu/energy/infrastructure/exemptions/doc/exemption_decisions.pdf.

European Commission, Eurostat, 2010a: EU energy and transport in figures. http://ec.europa.eu/energy/publications/statistics/doc/2010_energy_transport_figures.pdf.

European Commission, Eurostat, 2010b: Consumption of energy. http://epp.eurostat.ec.europa.eu/statistics_explained/index.php/Consumption_of_energy.

European Commission, Eurostat, 2011: File: Half-year electricity and gas prices, first half of year, 2009–2011 (EUR per kWh). http://epp.eurostat.ec.europa.eu/statistics_explained/index.php?title=File:Half-yearly_electricity_and_gas_prices,_first_half_of_year,_2009-2011_(EUR_per_kWh).png&filetimestamp=20111124164017.

European Neighbourhood and Partnership Instrument (ENPI). Regional strategy paper (2007–2013) and regional indicative programme (2007–2010) for the Euro-Mediterranean Partnership. http://ec.europa.eu/world/enp/pdf/country/enpi_euromed_rsp_en.pdf.

European Parliament, 2009: 3rd Energy Package gets final approval from MEPs. http://www.europarl.europa.eu/sides/getDoc.do?type=IM-PRESS&reference=20080616FCS31737&language=EN.

EU–Russia energy dialogue synthesis report. Brussels/Moscow 2001.

EU–Russia energy dialogue, the second progress report, 2002. http://ec.europa.eu/energy/international/bilateral_cooperation/russia/doc/reports/progress2_en.pdf.

EU–Russia energy dialogue, third progress report. Brussels/Moscow, 2002.

EU–Russia energy dialogue, fifth progress report. Moscow/Brussels 2004.

Everis; Mercados, Energy Markets International, 2010: From regional markets to a single European market. Final report, 28 April. http://ec.europa.eu/energy/gas_electricity/studies/doc/2010_gas_electricity_markets.pdf.

Everson, Michelle; Majone, Giandomenico; Metcalfe, Les; Schout, Adriaan, 1999: The role of specialized agencies in decentralising EU governance: Report to the Commission. http://ec.europa.eu/governance/areas/group6/contribution_en.pdf.

Federal Foreign Office: Strategic partnership with Russia. http://www.auswaertiges-amt.de/EN/Aussenpolitik/RegionaleSchwerpunkte/Russland/Russland_node.html.

Fifield, Anna, 2009: Political turmoil turns Iran's energy sector towards China. Financial Times, 11 April, p. 4.

Financial Times, 2009: Energy special report. Financial Times, 26 May.

Fischer, Severin; Geden, Oliver, 2012: The EU's energy roadmap 2050: Target without governance. Stiftung Wissenschaft und Politik. German Institute for International and Security Affairs. SWP Comments 8, March. http://www.swp-berlin.org/fileadmin/contents/products/comments/2012C08_fis_gdn.pdf.

Floudas, Demetrius A. M. A. (ed.), 2007: The extraterritorial application of European competition law. http://www.andrianopoulos.gr/xms Assets/File/Home/THE-EXTRATERRITORIAL-APPLICATION. pdf.

France – Energy mix fact sheet. http://ec.europa.eu/energy/energy_policy/doc/factsheets/mix/mix_fr_en.pdf.

Gao, Anton Ming-Zhi, 2008: Legal position of gas storage subject to the open access regime in the European Gas Directive and the 3rd Gas Liberalization Proposal. http://papers.ssrn.com/sol3/papers.cfm?abstract_id=1167793.

Gas Infrastructure Europe, 2011: Contributing to a sustainable society: Gas as an integral part of the Roadmap 2050. http://www.gasnaturally.eu/uploads/Modules/Publications/gie_roadmap2050.pdf.

Gazprom export: France. http://www.gazpromexport.ru/en/partners/france/.

Gazprom in questions and answers, book 2011. http://eng.gazprom questions.ru/index.php?id=4.

Geden, Oliver, 2008: Die Energie- und Klimapolitik der EU – zwischen Implementierung und strategischer Neuorientierung. Stiftung Wissenschaft und Politik.

Geden, Oliver; Marcelis, Clémence; Maurer, Andreas, 2006: Perspectives for the European Union's External Energy Policy: Discourse, ideas and interests in Germany, the UK, Poland and France. Working Paper FG 1, 2006/17, SWP Berlin. http://www.swp-berlin.org/fileadmin/contents/products/arbeitspapiere/External_KS_Energy_Policy__Dez_OG_.pdf.

Gehring, Thomas, 1996: Integrating integration theory: Neo-functionalism and international regimes. Global Society, Vol. 10, No. 3, pp. 225–253.

Geldhof, Wouter; Tormans, Stefan, 2008: Case C-439/06 – ECJ leaves limited scope for exemptions from third party access obligations for so-called operational networks. European Energy and Environmental Law Review, Vol. 17, No. 4, pp. 256–259.

Geldhof, Wouter; Vandendriessche, Frederik, 2008: European electricity and gas market liberalization. Background, status, developments, pp. 41–42. In Delvaux, Bram; Hunt, Michael; Talus, Kim, EU Energy Law and Policy Issues, 1st edn. ELRF Collection.

Geradin, Damien, 2004: Efficiency claims in EC competition law and regulation. http://papers.ssrn.com/sol3/papers.cfm?abstract_id=617922.

Geradin, Damien, 2005: Limiting the scope of Article 82 of the EC Treaty: What can the EU learn from the U.S. Supreme Court's Judgment in *Trinko* in the wake of *Microsoft, IMS* and *Deutsche Telekom*? Common Market Law Review, December.

Geradin, Damien, 2006: Twenty years of liberalization of network industries in the European Union: Where do we go now? http://papers.ssrn.com/sol3/papers.cfm?abstract_id=946796.

Geradin, Damien; Petit, Nicolas, 2004: The development of agencies at EU and national levels: Conceptual analysis and proposals for reform. Jean Monnet Working Paper 01/04. NYU School of Law, New York.

Geradin, Damien; O'Donoghue, Robert, 2005: The concurrent application of competition law and regulation: The case of margin squeeze abuses in the telecommunications sector. Global Competition Law Centre Working Paper 04/05.

Geradin, Damien; Muñoz, Rodolphe; Petit, Nicolas, 2005: Regulation through Agencies in the EU: A New Paradigm of European Governance. Edward Elgar, Cheltenham, Vol. 17, No. 4, pp. 256–259.

Germany – Energy Mix Fact Sheet. http://ec.europa.eu/energy/energy_policy/doc/factsheets/mix/mix_de_en.pdf.

Geveke, Arjan, 2003: Improving implementation by national regulatory authorities. Eipascope 2003/3, http://www.eipa.nl.

Glachant, Jean-Michel, 2009: Regulating networks in the new economy. EUI Working Papers. RSCAS 2009/05. http://cadmus.eui.eu/dspace/bitstream/1814/10622/1/EUI_RSCAS_2009_05.pdf.

Glachant, Jean-Michel; Lévêque, François, 2006: Electricity internal market in the European Union: What to do next? http://tisiphone.mit.edu/RePEc/mee/wpaper/2005-015.pdf.

Glachant, Jean-Michel; de Hauteclocque, Adrien, 2009: Long-term energy supply contracts in European competition policy: Fuzzy not crazy. EUI Working Papers. RSCAS 2009/06. http://cadmus.eui.eu/dspace/bitstream/1814/10623/1/EUI_RSCAS_2009_06.pdf.

Goltsblat BLP, 2012: President Decree dated 11 September 2012. Legal update No. 387, 14 September 2012. http://gblplaw.com/news/legal/72587/.

Gorst, Isabel, 2009: Russia and Italy sign gas supply deal. Financial Times, 16 May, p. 3.

Gorst, Isabel; Crooks, Ed, 2008: Fears of cartel rise after exporters meet. Financial Times, 24 December. http://www.ft.com/intl/cms/s/0/0925d9ae-d15c-11dd-8cc3-000077b07658.html#axzz1n2Gv1SL8.

Goyder, D. G.; Goyder, Joanna; Albors-Llorens, Albertina, 2009: Goyder's EC Competition Law, 5th edn. Oxford University Press, Oxford.

Gramlich, Ludwig, 2012: Regulating energy supranationally: EU energy policy. In Herrmann, Christoph; Terhechte, Jörg Philipp (eds), European Yearbook of International Economic Law. Springer, Berlin.

Grassani, Matteo, 2007: The "f-o-u-l-s" approach to unbundling in the electricity sector: pitfalls and legal risks of structural regulatory intervention. European Review of Energy Markets, No. 4. http://www.eeinstitute.org/european-review-of-energy-market/eei-review-issue-4.
Grätz, Jonas, 2009: Energy relations with Russia and gas market liberalization. Internationale Politik und Gesellschaft, pp. 66–80.
Green, Matthew, 2009a: Industry threatened after militants resume delta pipeline attacks. Financial Times, 26 May, p. 4.
Green, Matthew, 2009b: Into Africa. Financial Times, 25 May, p. 15.
Green, Richard, 2007: EU regulation and competition policy among the energy utilities. Institute for Energy Research and Policy.
Grigoriadis, Theocharis N., 2008: State responsibility and antitrust in the Energy Charter Treaty: Socialization vs. liberalization in bilateral investment relations. Texas International Law Journal, Vols 44–45, pp. 45–64.
Grigoriadis, Theocharis N.; Torgler, Benno, 2006: Energy regulation and legislative development in the state Duma of Russia: A spatial analysis on roll call votes with the optimal classification model, Russia 1994–2003. Center for Research in Economics, Management and the Arts.
Griller, Stefan; Orator, Andreas, 2004: Meroni revisited – Empowering European agencies between efficiency and legitimacy. New Modes of Governance Project.
Griller, Stefan; Orator, Andreas, 2007: Meroni revisited – Empowering European agencies between efficiency and legitimacy. http://www.eu-newgov.org/database/DELIV/D04D40_WP_Meroni_Revisited.pdf.
Griller, Stefan; Orator, Andreas, 2010: Everything under control? The "way forward" for European agencies in the footsteps of the Meroni doctrine. European Law Review, Vol. 35, No. 1.
Gromitsaris, Athanasios, 2011: Expropriation, takings annual report – 2011 – Germany. IUS Publicum Network Review. http://www.ius-publicum.com/repository/uploads/21_03_2012_12_05_Gromitsaris_Expropriation.pdf.
Growitsch, Christian; Stronzik, Marcus, 2008: Ownership unbundling of gas transmission networks – Theoretical background and empirical evidence. Competition and Regulation in Network Industries. EWI Working Paper, No. 11/7. http://www.ewi.uni-koeln.de/fileadmin/user_upload/Publikationen/Working_Paper/EWI_WP_11-07_unbundling_gas.pdf.
Growitsch, Christian; Stronzik, Marcus, 2009: Ownership unbundling of gas transmission networks – Empirical evidence. EWI Working Paper, No. 11/7. http://www.ewi.uni-koeln.de/fileadmin/user_upload/Publikationen/Working_Paper/EWI_WP_11-07_unbundling_gas.pdf.

Haar, Laura N.; Marinescu, Nicolae, 2011: Energy policy and European utilities' strategy: Lessons from the liberalization and privatisation of the energy sector in Romania. Energy Policy, Vol. 39, No. 5, pp. 2245–2255.

Haase, Nadine, 2008: European gas market liberalization: Are regulatory regimes moving towards convergence? Oxford Institute for Energy Studies.

Haase, Nadine, 2009: European gas market liberalization: Competition versus security of supply? Energy Delta Institute/Castel International Publishers, Gröningen.

Haase, Nadine; Bressers, Hans, 2008: In search of an optimal design for European gas markets. Network Industries Quarterly, Vol. 10, No. 3, pp. 15–17.

Haghighi, Sanam S. 2007: Energy Security: The External Legal Relations of the European Union with Major Oil- and Gas-supplying Countries. Hart, Oxford.

Hancher, Leigh; Ottervanger, Tom; Slot, Piet Jan (eds), 2006: EC State Aid. Sweet and Maxwell, London.

Haney, Aoife Brophy; Pollitt, Michael G. 2010: Exploring the determinants of "best practice" in network regulation: The case of the electricity industry. EPRG Working Paper 1012. Cambridge Working Paper in Economics 1020.

Hanf, Kenneth; Soetendorp, Ben, 1998: Small states and the Europeanization of public policy. In Hanf, Kenneth; Soetendorp, Ben (eds), Adapting to European Integration Small States and the European Union. Longman, London.

Harris, Nigel, 2006: EU Gas market – Trading and infrastructure development. Commodities Now, December.

Harris, Nigel, 2008: Towards a true EU gas market. In Fundamentals of the World Gas Industry. Kingston Energy Consulting.

Hayes, Mark H., 2007: Institutions and Gas Market Security. Leland Stanford Junior University.

Hellström, Per; Maier-Rigaud, Frank; Bulst, Friedrich Wenzel, 2009: Remedies in European antitrust law. Antitrust Law Journal, Vol. 76, No. 1, pp. 43–63.

Héritier, Adrienne, 2005: Managing regulatory developments in rail: Compliance and access regulation in Germany and the UK. In Héritier, Adrienne; Coen, David (eds), Refining Regulatory Regimes. Utilities in Europe. Edward Elgar, Cheltenham.

Hernández, Fabrizio; Gandolfi, Monica, 2005: EU Exemptions to TPA for New Gas Infrastructure. http://www.nera.com/NewsletterIssue/ER_Insights_Issue24_7.2005.pdf.

Himmelreich, Jörg, 2009: European gas policy in trouble. The German Marshall Fund of the United States. Policy Brief. http://www.gmfus.org/publications/article.cfm?id=544.

Hobe, Stephan, 2009: Energiepolitik. Europa Recht, Beiheft 1, 2009.

Hofer, Edith, 2011: The future role of regulation – More Europe. In Buschle, Dirk; Hirsbrunner, Simon; Kaddous, Christine (eds), European Energy Law Dossier de droit européen No. 22. Helbing Lichtenbahn, Bâle.

Hofmann, Herwig C. H., 2006: Negotiated and non-negotiated administrative rule-making: The example of EC competition policy. Common Market Law Review, Vol. 43, pp. 153–178.

Hofmann, Herwig C. H.; Morini, Alessandro, 2012: Constitutional aspects of the pluralisation of the EU executive through "agencification". http://papers.ssrn.com/sol3/papers.cfm?abstract_id=2031499.

Hofmann, Herwig C. H.; Türk, Alexander H. (eds), 2006: EU Administrative Governance. Edward Elgar, Cheltenham.

Hofmann, Herwig C. H.; Türk, Alexander H. (eds), 2009: Legal Challenges in EU Administrative Law Towards an Integrated Administration. Edward Elgar, Cheltenham.

Hofmann, Herwig C. H.; Rowe, Gerard C.; Türk, Alexander, 2011: Administrative Law and Policy of the European Union. Oxford University Press, Oxford.

Hope, Kerin; Troev, Theodor, 2009: Bulgaria taps into gas pipeline to end reliance on Russia. Financial Times, 15 July, p. 2.

Hough, David; Gammons, Sean; Whithington, Brian, 2005: Gas balancing rules in Europe: A report for Creg. Nera and TPA Solutions. Nera Economic Consulting.

House of Commons Energy and Climate Change Committee, 2011: UK energy supply security or independence: Eighth Report of Session 2010–2012, Vol. 1. http://www.publications.parliament.uk/pa/cm201012/cmselect/cmenergy/1065/1065.pdf.

House of Commons and European Union Committee, 2010: Impact assessments in the EU: Room for improvement? Chapter 2: Production of Impact Assessments. http://www.publications.parliament.uk/pa/ld200910/ldselect/ldeucom/61/6105.htm.

Hutcheon, Andrew; Spencer, James, 2008: Provisional application of the energy charter treaty. The European and Middle Eastern Arbitration Review Section 3: Regional Overview.

Hyndle-Hussein, Joanna, 2012: Estonia liberalises its gas market. Centre for Eastern Studies. CEWeekly. http://www.osw.waw.pl/en/publikacje/ceweekly/2012-01-11/estonia-liberalises-its-gas-market.

International Energy Agency, 2011: Key world energy statistics. http://www.iea.org/textbase/nppdf/free/2011/key_world_energy_stats.pdf.

International Energy Agency, 2010: World energy outlook 2010 factsheet. http://www.worldenergyoutlook.org/docs/weo2010/factsheets.pdf.

Italy – Energy mix fact sheet. http://ec.europa.eu/energy/energy_policy/doc/factsheets/mix/mix_it_en.pdf.

Jamasb, Tooraj; Pollitt, Michael, 2008: Security of supply and regulation of energy networks. Energy Policy, Vol. 36, No. 12, pp. 4584–4589.

Jégo, Marie, 2009: Pas de consensus sur le gaz entre la Russie et l'UE. Le Monde, 25 May, p. 6.

Joerges, Christian; Dehousse, Renaud, 2002: Good Governance in Europe's Integrated Market. Oxford University Press, Oxford.

Johnson, Debra, 2005: EU–Russian energy links: A marriage of convenience?, p. 273. Government & Opposition, Vol. 40, No. 2, pp. 256–277.

Jones, Alison; Sufrin, Brenda, 2004: EC Competition Law Text, Cases and Materials, 2nd edn. Oxford University Press, Oxford.

Jones, Christopher W.; Webster, William, 2006: EU Energy Law Volume I, 2nd edn. The Internal Energy Market. Claeys and Casteels, Leuven.

Jones, Clifford A., 2006: After the Green Paper: The third devolution in European competition law and private enforcement. The Competition Law Review, Vol. 3, No. 1, pp. 1–3.

Jordana, Jacint; Levi-Faur, David, 2004: The politics of regulation in the age of governance. In Jordana, Jacint; Levi-Faur, David (eds), The Politics of Regulation Institutions and Regulatory Reforms in the Age of Governance. Edward Elgar, Cheltenham.

Joskow, P. L.; Tirole, J., 2000: Transmission rights and market power on electric power networks. RAND Journal of Economics. Vol. 31, No. 3, pp. 450–487.

Kaderják, Péter; Cameron, Peter; Tóth, András István, 2007: Unilateral natural gas import dependence: a new supply security issue for Europe. European Review of Energy Markets, Vol. 2, No. 2, pp. 57–91.

Kahl, Wolfgang, 2009: Die Kompetenzen der EU in der Energiepolitik nach Lissabon. Europarecht 2009, Heft 5, 44. Jahrgang. Nomos Verlag, pp. 601–621.

Kanapinskas, Virginijus; Urmonas, Algimantas, 2011: Changes of legal regulation on natural gas market in the context of the Third European Union Energy Package, p. 245. Jurisprudencija Research Papers, 20. Mykolas Romeris University. Vilnius.

Kassim, Hussein; Menon, Anand, 2003: The principal–agent approach and the study of the European Union: promise unfulfilled? Journal of European Public Policy, Vol. 10, No. 1, pp. 121–139.

KEMA International B.V., 2009: Report study on methodologies for gas transmission network tariffs and gas balancing fees in Europe, p. 49. Arnhem, the Netherlands.

Keske, Sonja E., 2009: Group litigation in European competition law. A law and economics perspective. Erasmus Universiteit Rotterdam.

Keyaerts, Nico; Meeus, Leonardo; D'haeseleer, William, 2008: Natural gas balancing: Appropriate framework and terminology. TME Working Paper – Energy and Environment. K. U. Leuven Energy Institute. http://www.mech.kuleuven.be/tme/research.

Kiriazis, Georgios, 2001: Jurisdiction and cooperation issues in the investigation of international cartels. http://ec.europa.eu/competition/speeches/text/sp2001_010_en.pdf.

Koch, Charles H. Jr, 2009: Collaborative governance: Lessons for Europe from US electricity restructuring. College of William & Mary Law School. Faculty Publications. Paper No. 1270. http://scholarship.law.wm.edu/facpubs/1270.

Koch, Oliver; Gauer, Céline, 2011: Energy liberalisation and competitive law – The Commission's recent antitrust case practice. In Buschle, Dirk; Hirsbrunner, Simon; Kaddous, Christine, European Energy Law. Dossier de droit européen No. 22. Helbing Lichtenbahn, Bâle.

Koch, Oliver; Schnichels, Dominik, 2008: The E.ON seals case – €38 million fine for tampering with Commission seals. Competition Policy Newsletter 2008-2.

Koch, Oliver; Nagy, Károly; Pucinskaite-Kubik, Ingrida; Tretton, Walter, 2009: The RWE gas foreclosure case: Another energy network divestiture to address foreclosure concerns. Competition Policy Newsletter, Vol. 2.

Kourtikakis, Kostas, 2005: Networking with national authorities: Information sharing and bureaucratic accountability in the European Union. European Union Studies Association (EUSA) Biennial Conference, 31 March to 2 April, Austin, TX.

Kroes, Neelie, 2005: Damages actions for breaches of EU competition rules: Realities and potentials. Opening Speech at the conference "La réparation du préjudice cause par une pratique anticoncurrentielle en France et à l'étranger: bilan et perspectives", Cour de Cassation, 17 October 2005.

Kroes, Neelie, 2007: Improving competition in European energy markets through effective unbundling. Fordham International Law Journal, Vol. 31, No. 5, article 11, pp. 1387–1441.

Kubcik, Andrzej; Bielecki, Tomasz, 2012: Gazprom uważa, że jest dla UE nietykalny. http://wyborcza.biz/bizness/51,100896,1241720.html?i=0.

Laffont, Pascal; Simonet, Loïc, 2005: La Charte de l'Energie et le Transit des Matières Premières: Trop Loin? Trop Tôt? Annuaire Français de Droit International. LI-2005, pp. 524–541. CNRS Editions, Paris.

Larouche, Pierre, 2006: Contrasting legal solutions and the comparability of EU and U.S. experiences. http://papers.ssrn.com/sol3/papers.cfm?abstract_id=943615.

Larsen, Anders; Pedersen, Lene Holm; Sorensen, Eva Moll; Olsen, Ole Jess, 2004: Independent regulatory authorities in Europe. http://sessa.eu.com/documents/wp/D73.1-Larsen.pdf.

Lévêque, François; Glachant, Jean-Michel; Saguan, Marcelo; de Muizon, Guildas, 2009: How to rationalize the debate about "EU Energy Third Package"? Revisiting criteria to compare electricity transmission organizations. European University Institute.

Levy, Brian; Spiller, Paolo, 1996: Regulation, Institutions and Commitment. Cambridge University Press. Cambridge.

Lindroos, Maarit; Schnichels, Dominik; Svane, Lars Peter, 2002: Liberalization of European gas markets – Commission settles GFU case with Norwegian gas producers. Competition Policy Newsletter, No. 3, pp. 50–52.

Lindseth, Peter, 2002: Delegation is dead, long live delegation: Managing the democratic disconnect in the European market-policy. In Joerges, Christian; Dehousse, Renaud, 2002: Good Governance in Europe's Integrated Market. Oxford University Press, Oxford.

Locatelli, Catherine, 2008: EU gas liberalization as a driver of Gazprom's Strategies? IFRI. Russia/NIS Center.

Louis, Frédéric; Vallery, Anne, 2003: Court of Justice of the European Union clarifies application of State Aid rules to State financing of public service publications. http://www.wilmerhale.com/files/Publication/4d8b76c7-422c-4e48-9075-d32f04f9c9f5/Presentation/PublicationAttachment/f2a30d3d-f1f1-4cb9-bb21-26cef59f7b21/EC%20Bulletin072903.pdf.

Lowe, Philip, 2008: Can EU competition policy create competition in the energy sector? http://ec.europa.eu/competition/speeches/text/sp2008_09_en.pdf.

Lowe, Philip; Pucinskaite, Ingrida; Webster, William; Lindberg, Patrick, 2007: Effective unbundling of energy transmission networks: Lessons from the Energy Sector Inquiry. Competition Policy Newsletter.

Ludwigs, Markus, 2011: Das verändert Machtgefüge der Institutionen nach dem Dritten EU-Binnenmarktpaket. Deutsches Verwaltungsblatt, Heft 2/2011-126, pp. 61–124. Jahrgang.

Maass, Matthias, 2007: Small States in the International Society of States. Paper presented at the workshop "Small State Capacity Building", Birmingham.

Mackenzie, Kate, 2009: Use of power faces first fall since 1945. Financial Times, 22 May, p. 3.

Majone, Giandomenico, 1994: The Rise of the Regulatory State in Europe. West European Politics, Vol. 17, pp. 77–101.
Majone, Giandomenico, 1996: Regulating Europe. London. Routledge.
Majone, Giandomenico, 1997: From the positive to the regulatory state: Causes and consequences of changes in the mode of governance. Journal of Public Policy, Vol. 17, No. 2, pp. 139–167.
Majone, Giandomenico, 2001: Two logics of delegation. Agency and fiduciary relations in EU governance. European Union Politics, Vol. 2, pp. 103–122.
Majone, Giandomenico, 2002: Delegation of regulatory powers in a mixed polity. European Law Journal, Vol. 8, No. 3, pp. 319–339.
Majone, Giandomenico, 2003: The internationalization of regulation: Implications for developing countries. Paper presented at the International Conference on "Innovation and Change in Regulation and Competition", 13–15 October 2003, Mandaluyong City, Metro Manila, Philippines. http://www.competition-regulation.org.uk/conferences/Philippines03/majone.pdf.
Majone, Giandomenico, 2005: Dilemmas of European Integration. The Ambiguities and Pitfalls of Integration by Stealth. Oxford University Press, Oxford.
Majone, Giandomenico, 2009: Liberalization, re-regulation, and mutual recognition: Lessons from three decades of EU experience. Scottish Jean Monnet Centre Working Paper Series, Vol. 1, No. 1.
Makholm, Jeff D., 2007: Seeking competition and supply security in natural gas: The US experience and European challenge. Working Paper. Nera Economic Consulting. http://www.nera.com/Publication.asp?p_ID=3198.
McGowan, Francis, 2008: Can the European Union's market liberalism ensure energy security in a time of "economic nationalism"? Journal of Contemporary European Research, Vol. 4, No. 2, pp. 90–106.
McHarg, Aileen; Barton, Barry; Bradbrook, Adrian; Godden, Lee, 2010: Property and the Law in Energy and Natural Resources. Oxford University Press, Oxford.
Med-TSO – Mediterranean transmission system operators multilateral cooperation for the integration of the Mediterranean electricity systems. Rome, 4 May 2012.
Mercados, 2008: From regional markets to a Single European Market, p. 67. Final Report 28/04/2010. http://ec.europa.eu/energy/gas_electricity/studies/doc/2010_gas_electricity_markets.pdf.
Mileczarski, Władysław, 2007: Security of energy supply from the Polish perspective. European Review of Energy Markets, Vol. 2, No. 2. http://www.eeinstitute.org/european-review-of-energy-market/erem5-article-mielczarski.

Ministry of Employment and the Economy: Services of general economic interest. http://www.tem.fi/index.phtml?l=en&s=2017.

Molis, Arunas, 2011: Rethinking EU–Russia energy relations: What do the Baltic States want? SPES Policy Papers. Institut für Europäische Politik.

Monaghan, Andrew; Montaro-Jankovski, Lucia, 2006: EU–Russia energy relations: The need for active engagement. EPC Issue Paper No. 45.

Monopolkommission, 2009: Strom und Gas 2009 Energiemärkte im Spannungsfeld von Politik und Wettbewerb. http://www.monopolkommission.de/sg_54/s54_volltext.pdf

Monti, Giorgio, 2002: The single energy market: the relationship between competition policy and regulation. SPEECH/02/101 of 7 March 2002.

Monti, Giorgio, 2008: Managing the intersection of utilities regulation and EC competition law. Competition Law Review, Vol. 4, No. 2, pp. 123–145.

Monti, Mario, 2003: Applying EU competition law to the newly liberalized energy markets. World Forum on Energy Regulation, 6 October. http://europa.eu/rapid/pressReleasesAction.do?reference=SPEECH/03/447&format=P.

Monti, Mario, 2004: A reformed competition policy: achievements and challenges for the future. Center for European Reform. Brussels, 28 October.

Moodaliyar, Kasturi; Reardon, James F.; Theuerkauf, Sarah, 2010: The relationship between public and private enforcement in competition law – A comparative analysis of South Africa, the European Union and Swiss Law. South African Law Journal, Vol. 127, No. 1. pp. 141–162.

Moselle, Boaz; White, Martin, 2011: Market design for natural gas: The target model for the internal market. A report for the Office of Gas and Electricity Markets. LECG,March 2011. http://www.energy-regulators.eu/portal/page/portal/EER_HOME/EER_CONSULT/CLOSED%20PUBLIC%20CONSULTATIONS/GAS/GTM_CfE/Tab1/LECG%20Gas_Target_Model_0700311.pdf.

Natural Gas Market Report Glossary. http://www.spragueenergy.com/pages/content.aspx?p=Natural%20Gas%20Marketwatch%20Glossary.

Neumann, Anne; von Hirschhausen, Christian, 2004: Less long-term gas to Europe? A quantitative analysis of European long-term gas supply contracts. Zeitschrift für Energiewirtschaft, Vol. 28, No. 3, pp. 175–182.

Nizzo, Carlo, 1999: National public administrations and European integration. http://www.oecd.org/dataoecd/46/11/1850561.pdf.

Noël, Pierre, 2008: Beyond dependence: How to deal with Russian gas. European Council on Foreign Relations. Policy Brief. http://ecfr.3cdn.net/c2ab0bed62962b5479_ggm6banc4.pdf.

Nyssens, Harold; Schnichels, Dominik, 2007: Energy. In Faull, Jonathan; Nikpay, Ali (eds), The EC Law of Competition, 2nd edn. Oxford University Press, Oxford.

OECD, 2006: Remedies and sanctions in abuse of dominance cases. http://www.oecd.org/competition/abuseofdominanceandmonopolisation/38623413.pdf.

Oettinger, Günther, 2011: The importance of ACER for the internal energy market speech of Commissioner Oettinger at the opening ceremony of the Agency for the Cooperation of Energy Regulators (ACER), Ljubljana, March. http://europa.eu/rapid/press-release_SPEECH-11-144_en.htm?locale=en.

OFGEM, 2010: Notice of proposed modification in relation to the disconnection of vulnerable consumers to the standard conditions of: the electricity supply licence under section 11A of the Electricity Act 1989; the gas supply licence under section 23 of the Gas Act 1986. http://www.ofgem.gov.uk/Sustainability/SocAction/Publications/Documents1/SLC%2027%20Formal%20con%20doc%20-%20150710.pdf.

Orttung, Robert W.; Overland, Indra, 2011: Russia and the formation of a gas cartel. Problems of Post-Communism, Vol. 58, No. 3, pp. 53–66.

Padgett, Stephen, 1992: The single European energy market: The politics of realization. Journal of Common Market Studies, Vol. XXX, No. 1, p. 55.

Paillard, Christophe-Alexandre, 2010: Rethinking Russia: Russia and Europe's mutual energy dependence. Journal of International Affairs, Vol. 63, No. 2, pp. 65–84.

Pelkman, Jacques, 2001: Making EU network markets competitive. Oxford Review of Economic Policy, Vol. 17, No. 3, pp. 432–456.

Petrov, Konstantin; Hewicker, Christian; Boisseleau, François; Nabuurs, Pier, 2007: European internal electricity market – What next? European Review of Energy Markets, No. 4. http://www.eeinstitute.org/european-review-of-energy-market/eei-review-issue-4.

Philippe and Partners, 2011: Final report on unconventional gas in Europe. http://ec.europa.eu/energy/studies/doc/2012_unconventional_gas_in_europe.pdf.

Pielow, Johann-Christian; Ehlers, Eckhart, 2008: Ownership unbundling and constitutional conflict: A typical German debate? European Review of Energy Markets, Vol. 2, No. 3, pp. 34–46.

Pollack, Mark A., 2003: The Engines of European Integration. Delegation, Agency and Agenda-Setting in the EU. Oxford University Press, Oxford.

Pollitt, Michael G., 2007: Liberalisation and regulation in electricity systems: How can we get the balance right? In Sioshansi, F. P. (ed.), Competitive Electricity Markets: Design, Implementation, Performance. Oxford: Elsevier.

Pollitt, Michael G., 2008a: The arguments for and against ownership unbundling of energy transmission networks. Energy Policy, Vol. 36(2), pp. 704–713.

Pollitt, Michael G., 2008b: The future of electricity (and gas) regulation in a low-carbon policy world. The Energy Journal, Special Issue: The Future of Electricity: Papers in Honour of David Newberry, Vol. 29, pp. 62–94.

Prosser, Tony, 2005: The Limits of Competition Law: Markets and Public Services. Oxford University Press, New York.

Provisional central product classification of the United Nations. http://unstats.un.org/unsd/cr/registry/regcst.asp?cl=9&lg=1.

Rab, Suzanne, 2012: Czech energy commitments. Energy Newsletter. King & Spalding. http://www.kslaw.com/library/newsletters/EnergyNewsletter/2012/September/article2.html.

Rakovsky, Claude; Godhino de Matos, Manuel; Kopke, Alexander; Ohrlander, Peter; Shiels, Paul, 2009: EC merger regulation contributes to more efficient merger control in EU. Competition Policy Newsletter 2009-2.

Ramboll Oil and Gas, 2009: Future development of the energy gas market in the Baltic Sea Region. Final Report prepared for the EU Commission DG TREN C1. http://ec.europa.eu/energy/infrastructure/doc/2009_bemip_ramboll_bemip_final_report.pdf.

Ratner, Michael; Belkin, Paul; Nichol, Jim; Woehrel, Steven, 2013: Europe's energy security: Options and challenges to natural gas supply diversification. Congressional Research Service. http://www.fas.org/sgp/crs/row/R42405.pdf.

Renda, Andrea, 2006: Impact assessment in the EU: The state of the art and the art of the state. Centre for European Policy Studies. Brussels.

Report of the World Commission on Environment and Development: Our Common Future. http://www.un-documents.net/ocf-02.htm.

Représentation permanente de la France auprès de l'Union européenne, 2008: Effective and efficient unbundling of transmission system operators: Proposal for a Directive of the EP and of the Council amending Directive 2003/54/EC concerning common rules for the internal market in electricity and repealing Directive 96/92/EC. http://www.euractiv.com/sites/all/euractiv/files/3rdoptionletter.pdf.

Riley, Alan, 2007: Russian gas and EU energy security. Directorate General External Policies of the Union. Briefing Paper.

Riley, Alan, 2008: EU energy liberalization: Coming to a Member State near you! Competition Law Review, Vol. 4, No. 2, pp. 73–75.

Roggenkamp, Martha; Redgwell, Catherine; Rønne, Anita; del Guayo, Iñigo, 2007: Energy law in Europe national, EU and international regulation, 2nd edn. Oxford University Press, Oxford.

Rowe, Gerard C., 2002: Comment: Servants of the people – Constitutions and states from a principal–agent perspective. In Voigt, Stefan; Wagener, Hans-Jürgen, 2002: Constitutions, Markets and Law. Edward Elgar, Cheltenham.

Sadowska, Małgorzata M., 2011: Energy liberalization in antitrust straightjacket: A plant too far? World Competition: Law and Economics Review, Vol. 34, No. 3, pp. 449–76.

Salerno, Francesco Maria, 2008: The competition law-ization of enforcement: The way forward for making the energy market work? EUI Working Papers. RSCAS 2008/07. http://cadmus.eui.eu/dspace/bitstream/1814/8108/1/RSCAS_2008_07.pdf.

Sato, Chie, 2010: Extraterritorial application of EU competition law – Is it possible for Japanese companies to steer clear of EU competition law? Journal of Political Science and Sociology, No. 11, pp. 33–49.

Schaffer, Brenda, 2009: Energy Politics. University of Pennsylvania Press, Philadelphia, PA.

Scharpf, Fritz W., 1993: Co-ordination in hierarchies and networks. In Scharpf, Fritz W. (ed.), Games in Hierarchies and Networks: Analytical and Empirical Approaches to the Study of Governance Institutions. Campus, Frankfurt, pp. 125–165.

Scharpf, Fritz W., 2003: Problem-solving effectiveness and democratic accountability in the EU. http://www.mpifg.de/pu/workpap/wp03-1/wp03-1.html.

Schaub, Alexander, 2000: Competition policy and liberalization of energy markets. Brussels. http://ec.europa.eu/competition/speeches/text/sp2000_023_en.pdf.

Schipper, Frank; van der Vleuten, Erik, 2008: Trans-European network development and governance in historical perspective. Vol. 10, No. 3, pp. 5–7.

Schmitt, Stephan; Rammerstorfer, Margarethe, 2010: Ownership unbundling, investments and consumer prices – A panel data analysis. Paper for the third Annual Conference on Competition and Regulation in Network Industries, Brussels, 19 November. http://www.cne.es/cgi-bin/BRSCGI.exe?CMD=VEROBJ&MLKOB=543950160101.

Scholtz, Ulrich; Purps, Stephan, 2010: The application of EC competition law in the energy sector. Journal of European Competition Law and Practice, Vol. 1, No. 1, pp. 37–51.

Schütze, Richard, 2009: From Dual to Cooperative Federalism: The Changing Structure of European Law. Oxford University Press, Oxford.

Seliverstov, Sergey, 2009: Energy security of Russia and the EU: Current legal problems. IFRI. Gouvernance européenne et géopolitique de l'énergie.

Séréni, Jean-Pierre, 2007: Bataille du gaz en Europe. Le monde diplomatique. http://www.monde-diplomatique.fr/2007/03/SERENI/14505.

Shaw, Malcolm N., 2008: International Law, 5th edn. Cambridge University Press, Cambridge.

SIPRI Yearbook, 2007: Armaments, Disarmament and International Security. Stockholm International Peace Research Institute. Oxford University Press, Oxford.

Slabá, Monika, 2008: Liberalisation of Natural Gas Market – EU Vision vs. Reality. Energy Studies Review, Vol. 16, No. 1, pp. 199–212.

Smith, Herbert, 2005: EU to punish restrictive business practices – An EU competition law perspective on destination clauses in LNG contracts. http://www.herbertsmith.com/NR/rdonlyres/57514EFE-4A32-42F6-AF96-54F24845BF16/1149/euc_280605.html.

Sodupe, Kepa; Benito, Eduardo, 2001: Pan-European energy co-operation: Opportunities, limitations and security of supply to the EU, p. 170. Journal of Common Market Studies, Vol. 39, No. 1, pp. 165–177.

Soligo, Ronald; Jaffe, Amy Myers, 2004: Market structure in the new gas economy: Is cartelization possible? Geopolitics of Gas Working Paper Series. James A. Baker III, Institute for Public Policy Energy Forum.

Soriano, Leonor Moral, 2001: The role of the European Court of Justice in the design of regulatory space. http://aei.pitt.edu/2187/1/002676_1.pdf.

Spain – Energy mix fact sheet. http://ec.europa.eu/energy/energy_policy/doc/factsheets/mix/mix_es_en.pdf.

Spanjer, Abdelkader Rainaldo, 2008: Structural and Regulatory Reform of the European Natural Gas Market. Ponsen & Looijen BV, Wageningen.

Spanjer, Aldo, 2007: Do TPA exemptions solve the hold-up problem on European gas markets? http://media.leidenuniv.nl/legacy/as-2007-02.pdf.

Spector, David, 2008: Competition policy in the energy sector: Squaring the circle? Concurrences, No. 2-2008, pp. 50–56.

Stacey, Kiran, 2009: Gloom for energy and material amid cautious sentiment. Financial Times, 8 July, p. 30.

Stepina, Inese, 2008: Interests shaping EU energy policy at the Court of Justice of the European Union, p. 1. Paper presented at the Fourth Pan-European Conference on European Politics of the European Consortium for Political Research, 25–27 September 2008, University of Latvia, Riga, Latvia.

Stern, Jonathan, 2005: Gas pipeline cooperation between political adversaries: examples from Europe. http://www.chathamhouse.org.uk/files/3222_jsjan05.pdf.

Stern, Jonathan, 2006: The new security environment for European gas: Worsening geopolitics and increasing global competition for LNG. Oxford Institute for Energy Studies.

Stevens, Paul, 2010: The history of gas, pp. 4–5. Polinares Working Paper No. 5, September.

Stigler, George, 1975: Citizens and the State: Essays on Regulation, p. 183. University of Chicago Press, Chicago, IL.

Strange, Susan, 1996: The Retreat of the State, the Diffusion of Power in the World Economy, p. 124. Cambridge University Press, Cambridge

Strauss, Delphine; Crooks, Ed, 2009: Iraq offers to supply half of Nabucco's gas. Financial Times, 14 July, p. 5.

Study on Unbundling of Electricity and Gas Transmission and Distribution System Operators. Annexes Country Overview. http://ec.europa.eu/energy/gas_electricity/studies/doc/electricity/2006_03_unbundling_electricity_gas_annex.pdf.

Subiotto, Romano; Malone, Farrell; Little, David R.; de Brosses, Clémentine; Suciu, Simina, 2010: Recent EU case law developments: Article 102 TFEU. Journal of Competition Law and Practice. http://jeclap.oxfordjournals.org/content/early/2010/12/24/jeclap.lpq077.full.

Szwedo, Piotr, 2010: Case comment: (former) Yukos v. Russian Federation before the Permanent Court of Arbitration. Journal of International Cooperation Studies, Vol. 18, No. 2, pp. 57–69.

Szydło, Marek, 2009: Regulatory exemptions for new gas infrastructure. A key challenge for European Energy Policy. European Energy and Environmental Law Review, Vol. 18, No. 5, pp. 254–269.

Talus, Kim, 2007a: Role of the European Court of Justice in the Opening of Energy Markets. Springer Berlin/Heidelberg, Vol. 8, No. 3, pp. 435–448.

Talus, Kim, 2007b: Long-term gas agreements and security of supply – between law and politics. European Law Review, No. 23, pp. 535–548.

Talus, Kim, 2008: Security of supply – An increasing political notion, p. 125. In Delvaux, Bram; Hunt, Michael and Talus, Kim (eds), EU Energy Law and Policy Issues. Euroconfidentiel, Brussels.

Talus, Kim, 2010: Role of long-term contracts on EU–Russia Energy Relations. Paper Presented at Warwick Workshop Governing Energy in Europe and Russia, 3–4 September, Radcliffe House, University of Warwick.

Talus, Kim; Hunt, Michaël, 2011: Ownership unbundling: What end to the saga? In Buschle, Dirk; Hirsbrunner, Simon; Kaddous, Christine (eds), Dossier de droit européen, No. 22.

Tapia, Javier; Mantzari, Despoina, 2012: The regulation/competition interaction. In Geradin, Damien; Lianos, Ioannis (eds), Research Handbook on European Competition Law. Edward Elgar, Cheltenham.

The Economist, 2007: Gas cartels: A snag in the pipeline. http://www.economist.com/node/9005311.

Tonra, Ben, 2001: The Europeanization of National Foreign Policy: Dutch, Danish and Irish Foreign Policy in the European Union. Ashgate, Aldershot.

Torriti, Jacopo, 2008: Does the impact assessment on the "Third Package" provide the correct economic forecast for the liberalization of the EU energy markets? EUI Working Papers. RSCAS 2008/14. http://cadmus.eui.eu/dspace/bitstream/1814/8687/1/RSCAS_2008_14.pdf.

Truli, Emmanuela, 2009: White paper on damages actions for breach of the EC antitrust rules: The binding effect of decisions adopted by competition authorities. http://papers.ssrn.com/sol3/papers.cfm?abstract_id=1467808.

United Nations Economic Commission for Europe Committee on Sustainable Energy Working Party on Gas Study, 2012: The impact of liberalization of natural gas markets in the UNECE region – Efficiency and security. http://www.unece.org/fileadmin/DAM/energy/se/pdfs/wpgas/pub/ImpactLibNGM_UNECE_EffSec.pdf.

van der Linde, Coby; Correljé, Aad; de Jong, Jacques; Tönjes, Christoph, 2006: The paradigm change in international natural gas markets and the impact on regulation, pp. 53–54. Clingendael International Energy Programme.

Van Hoorn, Victor, 2009: "Unbundling", "reciprocity" and European energy market: WTO consistency and broader implications for Europe. European Energy and Environmental Law Review, Vol. 18, No. 1, pp. 51–76.

Vasconcelos, Jorge, 2005: Towards the Internal energy market: How to bridge a regulatory gap and build a regulatory framework. European Review of Energy Markets, Vol. 1, No. 1. EREM. Claeys & Casteels [and contributors].

Verde, Stefano, 2008: Everybody merges with somebody – The wave of M&As in the energy industry and the EU merger policy. Energy Policy, Vol. 36, No. 3, pp. 1125–1133.

Victor, David G., 2006: Natural gas and geopolitics. Paper Presented at the 23rd World Gas Conference, 9 June.

Vives, Xavier, 2008: Competition policy in Europe. Occasional Paper OP-161. http://www.iesep.com/Descargas/spdf/Gratuitos/R141.pdf.

Vollebergh, Herman R. J.; de Vries, Jan L.; Koutstaal, Paul R., 1997: Hybrid carbon incentive mechanisms and political acceptability. Environmental and Resource Economics, Vol. 9, pp. 43–63.

von Lichtenstein, Hildegard, 2007: "From Russia with gas": Forthcoming political disasters for Europe in energy supply. Working Program European Identities-Policies section, No. 10. http://www.institut-thomas-more.org/pdf/136_fr_WP-ITM10Eng.pdf.

Wäktare, Eleonora; Kovács, Kristóf; Gee, Alexander, 2007: The energy sector inquiry: Conclusions and way forward. Competition Policy Newsletter, No. 1.

Walters, Stephen; Meledo, Olivier, 2010: Acquisitions in the gas sector: The impact of sector regulations. Thomas Reuters (Legal), London.

Weatherill, Stephen, 2011: The limits of legislative harmonization ten years after tobacco advertising: How the court's case law has become a "drafting guide". German Law Journal, Vol. 12, No. 3, pp. 827–864.

Weihe, Thomas, 2007: Das Schwarze Meer zwischen der EU und Russland Sicherheit. Energie. Demokratie/134. Bergedorfer Gesprächkreis. Körber-Stiftung, Hamburg, 2007.

Weiss, Wolfgang, 2009: Agencies versus Networks: From divide to convergence in the administrative governance in the EU. Administrative Law Review, Vol. 61, Special Issue, pp. 45–70.

Whist, Bendik Solum, 2008: Nord Stream: Not just a pipeline. An analysis of the political debates in the Baltic Sea region regarding the planned gas pipeline from Russia to Germany. FNI Report 15/2008. Fridtjof Nansens Institutt.

Wilkinson, David, 1992: Maastricht and the environment: The implications for the EC's environment policy of the Treaty on European Union. Journal of Environmental Law, Vol. 4, No. 2, pp. 221–239.

Willis, Peter; Hughes, Paul, 2008: Structural remedies in Article 82 energy cases. Competition Law Review, Vol. 4, No. 2, pp. 147–174.

WTO Analytical Index: General Agreement on Trade in Services, http://www.wto.org/english/res_e/booksp_e/analytic_index_e/gats_01_e.htm#article1B5.

WTO Members and Observers. http://www.wto.org/english/thewto_e/whatis_e/tif_e/org6_e.htm.

Yataganas, Xénophon A., 2001: Delegation of regulatory authority in the European Union. The Jean Monnet Center for International and Regional Economic Law and Justice.

Youngs, Richard, 2007: Europe's external energy policy: Between geo-politics and the market. CEPS Working Document No. 278/November. www.ceps.eu/ceps/download/1390.

Zimmermann, Charles F.; Talus, Kim, 2008: Regulation of Electricity Markets at the EU level. European Energy and Environmental Law Review, Vol. 17, No. 1, pp. 12–22.

Annex I: Legislation applicable to the EU gas market

	Unbundling	Transmission	Transport (pipelines)	Storage	Cross-border trade
EU Law					
EU Treaties					
Adjacent Provisions			Article 56 TFEU prohibition of restrictions on freedom to provide services within the Union (this could be linked to territorial restriction clauses); article 58 (1) TFEU freedom to provide services in the field of transport shall be governed by the provisions of the Title VI relating to transport: articles 90–100 TFEU, especially art 91(1) EP and Council though the ordinary legislative procedure may lay down specific transport rules regarding international transport involving a MS, nonresident carriers, transport safety.		

Antitrust		GDF–ENI/ GDF–ENEL decisions, 26.10.2004: clauses that restrict the freedom of one of the contracting parties to use the delivered merchandise according to its own economic interests pose limits to competition (article 101 TFEU). Article 101(1)(d) prohibition of undertakings to apply dissimilar conditions to equivalent transactions with other trading parties (can be linked to non-discriminatory third-party access).	Article 101(1)(d) TFEU as in the case of transport.	Article 101(1)(a) TFEU incompatible with the internal market all agreements between undertakings and concerted practices that may affect the trade between the MS such as the direct or indirect fixing of purchase or selling prices or other trading conditions.
State Aid	**Article 107(1) TFEU:** if a public authority undertakes to provide infrastructure to impove connections to a new or upgraded site on which energy production facilities are to be sited, it may be concluded that this could amount to a selective benefit in the sense or article 107(1) TFEU. exception, if the use of the facility is subject to open and non-discriminatory access and the owners or operators of the infrastructure do not receive a selective benefit.			
Trans-European Networks		Articles 170–172 TFEU: interconnection and interoperability of national networks, as well as access to such networks;		Article 170(2) TFEU promotion of the interconnection and interoperability of national networks

	Unbundling	Transmission	Transport (pipelines)	Storage	Cross-border trade
International Agreements					
Specific Energy Provisions article 194 TFEU			Article 194 (1) (d) TFEU. Interconnection of energy networks		
EU Legislation					
Secondary Law governing the Gas Market					
Directive 2009/73/EC	Recitals (5)–(16): the need for efective separation of networks from activities of production and supply, the failure of legal and functional unbundling to prove effective; three possible ways: ownership unbundling, independent system operator ISO and independent transmission operator ITO; (18) the unbundling options are mutually exclusive; (20) implementation of effective unbundling should respect the principle of non-discrimination between the public and the private sector; (21) reciprocity: fully effective separation of network activities from supply and production should apply throughout the Community to both	Recital (25) and article 32: third party access to distribution and transmission systems.	Recital (23) transparent and non-discriminatory tariffs for access to transport; article 34 access to upstream pipeline networks.	Recital (23) and article 33 access to storage facilities; linepack and ancillary services; (24) independence of storage system operators in order to improve third-party access to storage.	

	Community and non-Community undertakings. Article 9 unbundling of transmission systems and transmission system operators. Article 14 independent system operators. Article 15 unbundling of transmission system owners and storage system operators. Articles 17–23 independent transmission operator. Article 26 unbundling of DSOs. Article 31 unbundling of accounts.				
Regulation (EC) No. 715/2009		Article 1 (a) nondiscriminatory rules for access conditions to: natural gas transmission systems; harmonizes principles for tariffs or the methodologies underlying their calculation (i.e. article 11 inclusion of the costs related to ENTSO-G activities, borne by TSOs), for access to the network (article 13 on tariffs for access to the network), third-party access services (article 14 on third-party access services concerning TSOs), harmonized principles for capacity-allocation and congestion-management (article 16). Article 2 definition of transmission annex I: guidelines on third-party access concerning TSOs, principles of capacity-allocation mechanisms and congestion-management	Article 2 definition of transport contract.	Article 1(b) nondiscriminatory rules for access conditions to: LNG facilities and storage facilities (see article 15 concerning third-party access services concerning storage and LNG facilities and article 17 regarding principles of capacity-allocation mechanisms and congestion-management procedures concerning storage and LNG facilities).	Article 1(c) harmonization of network access rules for cross-border exchanges in gas; determination of balancing rules and imbalance charges, facilitation of capacity trading; article 8 (7) the network codes developed by ENTSO-G shall be developed for market integration issues; article 21 balancing rules and imbalance charges (the latter shall avoid cross-subsidization between network users and shall not hamper the entry of new market entrants); article 21 (4) harmonization of balancing systems and streamline structures by TSOs in order to facilitate gas trade.

	Unbundling	Transmission	Transport (pipelines)	Storage	Cross-border trade
Regulation (EC) No. 715/2009 (cont.)		procedures concerning TSOs, definition of technical information necessary for network users to gain effective access to the system, the definition of all relevant points for transparency requirements and the information to be published at all relevant points and the time schedule according to which that information shall be published.			
Regulation (EU) No. 994/2010		Recital (7): priority infrastructure actions (tranmission network development) – the southern gas corridor (Nabucco and Interconnector Turkey Greece Italy), a diversified and adequate LNG supply for Europe, effective interconnection of the Baltic region, the Mediterranean Energy Ring and adequate north–south gas interconnections within central and south-east Europe; (14) sufficient and diversified gas infrastructure also connecting current isolated systems is important for the security of supply.			

TEN-E Regulation			
Regulation (EC) No. 713/2009 of the European Parliament and of the Council of 13 July 2009 establishing an Agency for the Cooperation of Energy Regulators	Article 6 Cooperation of the TSOs: (4) ACER: shall participate in the development of network codes, submit a nonbinding framework guideline to the Commission; (6) ACER shall monitor and analyse the implementation of the network codes and the Guidelines adopted by the Commission and their effect on the harmonization of applicable rules aimed at facilitating market integration as well as nondiscrimination, effective competition and the efficient functioning of the market; (7) ACER: monitor the creation of new interconnector capacity; (8) ACER: monitor the implementation of EU-wide network-development plans; (9) ACER: monitor the regional cooperation of TSOs.		Recital (5) MS close cooperation to eliminate obstacles to cross border exchanges of natural gas with a view to achieving the objectives of Community energy policy. Article 8 terms and conditions for access to and operational security of cross-border infrastructure: ACER – which regulatory issues are to be solved by NRAs

	Investment	Security of supply	Public service obligations	Supply
International Law	Provisional application of treaties: article 25 (1) Vienna Convention; Interpretation of treaties: article 31 of the Vienna Convention.			
International Agreements				
Energy Charter	Definition of investment: article 1(6) ECT. Provisional application of treaties: article 45 (1) ECT; ICSID case No. ARB/05/18 Ioannis Kardassopoulos v Georgia; article 45(2) ECT exception from the provisional application.			
EU Law				
EU Treaties				
General Principles			Principle of proportionality C-265/08, Federutility: obligations imposed in the general EC interest must be necessary to achieve this objective and must be limited in time; article 14 TFEU services of general economic interest.	
Adjacent Provisions	Article 207(1) TFEU common commercial policy shall be based on uniform principles regarding among others the commercial aspects of foreign direct investment and will be conducted in the context of the principles and objectives of the Union's external action.			Article 37 TFEU provides that MS are obliged to adjust any state monopolies of a commercial character so as to ensure that no discrimination regarding the conditions under which goods are procured and marketed exists between nationals of Member States (this could be linked to the idea of competion among suppliers and producers).
Antitrust	Article 101(1)(b) prohibition of undertakings' practices limiting investment.			101(1) TFEU applied to gas cases: restrictions in vertical supply agreements either preventing the buyer from using the gas for other

				purposes than agreed upon (use restrictions) or from selling the gas outside an agreed supply area. Territorial and use restrictions are essential restrictions entering the scope of EU competition law. Similarly profit splitting mechanisms violate EU competition rules and have comparable effects to restriction clauses (IP/02/1869 Commission and NLNG).
State Aid		Article 107(3) TFEU discretionary exceptions from article 107(1) in the interest of security of supply.	Article 106 (2) TFEU stipulates that "undertakings entrusted with the operation of services of general economic interest… shall be subject to the rules contained in this Treaty, in particular to the rules on competition, insofar as the application of such rules does not obstruct the performance, in law or in fact, of the particular tasks assigned to them. The development of trade must not be affected to such an extent as would be contrary to the interests of the Community". The article contains an escape clause for undertakings entrusted with a service of general interest: Federutility case C-265/08; article 107 case-law: Altmark, which sets four conditions when state compensation does not constitute state aid; Ferring C-53/00 [2001] ECR. I-9067; Enirisorse joined cases: Gemo SA, case C-126/01; article 93 TFEU aids are compatible with the Treaty if they meet the needs of coordination of transport or if they represent the reimbursement of certain obligations in the category of PSOs.	(1) TFEU general prohibition of state aids (in order to fall within the scope of this article, the state measure in question must have the effect of conferring a selective benefit or advantage); selectivity criterium reflected in Case C-143/99 Adria-Wien Pipeline; article 107(2), (3) exemptions from the state aid rules.

	Investment	Security of supply	Public service obligations	Supply
Trans-European Networks		Article 170 (1) TFEU: the achievement of the objectives enumerated in article 26 TFEU: Union's strategic interests, matters with defence implications, common foreign and security policy scope.		
International Agreements				
Specific Energy Provisions article 194 TFEU		Article 194(1) TFEU b. Ensure security of supply in the Union		
EU Legislation				
Secondary Law governing the Gas Market				
Directive 2003/55/EC			Article 3 of Directive 2003/55/EC, recognizes that the MS may impose PSOs relating to securing (security of supply, regularity, quality and price of supplies) and environmental protection on gas undertakings. Recital 27 of the same Directive notes that common minimum standards, respected by all MS, should be specified in the Directive. In the Commission's "guiding notes" of 16 January 2004, it is observed that only services which cannot be provided by the market can be deemed to be PSOs. Therefore, supply of gas to eligible customers cannot be deemed PSO and as of 2007 supply to all classes of consumers falls outside the scope of article 106(2) TFEU. In case of security and reliability of supply/ consumer and environmental protection being invoked as PSOs, the application of article 106(2) cannot be ruled out as a justification for financial	Article 3 of the Directive 2003/55/EC strengthened the existing provisions regarding consumer protection.

			support in relation to such objectives. A Commission's interpretative note concludes that PSOs must be clearly defined, transparent, non-discriminatory and verifiable	
Directive 2009/73/EC	Recital (35) investments in major new infrastructure while ensuring the proper functioning of the internal market in natural gas; (45) necessary investment to achieve objectives of social and economic cohesion.	Recital (22): security of energy supply is an essential element of public security and inherently connected to the functioning of the internal market in gas and the integration of the isolated gas markets of Member States, (40) the balance between supply and demand should be monitored in order to preserve the security of supply.	Recital (44) respect for PSOs is fundamental and it is important that common minimum standards respected by all Member States are specified in the Directive, (47) more precisions on PSOs; article 3(1), (2), (7), (10), (11)	Recital (42) long-term contracts should be taken into account when planning the supply and transport capacity of natural gas undertakings.
Regulation (EU) No. 994/2010	Recital (15): promotion of investments in new infrastructure (prior environmental impact assessment necessary) to enhance the security of gas supply and ensure the proper functioning of the internal market in gas; (27) necessary investments in indigenous production and infrastructures (interconnections providing access to the gas network of the Union, equipment allowing physical bidirectional gas flows on pipelines) should be made on time by natural gas undertakings – priority to those projects which support the integration of the internal gas market and security of gas supply.	Recital (2) security of gas supply aspects in relation to the existence of gas islands; (4) the new energy legislative package has an explicit objective to enhance the Union's security of energy supply; (7) + (8) diversification of gas routes and sources of supply to improve the security of supply of the Union as a whole and its Member States individually; the security of supply will depend in the future on the evolution of the fuel mix, development of production in the Union and third countries, investments in storage facilities and diversification; (26) measures necessary to ensure the fulfilment of the supply standard include: additional storage capacities and volumes, linepack, supply contracts, interruptible contracts, necessary technical measures; annex II+: list of market-based and nonmarket-based security of gas supply measures	Recital (49) Member States enjoy a wide discretion in providing for, commissioning and organizing public service obligations.	Recital (46) important that conditions for the supply from third countries do not distort competition and are in accordance with internal market rules

	Investment	Security of supply	Public service obligations	Supply
Balancing Rules				
TEN-E Regulation				
Regulation (EC) No. 713/2009 of the European Parliament and of the Council of 13 July 2009 establishing an Agency for the Cooperation of Energy Regulators	Article 4 (8) ACER: ensuring implementation of investments in accordance with the EU-wide network-development plans.	Recital (15) ACER: enhancing security of supply.		

	Consumers/consumer protection	Environmental protection	Other exemptions	Market foreclosure	Internal market (market integration)
EU Law					
EU Treaties					
General Principles	Article 12 TFEU: consumer protection requirements shall be taken into account in defining and implementing other Union policies and activities.	Article 11 TFEU environmental protection must be taken into account when defining Union policies and activities.			Article 3(3) TEU the Union shall establish an internal market; article 20(1) Member States enhanced cooperation.
Adjacent Provisions	Article 169 TFEU promotion of the interest of the consumers and insurance of a great level of consumer protection; how do articles 227 and 228 TFEU interact with private enforcement of consumer rights? The articles provide the right of MS individuals to address a petion to the EP on a matter which comes within the Union's fields of activity and which affects them directly. Article 228 provides for the interposition of the Ombudsman.				Article 27 TFEU derogations (temporary, the least disturbance to the functioning of the internal market); article 197(1) TFEU effective implementation of Union law by the MS essential for the proper functioning of the Union and regarded as a matter of common interest; article 207(1) and (6) TFEU the common commercial policy enviews the achievement of uniformity in measures of liberalization, but only in so far as the Treaties do not exclude such harmonization.
Antitrust	101 (3) the provisions of (1) do not apply in case the agreement between undertakings or concerted practices results in the improvement of the production or distribution of goods or in the promotion of technical or economic progress, while allowing consumers a fair share of the resulting benefit; article 102 (b) TFEU acts also in favour of the consumer's protection.			Article 101 (1)(c) TFEU prohibition of the sharing of markets or sources of supply between undertakings; COMP/37.541 Gas Natural+ ENDESA the Commission concluded that the original supply agreement between the two undertakings would consitute an infringement of art 102 TFEU as it had the same	Article 102 TFEU prohibition of any abuse of one or more undertakings of a dominant position within the internal market.

	Consumers/consumer protection	Environmental protection	Other exemptions	Market foreclosure	Internal market (market integration)
Antitrust (cont.)				effect as enforcing the already dominant position of Gas Natural in the Spanish gas market.	
State Aid	C-56/93 Belgium v Commission [1996] ECR I-723; joined cases 67, 68 and 70/85 Van der Kooy [1988] ECR 219 (discrimination among classes of consumers); C-482/99 France v Commission (Stardust Marine) (implications on energy matters due to the fact that several MS still own or control parts of their countries' energy enterprises). Article 107 (1) selectivity criteria: exemptions for certain category of consumers (i.e. large consumers) to pay a form of stranded costs levy into a compulsory fund, are considered a form of selective benefit. Aid may be provided to a certain group of final consumers through energy companies owned or controlled by the state insofar as the resources of these companies can be considered state resources. Such aid may be granted through: (a) preferential tariffs for energy supply – certain types of long-term contracts might enter this category; (b) cross-subsidization: i.e. Case C-280/00 and C-94/01P Chronopost if the revenue which a company earns on its reserved activities exceeds a reasonable level of compensation as calculated on the basis of the Altmark criteria and at the same time it uses its revenue to	Article 107(3)(c) TFEU discretionary exceptions from article 107(1) in the interest of environmental protection: article 107(3)(b) TFEU aid might be compatible with the common market if it promotes the execution of an important project of Common European interest – such as an environmental priority.	Regional aids, article 107(3)(a), i.e. Commission's decision N/660/A/2000 – exemption from climate change levy for natural gas in Northern Ireland: special geographical and political circumstances which had delayed the arrival of gas to Northern Ireland until 1996, correlated with the low levels of consumption and the resultant high gas prices in that region contributed to the state of fact of the gas infrastructure in NI. It was decided that an exemption from the CCL for natural gas consumption was necessary to prevent the further obstruction of the extension of gas infrastructure in NI.		

	cover its costs in the competitive sector (Chronopost criteria), then this could give rise to state aid in the meaning of article 107(1) TFEU.		
Trans-European Networks			Integrated market: article 170 (1) TFEU: achieval of the objectives set in article 174 TFEU: overall harmonious development of the Union, strengthening of its economic, social and territorial cohesion. Article 170 (2) : integration of landlocked and peripheral regions.
International Agreements			
Specific Energy Provisions article 194 TFEU			Article 194(1) TFEU spirit of solidarity among MS required to (a) ensure the functioning of the energy market; (b) ensure security of energy supply in the Union; (c) promote energy efficiency and energy saving and the development of new and renewable forms of energy; (d) promote the interconnection of energy networks.
EU Legislation			
Secondary Law governing the Gas Market			
Directive 2003/55/EC	Federutility, C-265-08: regarding Directive 2003/55/EC article 23(1)(c) providing the customers freedom to buy natural gas from the supplier of their choice from 1 July 2007 on and article 3 (the same in the new Directive) providing that natural gas undertakings must be operated in view of achieving a competitive market in gas.		

	Consumers/consumer protection	Environmental protection	Other exemptions	Market foreclosure	Internal market (market integration)
Directive 2009/73/EC	Recital (17) protection of customers: free choice of suppliers; gas contracts with several suppliers; protection against exclusivity clauses, (47) all consumers including vulnerable ones should benefit from competition and fair prices, (48) consumers interests should be at the heart of the Directive and quality of service should be a central responsibility of natural gas undertaking, (50) energy poverty, (51) greater consumer protection ensured by the availability of effective means of dispute settlement for all consumers; article 3(3), (4), (5), (6), (9); article 37 market opening and reciprocity annex I measures on consumer protection.		Recitals (27) + (28) exemption of small DSOs from legal distribution unbundling requirements and unnecessary administrative burdens. Article 28 exemptions for closed distribution systems; article 35 motivated refusal of third-party access; article 36 new infrastructure: exemptions for a defined period of time from the provisions of articles 9, 32, 33 and 34 and 41(6), (8) + (10) under certain conditions (enhancement of competition, certain level of risk, at the minimum legal unbundling, the exemptions must not be detrimental to competition or the effective functioning of the market). Article 48 derogations in relation to take-or-pay commitments; article 50 exemption from certain obligations which are not proportionate to the objective pursued.		Recitals (1)–(5) internal market in gas, progress made by Directive 2003/55/EC, persisting obstacles (absence of nondiscriminatory third party access and lack of an equally effective level of regulatory supervision in each Member State), lessons from the 2007 documents "An Energy Policy for Europe" on the completion of the internal market and the creation of a level playing field for all natural gas undertakings; (19): under this directive different types of market organisations will exist in the internal market in natural gas; (56)–(58) the development of a true internal market in natural gas: MS should foster the integration of their national markets and the cooperation of system operators at Community level, and the incorporation of isolated systems creating gas-islands markets; involvement of NRAs and ACER in the creation of such a market; incentives for cross-border interconnection. Article 1(1) lays down the rules for the organization and functioning of the gas sector, access to the market etc. Article 6 regional solidarity; article 7 promotion of regional cooperation; article 49 emergent and isolated markets.

Regulation (EC) No. 715/2009			Article 30 derogations and exemptions: reference to Article 49, 36 (1) + (2), 48 of Directive 2009/73/EC		Article 8(7) the network codes developed by ENTSO-G shall be developed for market integration issues.
Regulation (EU) No. 994/2010	Recital (3) increased protection of customers to ensure them continuous supply, particularly in case of difficult climatic conditions and in the event of disruptions; (9) + (10) vulnerable groups of customers – households and customers providing essential social services; (29) gas supply maintained particularly as regards household customers and a limited number of additional customers (delivering essential social services) in cases where the market is unable to ensure the supply; article 2 (1) definition of “protected customer” article 8 (3) supply standards: maintenance of gas supply in particular for protected customers (role of competent authority and natural gas undertakings).		Article 6 (10) exemptions from infrastructure standard regarding Luxembourg, Slovenia and Sweden and obligations on behalf of the said countries to submit regular reports to the Commission regarding the state of the conditions that allowed them to benefit of exemptions. Article 7 procedure for enabling bi-directional capacity or seeking exemptions Article 16 derogation for Malta and Cyprus from the application of the Regulation		Recitals (11) + (12) connection between internal gas market and security of energy supply in the Union; competition should not be unduly distorted; (22) the completion of the internal gas market requires a comprehensive and effective common approach to security of supply, transparency, solidarity and non-discriminatory policies; (39) the sovereign rights of Member States over their own energy resources are not affected by this Regulation.
Merger Regulation:Council Regulation (EC) No. 139/2004 of 20 January 2004 on the control of concentrations between undertakings				Article 3 of Merger Regulation 139/2004 of the Council of 20 January 2004 defines the types of concentrations on the market.	

	Consumers/consumer protection	Environmental protection	Other exemptions	Market foreclosure	Internal market (market integration)
Regulation (EC) No. 713/2009 of the European Parliament and of the Council of 13 July 2009 establishing an Agency for the Cooperation of Energy Regulators			Article 9 (1) ACER may decide on exemptions as provided for in art 36(4) of Directive 2009/73/EC – new infrastructure with cross-border elements.		Recital (5) ACER for the effective functioning of the internal markets in electricity and gas. Recital (10) uniform application of the legislation on the internal market in natural gas throughout the EU – this will be achieved in an integrated framework provided by ACER to NRAs and facilitating their participation and cooperation.

	Transparency/information sharing	National governments'/national regulators' powers	European Regulators: European Commission, ACER, ENTSO-G, Gas Coordination Group	Shared competences/cooperation among regulators
EU Law				
EU Treaties				
General Principles	Article 15(1) TFEU transparency in the work of the Union's institutions, bodies, offices and agencies, (3) access to documents.		Article 17 TEU legislative initiative, oversight of the Union law's application under the control of the CJEU; ensure the Union's external representation (with the exception of CFSP and other cases provided for in the Treaties), accountable to the EP	Article 3(6) TEU the Union shall pursue its objectives by appropriate means conforming to the competences conferred upon it by the Treaty. Article 4 (1) competences not conferred upon the Union remain with the MS. Article 4(3) mutual assistment between EU and MS in carrying out tasks flowing from the Treaties. Article 5 principle of conferral, subsidiarity and proportionality. Article 2(1) TFEU exclusive competence of the Union; article 2(2) shared competence between Union and MS (both can legislate and adopt legally binding acts in that area); article 3(b) TFEU EU exclusive competence in establishing the competition rules necessary for the functioning of the internal market; article 4(2) TFEU shared competence over energy.
Adjacent Provisions	Article 197(2) TFEU improving the administrative capacity of MS to implement Union law through the facilitation of the exchange of information and of civil servants between them.		Article 258, the EC has the attribute of setting in motion an infringement procedure should a MS fail to fulfil an obligation under the Treaties. (IP/06/1391 Competition: Commission opens infringement procedure against Malta for maintaining import monopoly for petroleum products).	Article 26 TFEU the Union shall adopt measures with the aim of establishing or ensuring the functioning of the internal market in accordance with the Treaty provisions; article 197(3) TFEU administrative cooperation among the MS and between them and the Union; article 207 (6) the exercise of competences conferred by this article in the field of the common commercial policy shall not affect the delimitation of competences between the Union and the MS; article 291 TFEU Ms shall adopt all measures of national law necessary to implement legally binding Union acts, but where uniform conditions for implementing legally binding Union acts are

	Transparency/information sharing	National governments'/national regulators' powers	European Regulators: European Commission, ACER, ENTSO-G, Gas Coordination Group	Shared competences/cooperation among regulators
				needed, those acts shall confer implementing powers on the Commission.
Antitrust		Article 101 TFEU, E.ON Ruhrgas vs. Bundeskartellamt.	Power of the Commission to demand amendments to contracts which do not conform with the liberalization of the gas market and the respect of competition principles (i.e. when infringing Article 102 TFEU)	
State Aid		Energy taxes: specific or special charges are solely concerned with an MS's discretion in forming its tax system and do not constitute state aid in the meaning of article 87(1).	Article 108 (2) the Commission can decide if an aid granted by a state is compatible or not with the internal market and can demand that the aid is terminated.	Article 108(1) TFEU the Commission in cooperation with the MS shall keep under constant review all systems of aid existing in those States.
Trans-European Networks				Article 171 TFEU shared competency upon TEN.
International Agreements				Articles 216 and 219(4) both the EU and the MS are competent to conclude international agreements (this could be linked to bilateral agreements signed between MS and certain external gas providers).
Specific Energy Provisions article 194 TFEU				Article 194(2) TFEU shared competency: energy policy; 194(3) TFEU exclusive competency: fiscal.

EU Legislation				
Secondary Law governing the Gas Market				
Directive 2003/55/EC	Article 6 of Directive 2003/55/EC provides that all measures or changes to such measures, whether these may involve a derogation from the Directive or not, to carry out PSOs obligations shall be notified separately to the Commission.			
Directive 2009/73/EC	Recital (36) increase of regulatory oversight in the market in order to strengthen the trust in the market; (37) transparency in regards to the formation of prices, (43) notifications and publication of reports, (46) to the extent to which measures taken by Member States to fulfill public service obligations constitute State aid there is an obligation under article 108(3) TFEU to notify them to the Commission, (47) consumers should have access to their consumption data, (49) clear and comprehensible information should be made available to consumers concerning their rights in relation to the energy sector; article 16 confidentiality obligations of TSOs and transmission system owners; article 27 confidentiality obligations of DSOs; article 30 right of access to accounts.	Recitals (29) + (30) strenthening of the role of NRAs + reinforced independence; (31) NRAs active role in ensuring that balancing tariffs are nondiscriminatory and cost-reflective; (32)–(34) powers of NRAs regarding fixing or approval of tariffs or the methodologies accompanying them, issuing of binding decisions in relation to natural gas undertakings and the power to impose effective, proportionate and dissuasive penalties on natural gas undertakings, which fail to comply with their obligations, ensuring of high standards of public service and the protection of vulnerable consumers. Article 4 authorization procedure: MS or any competent authority may grant or decline an authorization for the construction or operation of natural gas facilities. Article 5 MS or regulatory authorities will monitor the security of supply. Article 8 NRAs shall define the technical rules establishing the minimum technical design and operational requirements for the connection to the system of LNG facilities, storage facilities, other transmission or distribution	Recital (59) the Commission assisted by NRAs should observe and monitor the internal market from the short to the long term; (60) Competence of the Community through the Commission to adopt measures in accordance with the subsidiarity and the proportionality principles; (61) the Commission may adopt guidelines to achieve the necessary degree of harmonisation; such guidelines would constitute binding implementing measures; (62) recall of the Commission's implementing powers vis-a-vis the present Directive; (63) minimum degree of harmonization required to achieve the aim of the Directive. Article 36(4) where the new infrastructure is located in the territory of one or more MS the Agency may submit an advisory opinion to the regulatory authorities of the MS concerned, which may be	Recital (55) framework for regional cooperation in a spirit of solidarity. Article 3 (12) Commission in consultation with relevant stakeholders will design a concise and clear energy consumer checklist of practical information relating to energy consumer rights. Articles 10 and 11 cooperation between Commission and NRAs in the designation and cerification of TSOs and the certification in relation to third countries.

	Transparency/information sharing	National governments'/national regulators' powers	European Regulators: European Commission, ACER, ENTSO-G, Gas Coordination Group	Shared competences/cooperation among regulators
Directive 2009/73/EC (cont.)		systems and direct lines. Article 12 MS designation of storage and LNG operators. Article 13 tasks of TSOs, storage and/or LNG system operators. Article 24 MS designation of DSOs. Article 25 tasks of DSOs. Article 35 (2) MS measures in relation to refusal of access. Article 38 MS or relevant authorities may authorise the construction of direct lines. Articles 39–44 NRAs. Article 46 MS competent to take safeguard measures. Article 54 MS transposition.	used as a basis for their decision.	
Regulation (EC) No. 715/2009	Article 18 Transparency requirements concerning TSOs; article 19 Transparency requirements concerning storage facilities and LNG facilities; article 20 record keeping by system operators; article 21(3) any calculation methodology for imbalance charges as well as final tariffs shall be made public by the competent authorities or the transmission system operator. Article 25 provision of information from the Member States to the Commission on the purposes of Article 23.	Article 1 Member States may establish an entity equivalent to TSO which will be submitted to certification Article 5 establishment of ENTSO-G. Article 22 Trading of capacity rights by TSOs, LNG system operators, storage system operators. Article 26 right of the Member States to provide for more detailed measures. Article 27 Member States shall lay down rules on penalties applicable to infringements of the provisions of the Regulation and shall implement those rules.	Article 3 Commission's competence: certification of transmission system operators Article 6(1), (2), (4), (5), (6), (9), (10), (11) Commission's responsibilities regarding the establishment of network codes. Article 7 amendment to network codes by ACER, ENTSO-G, TSOs, etc. Article 8 ENTSO-G tasks: elaboration of network codes, adoption of common network operation tools to ensure coordination of network operation, nonbinding Community-wide 10-year network development plan, recommendations relating to the coordination of technical cooperation between Community and 3rd country	Article 4 cooperation of TSOs through ENTSO-G (completion and functioning of the internal market in natural gas and cross-border trade, optimal management, coordinated operation, etc.). Article 6(3) consultation between ACER and ENTSO-G in regards to the framework guideline. Article 6(7) consultation between ACER and relevant stakeholders. Article 10 consultations between ENTSO-G and relevant stakeholders (in the development of network codes, the draft Community network development plan, the annual work programme). Article 12 regional cooperation of TSOs. Article 23 Commission after consultation with ACER and ENTSO-G shall adopt guidelines providing the minimum degree of harmonization required to achieve the aims of the Regulation (in regards to third-party access services, capacity-allocation mechanisms and congestion-management procedures, tariff methodology related to cross-border trade).

			TSOs, adoption of an annual programme, of an annual report, of annual summer and winter supply outlooks Article 9 monitoring by ACER (i.e. monitoring of the implementation by ENTSO-G of network codes elaborated under article 8(2). Article 29 the Commission shall monitor the implementation of the Regulation.	Article 24 cooperation of regulatory authorities.
Regulation (EU) No. 994/2010	Recital (43) measures taken by competent authorities to redress crisis situations should be notified to the Commission; (46) where reliable information of a situation outside the Union threatening security of supply, the Commission should inform GCG without delay and the Union should take measures to try to defuse the situation; (47) transparency and reliability should be increased through meaningful exchange of information between the Commission and Member States on energy relations with third countries, including long-term supply arrangements, while preserving commercially sensitive information. Article 3(3) + (6) obligation of notification to the Commission of the name of competent authorities and transparency requirements for preventive action plans and emergency plans. Article 10(5) + (6) emergency plans and crisis levels: when the comepetent authoritity declares any crisis levels or an emergency it shall immediately inform the Commission; article 13 information exchange.	Recital (5) Member States enjoy a large margin of discretion as to the choice of measures regarding the ensurance of security of supply – according to the new legislative package they are required to foster solidarity and coordination in response to supply crises (preventive action + reaction to concrete disruptions of supply); recital (16) competent authorities or Member States should ensure that the gas market is tested as one of the necessary steps in the course of the process leading to compliance with the infrastructure model; (21) new roles and responsibilities for Member States, national regulatory authorities, transmission system operators; (28) TSOs should be allowed to consider the benefits on security of supply exerted by investments enabling physical capacity to transport gas in both directions in cross-border interconnections with third countries. Article 2(2) definition of competent authority. Article 3(4) + article 4 role of the competent authority. Article 6	Recital (15) in the case of investments for new infrastructure involving cross-border elements, ACER and ENTSO-G should be consulted in their respective areas of competence; (21) new roles and responsibilities for the Agency; (44) Commission's competences – with respect to the coordination of the actions regarding third countries (ensuring supply and transit) involving arrangements to handle crisis situations and the ensurance of a stable gas flow to the Union; Commission entitled to deploy a task force to monitor gas flows into the Union in crisis situations in consultation with the third countries involved; Commission entitled to assume mediation and facilitation role where a crisis arises due to difficulties in a third country. Article 3(5)	Recitals (17) + (18) close cooperation between competent authorities and other relevant national authorities (i.e. NRAs) and between Member States concerned, competent authorities and NRAs in the case of new cross-border interconnections; (23) + (24) security of gas supply is a shared responsibility of natural gas undertakings, Member States (through competent authorities) and the Commission within their respective areas of activities and competence (three-level approach); (25) + (26) coordination of measures in a spirit of solidarity and harmonization of security of supply standards; (33) + (34) + (42) ENTSO-G and ACER (members of Gas Coordination Group) should be fully involved in the process of cooperation and consultations at Union level; the GCG is the main body to be consulted by the Commission in the context of establishment of Action Plans and Emergency Plans; GCG adviser to the Commission to facilitate the coordination of security of supply measures in the event of a Union emergency; (36) Member States should devise measures to exercise solidarity. Article 3(1) ensuring security of supply is a shared responsibility involving a high degree of cooperation

	Transparency/information sharing	National governments'/national regulators' powers	European Regulators: European Commission, ACER, ENTSO-G, Gas Coordination Group	Shared competences/cooperation among regulators
Regulation (EU) No. 994/2010 (cont.)		role of Member States, competent authorities, NRAs, TSOs with respect to infrastructure standard; article 10 role of competent authorities and Member States in regards to emergency plans and crisis levels. Article 12 Gas Coordination Group.	Commission is competent to coordinate the action of competent authorities on its own or through GCG. Article 11(1) Commission may declare (discretion) a Union emergency or a regional emergency for a specifically affected geographical region (at the request of a Competent Authority); the Commission shall declare (obligation) a Union emergency at the request of at least two Competent Authorities; article 11(6) if the Commission considers (on its own or at the suggestion of a Competent Authority or natural gas undertaking) that in a Union or regional emergency an action taken by a Member State or a Competent Authority is contrary to the provisions of (5) it shall request that they change their action accordingly. Article 14 monitoring by the Commission.	between natural gas undertakings, Member States (competent authorities) and the Commission; article 3(2) cooperation among competent authorities to prevent a supply disruption annex IV: regional cooperation.
Merger Regulation:Council Regulation (EC) No. 139/2004 of 20 January 2004 on the control of concentrations between undertakings		E.On/Endesa Case power of regulators to impose conditions on mergers with Community dimension brought under scrutiny.	Article 8 gives the Commission the power to decide whether a notified concentration is compatible or incompatible with the Treaty. In the latter case, article 8(4) (b) provides that	Article 1 sets the division of competences between the Commission and the NCAs basing on the application of the turnover thresholds, which take into consideration the combined aggregate worldwide turnover (more than EUR 5000 million) and the aggregate Community-wide turnover of each

			the Commission might either require the dissolving of the concentration or impose any other measures to ensure the termination of the concentration or take other restorative measures (discretion). Articles 14 and 15 provide for pecuniary measures: fines and periodical penalties. Article 21 of the Merger Regulation: exclusive jurisdiction of the Commission to review concentrations that have a community dimension (and assess the competitive impact of such concentrations).	of at least two of the undertakings concerned (more than EUR 250 million).
Regulation (EC) No. 713/2009 of the European Parliament and of the Council of 13 July 2009 establishing an Agency for the Cooperation of Energy Regulators	Recital (6) ensuring ACER's transparency, amenability to democratic control and efficiency. Recital (7) transparency of TSOs cooperation. Recital (23) general rules regarding public access to documents held by EU bodies, protection of commercially sensitive data and personal data. Article 10 consultation and transparency: extensive consultation with market participants, TSOs, consumers, end-users, competition authorities Article 30 access to documents: provisions on the public access to documents shall apply to the documents held by the Agency	Recitals (3) + (5) Voluntary cooperation among national regulators should take place within a Community structure with clear competences and with the power to adopt individual regulatory decisions in a number of specific cases.	Recital (4) Commission – proposal for the set up of an independent mechanism for national regulators to cooperate. Recital (27) the Commission should be empowered to adopt the Guidelines necessary in situations when ACER becomes competent to decide upon the terms and conditions for access to and operational security of cross-border infrastructure. Article 4(4) Commission – discretion when deciding on adopting Guidelines on the situations in which the Agency becomes competent to decide upon the terms and conditions for access to and operational security of cross-border	

	Transparency/information sharing	National governments'/national regulators' powers	European Regulators: European Commission, ACER, ENTSO-G, Gas Coordination Group	Shared competences/cooperation among regulators
Regulation (EC) No. 713/2009 (cont.)			infrastructure. Recital (6) ACER: coordination of the regulatory functions performed by the NRAs in accordance with Directive 2009/73/EC and to that purpose ensuring the independence of the Agency from the producers, TSOs and DSOs and consumers. Recital (7) ACER: monitoring of the cooperation between TSOs as well as the execution of tasks of ENTSOs. Recital (9) ACER: important role in developing framework guidelines (which are nonbinding by nature with which network codes must be in line) and in reviewing network codes. Recital (10) In cross-border situations, ACER should be granted the power to adopt individual decisions covering technical issues, the regulatory regime for natural gas infrastructure and, as a last resort, exemptions from the internal market rules for new gas infrastructure located in more than one MS. Recital (11) ACER: advisory role to the Commission, other Community institutions and NRAs regarding the issues	

			related to the purpose for which it was established. Recital (12) recommendations for the sharing of good practices among regulatory authorities and market players. Recital (14) ACER: contribution to the implementation of the guidelines on trans-European energy networks. Recital (16) guarantee the independence of NRAs. Recitals (17) + (18) + (19) independence of ACER and its administrative and appeal boards. Article 1 (2) ACER: assistance to and coordination of the regulatory authorities in exercising at EU level, the regulatory tasks performed in the Member States. Article 7 ACER tasks regarding NRAs: individual decisions on technical issues, good practice, framework for cooperation.	Recital (8) Monitoring activities of ACER and the Commission should not overlap. Recital (26) shared implementing powers between ACER and the Commission. Recital (29) reference to subsidiarity and proportionality principles.
Commission Decision of 10 November 2010 amending Chapter 3 of Annex I to Regulation (EC) No. 715/2009 of the European Parliament and of the Council on conditions for access to the natural gas transmission networks	Definition of the technical information necessary for network users to gain effective access to the system: all relevant points for transparency reauirements, the information to be published at all relevant points, time schedule according to which this information shall be published.			

Annex II: Case law

Case law (gas market)	Unbundling	Transmission	Transport (pipelines; trade)	Storage
Commission v Spain, Case C-196/07; E.ON/ ENDESA, Case COMP/M.4197; Commission v. Spain, Case C – 207/07			Spain: pipelines for import of natural gas lack spare capacity and are blocked by long-term contracts. Current gas pipelines linking Spain to the rest of the EU and its suppliers: Lacq-Calahorra (Norway), Maghreb (Algeria), Medgaz (Algeria).	
Decision of the Commission 2006/622/CE of 21.12.2006, Case COMP/M. 3696	Ownership unbundling			
Distrigaz, COMP/B-1/37.966			Third-party access to network; improved access to wholesale gas, transport and customers	
DONG/DUC, Case COMP/38.187			DONG undertook to introduce an improved access regime for its off-shore pipelines linking the Danish gas fields with the Danish main land and to introduce interruptible transport contracts.	
Eni SpA v Commission, Case T-558/08; ENI SpA, Case COMP/39.315			**Capacity hoarding**: refusal to grant access to capacity available on the transport network; **capacity degradation**: granting of access in an allegedly less useful manner.	
E.ON electricity, Case COMP/39.389; RWE gas foreclosure, Case COMP/39.402	E.ON: divestiture of 5000 MW of generation capacity and its high voltage transmission grid for electricity supply in Germany. RWE offered a structural remedy, regarding the selling of its existing			

	German high-pressure gas transmission network – including the necessary personnel and ancillary assets and services necessary for a viable gas transport business – to an independent purchaser, who did not give rise to competition concerns.			
E.ON gas foreclosure, Case COMP/39.317; GDF gas foreclosure, Case COMP/39.316			Third-party access refusal and long-term reservation of transport capacity.	
E.ON/MOL, COMP/M.3696	MOL commitments: to remain a minority shareholder in MOL WMT and MOL Storage, to divest its remaining shareholdings on 25% +1 share in MOL Storage and MOL WMT within 6 months following the transaction. Ownership unbundling in between the activities of production and transport conserved by MOL and the commercial activities (wholesale and storage) conserved by E.ON. As a consequence of the ownership unbundling the Hungarian gas market became more liquid.			E.ON: grant third-party access to storage at regulated prices and conditions and sufficient storage capacities.
Fearon v Irish Land Commission, Case 182/83; Klaus Konle v Republic of Austria, Case C-302/97	CJEU: national ownership concepts are subject to basic freedoms and to the principle of prohibition to discriminate.			

Case law (gas market)	Unbundling	Transmission	Transport (pipelines; trade)	Storage
Fédération française des sociétés d'assurances (FFSA), Union des sociétés étrangères a'assurances (USEA), Groupe des assurances mutuelles agricoles (Groupama), Fédération nationale des syndicats d'agents généraux d'assurances (FNSAGA), Fédération française des courtiers d'assurances et de réassurances (FCA) and Bureau international des producteurs d'assurances *et* de réassurances (BIPAR) v Commission of the European Communities, Case T-106/95	The decision in this case law produces similar effects to the introduction of effective unbundling in the gas sector.			
GDF gas foreclosure, Case COMP/39.316			Release of a large share of long-term reservations of gas import capacity into France	
Gaz de France, Case COMP/38.662			Territorial restriction clauses in gas transport contracts – common practice for an undertaking ensuring the transport of gas for another undertaking within the same territory where the former sells gas as well.	
Gaz de France/ Suez, COMP/M.4180; DONG/ Elsam/E2, Case M.3868	(1) Divestiture of Suez's gas supply business Distrigas, which gave up control over the Belgian gas transmission network operator Fluxys; divestiture of GDF's shareholding in the Belgian electricity supplier SPE. (2) Remedies included full unbundling of the ownership of the gas production and gas activities bestowed upon MOL and the wholesale and storage activities acquired by E.ON. (3) DONG offered to divest the larger of its two Danish gas storage facilities and to carry out a gas release programme through auctions, thus			

	making large quantities of gas available to competitors. The commitments offered by DONG constituted the first structural unbundling remedy in a merger case with Union dimension, which envisaged the separation of assets of the incumbent wholesale operator and infrastructure facilities.			
GFU – Norwegian Gas Negotiation Committee, Case COMP/ 36.072		Third-party access: effective, cost-reflective and nondiscriminatory access regime	Effective transport of Norwegian gas through the European gas pipelines without any artificial obstacles	
Marathon/Ruhrgas/GDF et alia, Case COMP/ 36.246		Thyssengas' commitments are supposed to contribute to a better functioning of the gas transmission market in Germany. Ruhrgas: extension of the entry/exit regime beyond its own network offering the same to other regional transmission companies in which it holds a minority stake.	Thyssengas GmbH committed itself to grant improved third-party access to its pipeline network. (The Commission's investigation regarded the prevention of access to the European gas network of Norwegian gas producer Marathon.) In order to improve third-party access, Thyssengas put forward five substantial proposals regarding: (1) balancing, (2) trade in capacity rights, (3) congestion management, (4) transparency, (5) handling of access requests. BEB originally refused to grant Norwegian gas producer Marathon access to its Northern German pipeline network. In the settlement BEB offered to improve third-party access to its network and to transform the transport system into a user-friendly system for its network: entry/exit system (better reflecting physical flows and costs incurred when	Improved third-party access to BEB's storage facilities. BEB will allow companies to use its storage facilities even if the technical minimum flow requirements are not fulfilled (with the condition that at the same time other shippers fulfil the minimum flow requirements) GdF: introduction of an optional daily balancing service as part of the offer of access to storage for third parties.

Case law (gas market)	Unbundling	Transmission	Transport (pipelines; trade)	Storage
Case COMP/36.246 (cont.)			transporting gas). BEB will improve its handling of access requests by introducing online screen-based booking procedures, particularly relevant for short-term trading. Additionally, the introduction of "use it or lose it" principle for capacity reservations on its own gas trading branch regarding third party access to unused transport capacity originally booked by BEB's trading branch. Access to pipeline network (refusal to grant access would have a negative impact on the single market project and therefore has been tackled as a potential abuse or restrictive concerted practice); linking of other pipelines to Gasunie's own pipeline system. Improve third party access to their transport networks in view of benefiting customers in France and Germany. Commitments by GdF: gradual reduction in the number of tariff and balancing zones on which the entry/exit transport system is based in France. Commitments of Ruhrgas: capacity reservation based on the booking of gas transport capacities separately at entry and exit points, across tariff and quality zones.	

Oscar Bronner GmbH & Co. KG v Mediaprint Zeitungs, Case C-7/97		CJEU enunciated three cumulative conditions under which access to facilities deemed essential had to be granted: (1) the refusal of access should be likely to eliminate all competition in the market of the entity requesting the service; (2) there is no objective justification for the refusal of access; (3) access to the facilities is indispensable for carrying on that entity's business.	
Territorial restrictions 1) Algeria gas transport contracts, 2) Expansion of TAG pipeline, Case COMP/37.811			ENI agreed to increase capacity on the pipeline that transports Russian gas to Italy via Austria and promote an improved third-party access regime facilitating the use of its pipeline for transit purposes.
Territorial restrictions – Austria (Gazprom/OMV), Case COMP/38.085			**OMV:** increase of pipeline capacity and improvement of third-party access.
UK Gas Interconnector, COMP/38.075			Alleged influence upon the flow direction of the Interconnector.
Sector Inquiry Lessons	MEMO/05/203: in a number of Member States TSOs unbundling as required by the 2003 Directive has not been implemented; discriminatory treatments in unbundled companies. IP/05/1421: many wholesale markets in many Member States are liquid because of long-term contracts or because companies are active both in production and in the retail market; inadequate level of unbundling of network and supply activities (reflected in a lack of competition on the supply market and few incentives to invest in networks – therefore resulting into major	Speech/07/4: lack of interconnection capacity leading to a selected group of TSOs acquiring significant congestion revenues. Memo/09/567: Commission concerned that certain E.ON practices may constitute a breach of EU rules on abuse of a dominant market position (article 102 TFEU). In specific, that E.ON may have closed off competitors from the market by booking the entire capacity at key entry points into the gas network on a long-term basis. E.ON remedies: reduction of its long-term gas capacity reservations which prevent access of competitors to infrastructure needed to supply gas to customers within E.ON's network.	IP/05/1421: inadequate use of existing infrastructure; gas market suffering from a lack of liquidity of both gas and transport capacity. IP/10/494: Almunia: "the notorious lack of transport capacity is currently one of the major obstacles to gas competition in Germany. With today's commitments we have achieved a far-reaching solution which will give competitors access to the transport capacities they need to enter the market."

Case law (gas market)	Unbundling	Transmission	Transport (pipelines; trade)	Storage
Sector Inquiry Lessons (cont.)	obstacles to new entry and threats to security of supply)	IP/10/494: Access to gas pipelines is vital for new market entrants. Insufficient access limits the ability to acquire customers, no matter how competitive the offer might be. E.ON: release of large capacity volumes at entry points to its gas network by October 2010 – around 15% of the pipeline capacity		

Case law (gas market)	Cross-border (trade; transit; transport)	Investment	Supply	Security of supply	Public service obligations
Adria Wien Pipeline GmbH v Finanzlandesdirektion für Kärtnen, Case C-143/99			An Austrian tax measure awarding a rebate on energy taxes charged on supply of natural gas and electricity to undertakings active (1) in the manufacture of goods, (2) in so far as energy taxes exceeded 0.35% of the production value, was in fact a selective measure, therefore entered in the scope of article 107(1) TFEU.		
Altmark Trans GmbH, Regierungsprasidium Magdeburg and Nahverkehrsgesellschaft Altmark GmbH Case C-280/00					Decision of the Court of Justice ruled on four conditions that exclude financial compensation for the costs of carrying out PSOs from the application of state aid rule: (1) the recipient undertaking must have PSOs to discharge, such obligations have to be clearly defined; (2) the parameters for the calculation of the compensation must be established in advance in a transparent and clear manner; (3) the compensation cannot exceed what is necessary to cover all or parts of the costs incurred while delivering PSOs; (4) where the undertaking is not chosen pursuant a public procurement procedure, the level of compensation needed must be determined on the basis of an analysis of the costs that a typical

Case law (gas market)	Cross-border (trade; transit; transport)	Investment	Supply	Security of supply	Public service obligations
					undertaking would incur in such a situation.
British Gas/British Gas Trading/Wingas, Case COMP/ 36.559			Long-term gas supply contracts in the wholesale sector may be compatible with European competition law.	Long-term contracts have played an important role for Europe's supply security in the past and will probably continue to do so in the future (as long as they do not contain clauses foreclosing the market).	
Campus Oil Limited and others v Minister for Industry and Energy and others, Case C-72/83				CJEU deemed the objective of security of energy supply as an integral part of the "public security" concept, covered by article 36 TFEU (ex-article 30 TEC).	
Criminal proceedings against Paul Corbeau, Case C-320/91; Municipality of Almelo and others v NV Energiebedrijf Ijsselmij, Case C-393/92	CJEU: restrictions on competition are allowed where the guarantee of the economic stability of an undertaking entrusted with PSOs would be prejudiced.				
Distrigaz, COMP/B-1/37.966			Lack of competititon among suppliers. Committment regarding the noninclusion of any use, resale or destination restriction in any new supply contract and the removal of such clauses where appropriate. No tacit renewal clause, no long-term contract with resellers longer than 2 years. Concerns about		

			competition in supply owing to long-term contracts signed by Distrigas with its gas customers. New commitments encourage competition among suppliers.		
DONG/DUC, Case COMP/38.187			The gas supply agreements concluded between the DUC partners and DONG contained certain provisions, which were considered to be anticompetitive.		
Elliniki Radiophonia, Case C-260/89; Klaus Höfner and Fritz Elser v. Macrotron GmbH, Case C-41/90; Merci convenzionali porto di Genova SpA v Siderurgica Gabrielli SpA, Case C-170/90					In these cases CJEU found that competitiveness prevails public service obligations in regulated sectors.
Eni SpA v Commission, Case T-558/08; ENI SpA, Case COMP/39.315		**Strategic underinvestment**: alleged strategic limitation of investment in capacity on transport networks	Weakening of supply competitiors on the Italian market, where ENI is the largest producer, importer and supplier of natural gas on the wholesale and retail markets.	Negatively affecting the security of gas supply in Italy as a result of creating bottlenecks on import capacity and owing to limited investment in transport capacity. The security of supply is also affected negatively by the lack of more suppliers on the market.	
E.ON gas foreclosure, Case COMP/39.317; GDF gas foreclosure, Case COMP/39.316		GDF: strategic underinvestment in import infrastructure.			
E.ON/GDF collusion, COMP/39.401			Ruhrgas supply area in Germany was de facto protected from competition by a system of so-called		

Case law (gas market)	Cross-border (trade; transit; transport)	Investment	Supply	Security of supply	Public service obligations
Case COMP/39.401 (cont.)			"demarcation agreements" with other German suppliers until such agreements became illegal in April 1998.		
E.ON/MOL, COMP/M.3696			E.ON: transfer of gas supply rights for a period of 8 years (1 BCM yearly) – increase of competition among suppliers		
Federutility, Case C-265/08; Altmark Trans GmbH, Regierungsprasidium Magdeburg and Nahverkehrsgesellschaft Altmark GmbH Case C-280/00; Ferring SA v Agence Centrale des Organismes de Sécurité Sociale, Case C-53/00; Enirisorse SpA v Sotacarbo SpA, Case C-237/04; Ministre de l'Economie, des Finances et de l'Industrie v Gemo SA, Case C-126/01					Competition rules should not obstruct in law or fact the accomplishment of public service obligations.
Gas Natural v ENDESA, Case C 122/06 P-DEP and GAS NATURAL + ENDESA, Case COMP/37.542			Restrictions in vertical supply agreements either preventing the buyer from using the gas for other purposes than agreed upon (use restrictions) or from selling the gas outside an agreed supply area. Territorial and use restrictions are essential restrictions entering the scope of EU competition law.		
GDF gas foreclosure, Case COMP/39.316			Concern that GDF Suez actions distorted competition among suppliers by preventing access to gas import capacity into France. The commitments allowed new entrants on the market to have access to gas import infrastructure and acquire new customers.		

Gaz de France, Case COMP/38.662				Directive 2003/55/EC read in conjunction with article 106 TFEU
GFU – Norwegian Gas Negotiation Committee, Case COMP/ 36.072			Fixing of price and quantities sold; the commitments facilitate the establishment of new supply relationships; no territorial sale restrictions and use restrictions in gas supply contracts.	
Marathon/Ruhrgas/GDF et alia, Case COMP/ 36.246	It is expected that the entry/ exit model will facilitate domestic as well as cross-border transports for third parties	**GDF:** investment programme to reduce congestion on the network and thus facilitate access for new entrants by reducing the cost of transport connected with the crossing of several zones.	As part of the settlement, BEB will assist shippers having a flexible supply source to avoid getting into a situation of imbalances, by introducing a free of charge online balancing system and a bulletin board that will allow shippers to get in contact with each other and optimise their transport and storage requirements. As part of the settlement, Gasunie will assist shippers with a flexible supply source to avoid getting into a situation of imbalances, by introducing an online balancing system to avoid situations where suppliers/ shippers are charged very high prices for the gas supplied by Gasunie. Ruhrgas: gradual reduction of tariff zones and online balancing services.	

Case law (gas market)	Cross-border (trade; transit; transport)	Investment	Supply	Security of supply	Public service obligations
Poland Aid to PGNiG for underground gas storage, Case N 660/2009				Exemption from state aid rules in cases regarding the improvement of the security of gas supply.	
Poland- Regional aid scheme for construction and of the gas distribution network, Case N 435/2008 reconstruction		The aid offered incentives for investments in the construction of gas distribution networks on territories, which at that moment were not supplied with gas.			
Swedish Interconnectors, Case COMP/39.351	The case underlines the need for a unified approach to national and cross-border congestion management. The trade-off between managing domestic congestion and limiting cross-border capacity trade is in opposition to the advancement of the European energy transmission network, which aims at interconnecting the EU energy market.				

Territorial restrictions 1) Algeria gas transport contracts, 2) Expansion of TAG pipeline, Case COMP/37.811	Territorial restrictions prevent cross-border trade.		Agreement to delete a clause preventing one of NLNG's European customers to re-sell gas outside its national borders. Territorial restriction clauses in supply contracts preventing wholesaler from reselling the gas outside the countries where they are established. Market foreclosure through territorial restriction clauses and profit sharing mechanisms in gas supply agreements.	
Territorial restrictions – Austria (Gazprom/OMV), Case COMP/38.085			Removal of clauses infringing Treaty rules on restrictive business practices (article 101 TFEU), which negatively affected competition among suppliers. This is a breakthrough case while it addresses Europe's largest external gas supplier – Gazprom – and the main transit route for this supplier's gas to Europe.	Territorial restriction clauses have negative impact on the security of supply.
Territorial restrictions in Germany (Ruhrgas/Gazprom), Case COMP/38.307	Germany is an important transit country for both Norwegian and Russian gas to neighbouring countries.		Removal of clauses infringing the Treaty rules on restrictive business practices (article 101 TFEU), which negatively affected competition among suppliers.	

Case law (gas market)	Cross-border (trade; transit; transport)	Investment	Supply	Security of supply	Public service obligations
Sector Inquiry Lessons	IP/05/716: cross-border flow seems insufficient to constrain price differences between most Member States. IP/05/1421: limited cross-border competition, absence of price convergence across the EU and low-level of cross-border trade. Speech/07/4: absence of cross-border integration and cross-border competition (no uniformity of energy prices, different market designs between Member States)		IP/05/1421: barriers to cross-border supply of gas preventing the development of an integrated EU gas market. Speech/07/4: insufficient tradeable supplies on energy markets due to long-term contracts which help locking-in the markets (prevent alternative suppliers from supplying customers on the retail market).		

Case law (gas market)	Consumers/consumer protection	Other exemptions (including environment)	Market foreclosure	Internal market/market integration
Austria – Aid to renewable energy installations, Case N 241/2004; HAMSA v Commission, Case T-152/99	Protection of vulnerable energy consumers exposed to energy poverty and certain cases of security of energy supply concerns.	Automatic exemptions from the rules of state aid in cases of measures having a social character and granted to individual consumers, if the aid is provided without discrimination in regards to the origin of the products or services concerned.		
Commission v Belgium, Case C-475/08				Failure to transpose Directive 2003/55/EC
Commission v Bulgaria, Case C-198/2012				Misapplication of TEP provisions, specifically with respect to market interconnection and adoption of national transposition measures.
Commission v Estonia, Case C-351/05				Failure to transpose Directive 2003/55/EC.
Commission v Estonia, Case C-241/2013; Commission v Poland, Case C-55/2013; Commission v Slovenia, Case C-9/2013				Member States have yet to transpose the provisions of TEP, contributing thus through their action to the continuation of failures in market interconnection.
Commission v France, Case C-259/01				Failure to transpose Directive 98/30/EC.
Commission v Germany, Case C-64/03				Failure to transpose Directive 98/30/EC.
Commission v Luxembourg, Case C-354/05				Failure to transpose Directive 2003/55/EC.
Commission v Spain, Case C-357/05				Failure to transpose Directive 2003/55/EC.
Commission v Spain, Case C-196/07; E.ON/ENDESA, Case COMP/M.4197; Commission v. Spain, Case C – 207/07				Concerning the Spanish gas market: The Commission has observed the existence of significant barriers to import gas through pipelines.

Case law (gas market)	Consumers/consumer protection	Other exemptions (including environment)	Market foreclosure	Internal market/market integration
Distrigaz, COMP/B-1/37.966	No long-term contract with customers longer than 5 years. Increasing consumers' choice by allowing new entrants to acquire consumers. With very few exceptions, gas customers have only one gas supplier at one time. As a result, competition in the gas supply market only takes place when a contract expires and other contract is concluded. Removal of any barrier to access to customers.		Prohibition on abuse of dominant position (article 102): the Distrigaz case raised concerns that the Belgian gas supply market could be foreclosed due to long-term gas supply agreements	
DONG/DUC, Case COMP/38.187	A significant number of customers inside and outside Denmark have actively looked for alternative sources of supply than that offered through the joint agreement both prior and upon the settlement.		Joint marketing activities of three Danish gas producers (Shell, Maersk, Chevron Texaco) as well as anticompetitive arrangements in their gas supply contracts with the Danish incumbent wholesaler DONG.	
EDP v Commission, Case T-87/05				Derogations from the complete opening of the gas market under Directive 2003/55/EC.
EDP/ENI/GDP, Case M.3440; Energias de Portugal (EDP) SA v Commission, Case T-87/05			Merger prohibition decision: the concentration would have caused an important potential competitor to disappear from all the electricity markets, while another's dominant position would have been reinforced to the detriment of effective competition.	
Enel/Endesa, Case M.5494; EdF/British Energy, Case M.5224; Vattenfall/Nuon, Case COMP/ M.5496; E.ON/ Endesa, Case M.4197			Examples of mergers between energy undertakings either within the confines of the same sector or cross-sector, which had the potential effect of eliminating or distorting market competition.	

ENEL/ENI/Gaz de France, Case COMP/38.662			Market sharing.	
Eni SpA v Commission, Case T-558/08; ENI SpA, Case COMP/39.315	Preventing proper access to gas infrastructure despite very significant short- and long-term demand from three party shippers. Obstacles to market integration.		Abuse of dominant market position article 102 TFEU	The 2009 gas crisis demonstrated that access to gas infrastructure in Europe is a requirement for market integration and the development of competition plays a crucial role in assuring the security of gas supply.
E.ON gas foreclosure, Case COMP/39.317; GDF gas foreclosure, Case COMP/39.316			Incumbents were suspected of concerted practices constituting an abuse of dominant market position in breach of article 102 TFEU	
E.ON/GDF collusion, COMP/39.401	Weakening of consumers' protection through depriving them of price competition and choice of supplier in two of the largest gas markets in the EU. **Action for damages:** any person or firm affected by anticompetitive behaviour as described in this case may bring the matter before the courts of the Member States and seek damages. (private enforcement)		Market-sharing in breach of EC Treaty rules on cartels and restrictive business practices (article 101 TFEU).	
E.ON/MOL, COMP/M.3696			The proposed concentration was found to have a significant impact on the gas and electricity sectors in Hungary, owing to MOL's position as almost exclusive controller over the procurement of gas – this allowed it to foreclose third-party access to resources and infrastructure.	
Exemption from climate change levy for natural gas in Northern Ireland, Case N/660a/2000		It was considered that the distortion of competition through the application of the tax exemption was necessary in order to realize the environmental benefits.		

Case law (gas market)	Consumers/consumer protection	Other exemptions (including environment)	Market foreclosure	Internal market/market integration
Federutility, Case C-265/08; Altmark Trans GmbH, Regierungsprasidium Magdeburg and Nahverkehrsgesellschaft Altmark GmbH Case C-280/00; Ferring SA v Agence Centrale des Organismes de Sécurité Sociale, Case C-53/00; Enirisorse SpA v Sotacarbo SpA, Case C-237/04; Ministre de l'Economie, des Finances et de l'Industrie v Gemo SA, Case C-126/01	Articles 3 and 23 (1)(c) of Directive 2003/55/EC, beginning from 1 July 2007 all customers have the right to freely choose their gas supplier, and all gas suppliers also have the right to nondiscriminatory access to customers			
France v Commission (Stardust Marine), Case C-482/99	Should a government seek to exert control or influence over the tariffs charged to certain groups of users or to ensure that capacity in pipelines, electricity networks, storage facilities or interconnectors is reserved to particular undertakings, state aid issues may well arise.			
Gas Insulated Switchgear, Case COMP/38.899	Switchgear: action for damages can be brought by any person or firm affected by anti-competitive behaviour before the courts of the Member States. (private enforcement)		Switchgear allegations: cartel agreement and related practices – bid rigging, exchange of sensitive price information, price fixing, market sharing.	
Gas Natural v ENDESA, Case C 122/06 P-DEP and GAS NATURAL + ENDESA, Case COMP/37.542	Discrimination of Spanish gas purchasers. In the original agreement some of the clauses favored ENDESA with a better treatment than other future clients of Gas Natural.	The original contract provided for the Spanish gas market segmentation (it contained a destination clause) in favour of the dominant firm.	Non-discriminatory third-party access to the Spanish gas market. Not allowing the dominant gas supplier to impede the entry of new suppliers on the market: the long-term contract signed between Gas Natural and ENDESA would have raised barriers to entry into the liberalized Spanish gas market.	
GDF gas foreclosure, Case COMP/39.316	Strengthening of consumers' protection by improving competition among suppliers and allowing access to new entrants on the market.			

Gaz de France, Case COMP/38.662			Clauses providing territorial restrictions in contracts between GDF and ENI and GDF and ENEL determined the limitation of competition inside the common market (article 101(1) TFEU)	Horizontal markings of different geographical markets
GFU – Norwegian Gas Negotiation Committee, Case COMP/ 36.072	Choice among 30 suppliers; the settlement included reservation of certain gas volumes for new customers, who in the past have not bought gas from Norwegian gas producers.		**GFU** has entered into a large number of long-term supply agreements with European gas operators which perpetuate the adverse effects of the joint selling schemes for several years and which have led to a significant rigidity and lack of liquidity in the European gas markets.	The commitments contribute to the creation of a single market for gas in Europe
Greece – Aid in favour of the gas supply companies in Anatoliki Makedonia Thraki; Kentriki Makedonia and Sterea Ellada, Case SA.31.225 (2011/N)		National measures that amounted to state aid were allowed because they promoted regional development in areas designated in accordance with articles 107(3)(a) and (c) TFEU.		
Irish Corrib, Case COMP/37.708	Commission expressed concerns regarding the effects of joint marketing upon the ongoing gas market liberalization process and specifically in respect of the eligibility of gas consumers to choose freely among their suppliers.	Application for an exemption from the European Commission to market gas produced at Corrib jointly for the first five years of production		
Kwekerij Gebroeders van der Kooy BV and others v Commission, joined cases 67, 68 and 70/85; Société CdF Chimie azote et Fertilisants et al v Commission, Case C-169/84; Belgium v Commission, Case C-56/93.	**Discrimination among classes of consumers**: state-aid; **preferential energy tariffs**: CJEU concluded that the decision by Gasunie of granting preferential tariffs to large consumers could be attributed to the Dutch government.			

Case law (gas market)	Consumers/consumer protection	Other exemptions (including environment)	Market foreclosure	Internal market/market integration
Marathon/Ruhrgas/GDF et alia, Case COMP/36.246	**Gaz de France:** offered a settlement which was supposed to allow the less well placed French customers (low quality of the gas they used or geographical location) to benefit from the liberalization of the gas market.		Refusal to grant access can be tackled as a potential abuse of a dominant position or a restrictive concerted practice.	Regional initiatives – cooperation of market incumbents for the harmonization of gas transport conditions: BEB is open to discuss with adjacent pipeline system operators possible cooperation to extend the entry/exit system to larger territories. Gas market's rules harmonization by incumbents: Ruhrgas offered to extend its entry/exit regime beyond its own network offering the same to other regional transmission companies in which it holds a minority stake.
PreussenElektra v Schleswag, Case C-379/98		CJEU underlined the importance of state measures for protecting the environment through their contribution to the reduction in emissions of greenhouse gases.		
Territorial restrictions 1) Algeria gas transport contracts, 2) Expansion of TAG pipeline, Case COMP/37.811	Better protection of consumer interests in the absence of territorial restriction clauses: ENI agreed to offer significant gas volumes to customers located outside Italy over a period of five years.		Market foreclosure through territorial restriction clauses and profit sharing mechanisms in gas supply agreements.	Territorial restriction clauses undermine the on-going creation of a European single gas market. The presence of territorial restriction clauses in supply contracts represents a breach of European competition law and undermines the accomplishment of the internal gas market. Major breakthrough case in relations with one of Europe's most important natural gas suppliers, while it eliminates an important obstacle for the creation of a single EU-wide market in gas.

Territorial restrictions – Austria (Gazprom/OMV), Case COMP/38.085		The case is important because it shows the EC's resolution to introduce competition in the gas sector and fight market partitioning agreements or other anticompetitive arrangements. Territorial restriction clauses prevent wholesalers from reselling gas in the neighbouring territories, which has negative impact on the creation of an integrated European gas market, fostering both competition and security of supply.
Territorial restrictions in Germany (Ruhrgas/Gazprom), Case COMP/38.307		Removal of territorial sale restrictions from gas supply contracts is a step towards creating a competitive and integrated European gas market.
UK Gas Interconnector, COMP/38.075	No evidence of cartel-like behaviour, despite rigidities in the agreements regarding flow reversal and short-term capacity transfers.	The priority status granted by the Commission to interconnector issues shows that they constitute a cornerstone for achieving an effective single market for energy.
Vattenfall/Nuon, Case COMP/ M.5496; E.ON/ Endesa, Case M.4197	Cross-sector mergers between gas and electricity undertakings with the potential effect of eliminating or distorting market competition.	

Case law (gas market)	Consumers/consumer protection	Other exemptions (including environment)	Market foreclosure	Internal market/market integration
Sector Inquiry Lessons	MEMO/05/203: significant price increases combined with the fact that often customers do not receive genuinely competitive offers affects negatively the benefits that the consumers could get out of the gas market liberalization; the sector inquiry aimed at ensuring developments in competition on the gas market for those customers who have not been eligible for many years. IP/10/494: commitments are expected to benefit domestic and industrial gas consumers.		IP/05/716: limited new market entry and very high market concentration (barriers to entry: long-term agreements and access to gas customers markets) IP/05/1421: gas markets in many Member States continue to be concentrated, creating scope for incumbent operators to influence prices.	IP/05/716: integration between national markets has been slow in many regions. IP/05/1421: Member States must implement more effectively the market opening measures required under the EU's gas and electricity Directives.

Case law (gas market)	Transparency/information sharing	National governments/regulators' powers	European Commission	Regulators' cooperation	Shared competences
Altmark Trans GmbH, Regierungsprasidium Magdeburg and Nahverkehrsgesellschaft Altmark GmbH Case C-280/00			When applying competition law to assess the legality of public monopolies and privileged undertakings the Commission must respect the principle of proportionality.		
Ahlström Osakeyhtiö and others v Commission of the European Communities, Cases C-89/85, C-104/85, C-114/85, C-116/85, C-117/85, C-125/85 [1988] ECR 5193, [1988] 4 CMLR 901, [1993] ECR I – 1307, [1993]			The Commission is competent to apply competition law rules to actions performed by foreign entities, which produce their effects on the territory of the Common Market.		
Alrosa Company Ltd v Commission of the European Communities, Case T-170/06			Commitments accepted by the Commission should not exceed what was appropriate and necessary for terminating the infringement of competition law.		
Béguelin Import Co. v S.A.G.L. Import Export, Case 22/71; Gencor Ltd v Commission, Case T-102/96; Imperial Chemical Industries Ltd. v. Commission of the European Communities, Case 48/69			The Commission is competent to apply competition law rules to actions performed by foreign entities, which produce their effects on the territory of the Common Market.		

Case law (gas market)	Transparency/information sharing	National governments/regulators' powers	European Commission	Regulators' cooperation	Shared competences
Campus Oil Limited and others v Minister for Industry and Energy and others, Case C-72/83		Member States are allowed to adopt complementary national measures required to guarantee public security, despite the application of EU rules to the same matter.			
Cementbouw Handel & Industrie BV v Commission of the European Communities, Case C-202/06			CJEU has ruled that the Commission's acceptance of commitments under the merger regulation must respect the principle of proportionality.		
Commission v Alrosa, Case C-441/07 P			CJEU maintained that article 9 did not require the Commission to find an infringement, but to examine and possibly accept commitments formulated by the parties. Hence, the principle of proportionality should be restricted to verifying that the commitments in question address the competition concerns formulated by the Commission and that the undertakings concerned have not previously formulated less onerous commitments satisfying the concerns.		

Commission v Spain, Case C-196/07; E.ON/ENDESA, Case COMP/M.4197; Commission v. Spain, Case C – 207/07	CNE's powers increased by the Spanish Council of Ministers. Merger bids require CNE approval, which in this case adopted 19 conditions to allow the bid; CNE has exceeded its attributions by imposing conditions on a merger with Community dimension. The conditions were declared illegal.	Art. 21 of Merger Regulation, gives the Commission exclusive competence to assess the competitive impact of concentrations with a Community dimension. The Commission is entitled to monitor national measures in order to ensure that Member States do not impose unjustified restrictions (which are not necessary and proportionate for the protection of a legitimate public interest) on cross-border mergers that are the exclusive competence of the Commission.
Distrigaz, COMP/B-1/37.966		Annual report on compliance presented to the Commission; discretionary power of the Commission to impose on the party which does not respect the commitments a fine amounting to 10% of total worldwide turnover; possibility to adopt decisions making legally binding commitments assumed by an undertaking.

Case law (gas market)	Transparency/information sharing	National governments/regulators' powers	European Commission	Regulators' cooperation	Shared competences
DONG/DUC, Case COMP/38.187	DONG committed to increase the transparency of the system by publishing information on available capacity.				
EDP v Commission, Case T-87/05			Court of Justice supports the Commission's position despite the latter's wrong qualification of increase of dominant position. The Commission is entitled to monitor national measures in order to ensure that MS do not impose unjustified restrictions on cross-border mergers that are the exclusive competence of the Commission.		
EDP/ENI/GDP, Case M.3440; Energias de Portugal (EDP) SA v Commission, Case T-87/05			The General Court has given the Commission its support in liberalizing the EU gas market through formal merger control.		
E.ON electricity, Case COMP/39.389; RWE gas foreclosure, Case COMP/39.402			Under Article 9 of Regulation (EC) 1/2003 the Commission is allowed to make use of its antitrust enforcement powers in a regulatory manner, rather than a corrective one.		

E.ON/ Endesa, Case M. 4110; ENEL/ ACCIONA/ENDESA, Case M.4685		The Commission has started infringement proceedings under article 258 TFEU, as a result of the failure of Member States' authorities to withdraw certain national measures that created unjustified restrictions to mergers with a Community dimension.	
E.ON/GDF collusion, COMP/39.401		Ground-breaking case: the Commission has a right to impose fines in case of an antitrust infringement. These were the first fines imposed by the Commission in the energy sector.	
E.ON/MOL, COMP/M.3696	HEO (Hungarian Energy Office: energy regulator) is entitled to monitor over any issues that might result in storage capacity constraints and to agree to E.ON's storage development plan.		Collaboration between the EC and national regulators in the framework of the market test on the proposed undertakings.
E.ON Ruhrgas v Bundeskartellamt, Case KVR 67/07	NCA has the power to impose conditions on future contracts.		
E.ON Seals, Case COMP/39.326		The Commission can fine an undertaking for tampering with seals applied by European officials during a surprise descent upon the premises of the undertaking.	

Case law (gas market)	Transparency/information sharing	National governments/regulators' powers	European Commission	Regulators' cooperation	Shared competences
Federal Republic of Germany v Council (Product Safety), Case C-359/92; UK v Parliament and Council (Smoke Flavourings), Case C-66/04			Product Safety: the Commission has the power to take decisions replacing those of national authorities if the partially harmonizing provisions of the EU secondary legislation could give rise to the necessity of temporary emergency intervention. Smoke Flavourings: the Commission enjoys discretion in regards to the "harmonization technique most appropriate for achieving the desired result, in particular in fields which are characterized by complex technical features."		
Fédération française des sociétés d'assurances (FFSA), Union des sociétés étrangères a'assurances (USEA), Groupe des assurances mutuelles agricoles (Groupama), Fédération nationale des syndicats d'agents généraux d'assurances (FNSAGA), Fédération française des courtiers d'assurances et de réassurances (FCA) and Bureau international des producteurs d'assurances *e*t de réassurances (BIPAR) v Commission of the European Communities, Case T-106/95			The European Commission is entitled to exercise its power of supervision discretionary when it comes to Member States liable for infringing the rules of the Treaty (article 106 – 3 TFEU). Such a discretion increases if the compliance of Member States with competition rules is at		

			stake. The Commission must be allowed a certain discretion in deciding on the most appropriate method for ensuring that the state aid does not involve any cross-subsidy for the benefit of the competitive activities of the undertaking concerned.		
Federutility, Case C-265/08; Altmark Trans GmbH, Regierungsprasidium Magdeburg and Nahverkehrsgesellschaft Altmark GmbH Case C-280/00; Ferring SA v Agence Centrale des Organismes de Sécurité Sociale, Case C-53/00; Enirisorse SpA v Sotacarbo SpA, Case C-237/04; Ministre de l'Economie, des Finances et de l'Industrie v Gemo SA, Case C-126/01		National regulatory authorities play an important role in balancing competitiveness and public service obligations. Federutility: The case-law sets the parameters within which NRAs may set rates for energy companies. The question raised in the case was whether national regulators could intervene in gas market prices, having in mind the total opening of the market from 1 July 2007. The European Court of Justice set three conditions for such an intervention: (1) it must further a general economic interest containing the maintaining of prices at reasonable levels for end consumers. (2) It must be limited in time and scope. (3) Its rules must be clearly defined, transparent, nondiscriminatory and verifiable.			
Gas Insulated Switchgear, Case COMP/ 38.899			Switchgear, IP/07/80 the possibility for the EC to impose fines on the members of gas insulated switchgear cartel as a result of their serious		EU's competences are considered specific and cannot be extended beyond the general framework of the Treaty.

Case law (gas market)	Transparency/information sharing	National governments/regulators' powers	European Commission	Regulators' cooperation	Shared competences
Case COMP/38.899 (cont.)			infringement of EU Treaty antitrust rules – the total fines imposed are the largest set of fines ever imposed on a single cartel.		
GDF gas foreclosure, Case COMP/39.316			Discretionary power of the Commission to impose on the party which does not respect the commitments a fine amounting to 10% of total worldwide turnover.		
Gaz de France, Case COMP/38.662			The first formal decisions adopted in the field of energy by the Commission and confirming that the clauses providing territorial restrictions are violating article 101 TFEU; the case outlines Commission's power to demand commitments from the parties.		
Gazprom – Upstream gas supplies in Central and Eastern Europe, Case COMP/39.816			The Commission can open formal investigation regarding third-country gas producers and suppliers in light of allegations that these might be hindering competition in EU gas markets.		

GFU – Norwegian Gas Negotiation Committee, Case COMP/ 36.072	NRAs focused on transport cases having a national dimension	Regards cross-border cases. The European Commission has the right to adopt a statement of objections, which is a legal step in proceedings under article 101 TFEU and article 53 of the EEA agreement. Both articles ban cartels and other damaging concerted business practices. The Commission is competent to accept commitments offered by gas producers and to close ongoing investigations. The Commission is also competent to initiate formal proceedings against gas companies, which behave in a way incompatible with European competition law.	
Italy – R&D in the energy sector, Case S.A. 15876, NN 27/2005	High involvement of national governments in the regulation of the gas sector.		
Kadi and al Barakaat International Foundation v Council, joined cases C-402/05 and C-414/05			CJEU held that the Union had competence to impose sanctions on individuals, despite the lex specialis of 75 and 301 TFEU (ex-articles 60 and 301 TEC).

Case law (gas market)	Transparency/information sharing	National governments/regulators' powers	European Commission	Regulators' cooperation	Shared competences
Marathon/Ruhrgas/GDF et alia, Case COMP/36.246	Improving transparency through regular publishing and update of the available transport capacity at all major entry and exit points of the transmission network. This will enable shippers to have easier access to information about available transmission and storage capacity. Gasunie has undertaken on behalf of its transportation arm to increase transparency as regards access to its network, to better manage the capacity available and to speed up its handling of access requests (online screen-based booking procedures). Transparency will be improved by publishing on its internet site the contracted transport capacity at all entry and all major exit points of its gas network. GdF and Ruhrgas: improvement of transparency	The Dutch energy regulator DTE plays an important role in the Dutch gas market and the monitoring of the access regime. The French Energy Regulator (CRE) plays a very active role in the French market and assumed the monitoring of GdF's access regime (e.g. regarding the transparent and non-discriminatory opportunity for operators to convert H gas into L gas; the implementantion of a gas release programme in southern France where at the moment of the settlement there was no competition).			
Meroni & Co, Industrie Metallurgiche SpA v High authority, Case 9/56; Draft Agreement establishing the European laying-up fund for inland vessels, Opinion 1/76; Giuseppe Romano v Institut national d'assurance maladie-invalidité, Case 98/80; The Queen, on the application of Alliance for Natural Health and Nutri-link Ltd v Secretary of State for Health, Cases C-154 and 155/04; Tralli v ECB, Case C-301/02 P.		The jurisprudence of the European Courts shows that a delegation of wide discretionary powers is clearly unlawful under present primary law and that the creation of regulatory agencies which typically would dispose of such powers, is violating the principle of institutional balance, and is therefore not permissible.			
Ralf Schräder v Community Plant Variety Office (CPVO), Case C-38/09; Schräder v OCVV – Hansson (Lemon Symphony), joined cases T-133/08, T-134/08, T-177/08, T-242/09.		The Courts accepted that an agency may be entrusted with the power of taking externally binding decisions under EU law, which entail the exercise of administrative discretion.			

Territorial restrictions – Austria (Gazprom/ OMV), Case COMP/38.085				In improving third-party access, OMV based its actions on the Guidelines for good practice developed by the European Commission, European regulators and European gas industry in the context of the Madrid Forum.
T-Mobile Deutschland/ O2 Germany, Case COMP/38.369; Deutsche Telekom AG, Case COMP/ C-1/37.451, 37.578, 37.579; Deutsche Telekom AG, Case T-271/03.				Regarding the regulation of utilities sectors both the Commission and the European Courts seem to have opted for a complementarity approach between the Commission and the NRAs.
Sector Inquiry Lessons	IP/05/716: during the inquiry the Commission may ask companies and trade associations to supply information, documents or statements; open dialogue with all stake-holders, the sector will be kept informed about the progress. IP/05/1421: lack of transparency on the markets benefiting incumbents and undermining the position of new entrants.		MEMO/05/203: legal basis for the EC's action – article 17 of Regulation 1/2003 provides that the Commission can undertake a sector inquiry when the trend of trade, price developments or other circumstances suggest that competition in a given sector might be distorted. Investigative powers: articles 18, 19, 20 and 22 of Regulation 1/2003: the authority to request information from undertakings, to request information	IP/05/716: both the Commission and national competition authorities may proceed to enforcement action in case of distortions of competition. MEMO/ 05/203: cooperation between the Commission and national competition authorities in the framework provided by Regulation 1/2003 (on applying articles 101 and 102 TFEU); cooperation with national energy regulators.

Case law (gas market)	Transparency/information sharing	National governments/regulators' powers	European Commission	Regulators' cooperation	Shared competences
Sector Inquiry Lessons (cont.)			from governments and national competition authorities, to take oral statements from natural or legal persons and to undertake inspections in the framework of the inquiry. The European Commission is entitled to ask non-European companies for information in case their activities affect trade in Europe (but it cannot enforce orders to provide information outside the EEA territory). If it finds competition restrictions leading to a formal infringement decision, the Commission can also impose fines on companies even if they are located outside the EU. IP/10/494: The Commission has adopted a decision rendering legally binding the commitments offered by E.ON to effectively open up access to the German gas market.		

Index